GREEN CORROSION INHIBITORS

WILEY SERIES IN CORROSION

R. Winston Revie, Series Editor

GREEN CORROSION INHIBITORS

Theory and Practice

V. S. SASTRI

A John Wiley & Sons, Inc., Publication

Library of Congress Cataloging-in-Publication Data:

Sastri, V. S. (Vedula S.), 1935-
    Green corrosion inhibitors : theory and practice / V.S. Sastri.
        p. cm.
    Includes bibliographical references and index.
    ISBN 978-0-470-45210-3 (cloth)
1. Corrosion and anti-corrosives. 2. Corrosion and anti-corrosives—Environmental aspects. 3.
Green products. 4. Environmental chemistry—Industrial applications. I. Title.

    TA462.S3187 2011
    620.1′1223—dc22

                                                                        2010043327

Printed in Singapore

oBook ISBN: 978-1-118-01543-8
ePDF ISBN: 978-1-118-01417-2
ePub ISBN: 978-1-118-01541-4

10 9 8 7 6 5 4 3 2 1

*To Sri Vighneswara, Sri Venkateswara, Sri Anjaneya,
Sri Satya Sai Baba, and my parents, teachers, wife Bonnie, and
children Anjali Eva Sastri and Martin Anil Kumar Sastri*

CONTENTS

PREFACE

The excellent first book on corrosion inhibitors—*Corrosion Inhibitors* by Professor J. I. Bregman—appeared in 1963. After a gap of 35 years, I wrote and published *Corrosion Inhibitors: Principles and Applications* in 1998. This book presented an account of modern principles and applications, and it has proved a useful resource for instructors, students, neophytes, practicing engineers, and researchers in corrosion.

Corrosion is more than just an inevitable natural phenomenon; it is very important from the points of view of economics and tragic accidents involving loss of life. Corrosion can be very expensive as well as unsafe. Corrosion chemistry is changing at a rapid pace and every chemical process is viewed very critically from the points of view of safety, environmental impact, and economics.

One of the methods of combating corrosion is the use of corrosion inhibitors that decrease the corrosion rates to the desired level with minimal environmental impact. The field of corrosion inhibitors is undergoing dramatic changes from the viewpoint of environmental compatibility. Environmental agencies in various countries have imposed strict rules and regulations for the use and discharge of corrosion inhibitors. Strict environmental regulations require corrosion inhibitors to be environmentally friendly and safe.

As the title itself reflects, the book highlights the attempts made in developing environmentally acceptable corrosion inhibitors. It covers the senior undergraduate/graduate level syllabus on corrosion inhibitors. The book is also useful for practicing corrosion scientists in industry and national research institutions. Its coverage facilitates the understanding of the principles of electrochemistry. The book does not provide encyclopedic coverage of the subject of "green" corrosion inhibitors.

Rather, the main thrust is on the principles of green corrosion inhibitors with respect to theory and practice.

Chapter 1 is an introduction dealing with historical developments in corrosion inhibition science and the general concepts such as passivity, role of oxygen, classification of inhibitors, corrosion inhibition mechanisms, selection of inhibitors, application of Sastri's inhibitor field theory in corrosion inhibition, photochemical corrosion inhibition, role of thermodynamics and kinetics in corrosion, economics of corrosion, and various forms of corrosion with examples.

Chapter 2 focuses on electrochemical principles and corrosion monitoring. The extensive list of topics include the nature of corrosion reactions, electrode potentials, Pourbaix diagrams, dynamic electrochemical processes, monitoring corrosion and corrosion inhibitors, and monitoring corrosion by physical techniques, electrochemical techniques, and indirect corrosion measurement techniques, such as corrosion products, corrosion potential, water chemistry parameters, residual inhibitor, and chemical analysis of process samples. This chapter also deals with direct intrusive corrosion monitoring techniques (physical techniques, electrical resistance, and inductive resistance probes); electrochemical techniques (linear polarization resistance, zero resistance ammetry, potentiodynamic–galvanodynamic polarization, electrochemical noise, and electrochemical impedance); direct nonintrusive techniques (ultrasonics, magnetic flux leakage, eddy current, radiography, thin-layer activation and gamma radiography, and electric field mapping); indirect online measurement techniques (hydrogen monitoring and corrosion potential); online water chemistry parameters (pH, conductivity, dissolved oxygen, redox potential, flow regime, and flow velocity); and process parameters (pressure, temperature, dew point, and fouling). In addition, indirect off-line techniques including water chemistry parameters—alkalinity, metal ion analysis, dissolved solids, gas analysis, residual oxidant, residual inhibitor, chemical analysis of process samples, sulfur content, total acid number, nitrogen content, and salt content of crude oil—have also been studied.

Chapter 3 deals with adsorption of corrosion inhibitors, Helmholtz double layer, types of inhibitors, adsorption isotherms, role of oxyanions and passive films, inhibition of localized corrosion, influence of environmental factors, and passivation of metals.

Chapter 4 is concerned with general aspects of inhibitors, factors pertaining to metal samples, sample preparation, environmental factors, concentration of inhibitors, process conditions, cooling systems, processing with acids, corrosion in the oil industry, reinforcing steel in concrete, corrosion inhibition in coal-water pipelines, mining operations, and atmospheric corrosion inhibition.

Chapter 5 studies the interface corrosion inhibition, structure, and stability of metal–inhibitor complexes, the Hammett equation, the Sastri equation, quantum chemical considerations, the hard and soft acid base principle, inhibitor field theory, photochemical corrosion inhibition, interphase and intraphase inhibition, passive oxide films, and interaction of anions with oxide films.

Chapter 6 discusses the industrial applications of corrosion inhibition such as reinforcing steel in concrete, coal-water slurries, cooling water systems, acid treatment, oxygen scavenging, coatings, and corrosion resistance of magnesium, aluminum, chromium, iron, nickel, copper, silver, zinc, cadmium, tin, lead, zirconium, and tantalum.

Chapter 7 finally deals with environmental testing, PARCOM guidelines, toxicity, chemical hazard assessment and risk management (CHARM) model, macrocyclic phthalocyanines and porphyrins, plant products, oleochemicals, rare earth metal compounds, barbiturates, hybrid coatings as green inhibitors, and sol-gel coatings in preventing corrosion in microelectronics.

ACKNOWLEDGMENTS

I wish to thank the following organizations for granting permission to reproduce the figures: National Association of Corrosion Engineers for Figs. 1.1–1.3, 1.5, 3.9–3.19, 5.11, 6.1, and 6.2; McGraw Hill for Figs. 1.10–1.12; Metal Samples Company for Fig. 2.23 (www.metalsamples.com); Common Ltd. for Fig. 2.26, and Kingston Technical Software for Fig. 2.29.

I wish to express my deep gratitude to my wife, Bonnie, for transcribing the text. I also wish to thank my daughter, Anjali, for imaging the figures and my son, Martin, for his moral support.

Finally, I wish to express my gratitude to the editorial and production staff of John Wiley & Sons, Inc. for their kind cooperation shown during writing, reviewing, and bringing out this book.

V. S. SASTRI
Sat Ram Consultants

Ottawa, Ontario, Canada
March 2011

1

INTRODUCTION AND FORMS OF CORROSION

1.1 DEFINITION

The term "corrosion" has its origin in Latin. The Latin word *rodere* means "gnawing," and *corrodere* means "gnawing to pieces." In daily life, corrosion manifests itself in many forms, such as corroded automobiles, nails, pipes, pots, pans, and shovels. Corrosion is a costly materials science problem. Metallic corrosion has been a problem since common metals were first put to use.

Most metals occur in nature as compounds, such as oxides, sulfides, silicates, or carbonates. Very few metals occur in native form. The obvious reason is the thermodynamic stability of compounds as opposed to the metals. Extraction of a metal from ore is *reduction*. The reduction of iron oxide with carbon as a reducing agent gives rise to metallic iron

$$2\,Fe_2O_3 + 3\,C \rightarrow 4\,Fe + 3\,CO_2,$$

while the oxidation of metallic iron to produce the brown oxide known commonly as "rust" is corrosion. The extraction of iron from the oxide must be conducted with utmost careful control of the conditions so that the reverse reaction is prevented.

Green Corrosion Inhibitors: Theory and Practice, First Edition. V. S. Sastri.
© 2011 John Wiley & Sons, Inc. Published 2011 by John Wiley & Sons, Inc.

1.2 DEVELOPMENTS IN CORROSION SCIENCE

During the Gupta Dynasty (320–480 CE), the production of iron in India achieved a high degree of sophistication, as attested by the Dhar Pillar, a 7-tonne (7000 kg), one-piece iron column made in the fourth century CE. The existence of this pillar implies that the production of iron from oxide ore was a well-established process, and the personnel involved in the production of the iron pillar were aware of the reverse reaction involving the oxidation of iron to produce iron oxide (the familiar rusting of iron) and of the need to minimize the extent of this reverse reaction.

Copper nails coated with lead were used by the Greeks in the construction of lead-covered decks for ships (1). The Greeks probably realized that metallic couples of common metals are undesirable in seawater. Protection of iron by bitumen and tar was known and practiced by the Romans.

TABLE 1.1 Timeline of Developments in Corrosion Science

L.J. Thénard	1819	Enunciated electrochemical nature of corrosion
Sir H. Davy	1829	Principle of cathodic protection
A. de la Rive	1830	Established best quality of zinc for galvanic batteries
M. Faraday	1834–1840	Established relations between chemical action and generation of electric currents based on what were later called "Faraday's laws"
S. Arrhenius	1901	Postulated the formation of microcells
W.R. Whitney A.S. Cushman	1903 1907	Confirmed the theory of microcells
W. Walker A. Cederholm L. Bent, W. Tilden	1907 1908	Established the role of oxygen in corrosion as a cathodic simulation
E. Heyn and O. Bauer	1908	Corrosion studies of iron and steel, both alone and in contact with other metals, leading to the concept that iron in contact with a nobler metal increased corrosion rate, while its contact with a base metal resulted in partial or complete protection
R. Corey T. Finnegan M. de Kay Thompson	1939 1940	Investigated attack of iron
A. Thiel Luckmann	1928	Investigated the attack of iron by dilute alkali with liberation of hydrogen
W. Whitman, R. Russell U. Evans G.V. Akimov	1924 1928 1935	Observed increased corrosion rate when a small anode is connected to a large cathode

Robert Boyle (1627–1691) published two papers, "Of the Mechanical Origin of Corrosiveness" and "Of the Mechanical Origin of Corrodibility" in 1675 in London (2). At the turn of the nineteenth century (3, 4), the discovery of the galvanic cell and Davy's theory on the close relationship between electricity and chemical changes (5) led to the understanding of some of the basic principles of corrosion.

Some of the developments in corrosion science are summarized in Table 1.1.

1.3 DEVELOPMENT OF SOME CORROSION-RELATED PHENOMENA

Some of the developments of corrosion-related phenomena are given in Table 1.2. The development of corrosion science in terms of published scientific literature through 1907–2007 is illustrated by the number of scientific papers given in Table 1.3.

Developments can be found in the scientific journals listed in Table 1.4. Some leading organizations championing corrosion science are detailed in Table 1.5.

TABLE 1.2 Development of Some Corrosion-Related Phenomena

J.S. MacArthur	1887	Process of cyanide dissolution of gold (gold is not soluble in hot acids)
P.F. Thompson	1947	Dissolution of gold in dilute cyanide solutions recognized as an electrochemical process
Concept of Passivity		
J. Keir	1790	Observed that iron in concentrated nitric acid altered its properties
C.F. Schönbein	1799–1868	Suggested the state of iron in concentrated nitric acid as passivity
W. Müller (K. Konopicky, W. Machu)	1927	Postulated the mathematical basis of the mechanism of anodic passivation
G.D. Bengough (J.M. Stuart, A.R. Lee, F. Wormwell)	1927	Systematic and carefully controlled experimental work on passivity
Role of Oxygen		
Marianini	1830	Indicated the electric currents were due to variations in oxygen concentration
Adie	1845	
Warburg	1889	
V.A. Kistiakowsky	1908	
	~1900	Hydrogen peroxide detected during the corrosion of metals
	~1905	The view that acids were required for corrosion to occur dispelled by observation of rusting of iron in water and oxygen
Aston	1916	Role of local differences in oxygen concentration in the process of rusting of iron
McKay	1922	Currents due to a single metal of varying concentrations of metal ion

(continued)

TABLE 1.2   (*Continued*)

U.R. Evans	1923	Differential aerations and their role in metallic corrosion
Evans and co-workers	1931–1934	Electric currents due to corrosion of metal in salt solutions were measured and a quantitative electrochemical basis of corrosion was propounded. The oxygen-rich region becomes cathodic and the metal is protected, while the lower oxygen region, being anodic, is attacked

Inhibitors

Roman civilization	–	Protection of iron by bitumen, tar, extracts of glue, gelatin, and bran were used to inhibit corrosion of iron in acid
Marangoni, Stefanelli	1872	Distinction between inhibitive paints and mechanically excluding paints made, based on laboratory and field tests Development of paints containing zinc dust
Friend	1920	Colloidal solution of ferric hydroxide acts as an oxygen carrier, passing between ferrous and ferric states
Forrest, Roetheli, Brown	1930	Protective property of coating varied and depended on the rate of supply of oxygen to the surface
Herzog	1936	Postulated that iron, on long exposure to water, becomes covered by magnetite overlaid by ferric hydroxide. Magnetite layer acts as cathode and ferric hydroxide is converted to hydrated magnetite. Hydrated magnetite may lose water and reinforce the preexisting magnetite or absorb oxygen from air to give ferric hydroxide.
Chyzewski	1938	Classified inhibitors as cathodic and anodic inhibitors
V.S. Sastri	1988	Classification of corrosion inhibition mechanisms as interface inhibition, intraphase inhibition, interphase inhibition, and precipitation coating (*Corrosion '88*, paper no. 155)
V.S. Sastri	1990	Modern classification of inhibitors as hard, soft, and borderline inhibitors (30)
K. Jüttner, W.J. Lorenz, F. Mansfeld	1993	*Reviews on corrosion inhibitor science and technology*, 1993
V.S. Sastri, J.R. Perumareddi, M. Elboujdaini	1994	Novel theoretical method of selection of inhibitors (*Corrosion*, **50**, 432, 1994)

TABLE 1.2 *(Continued)*

V.S. Sastri, J.R. Perumareddi, M. Elboujdaini	2005	Sastri equation relating percent inhibition to the fractional electronic charge on the donor atom in the inhibitor (6)
V.S. Sastri, J.R. Perumareddi, M. Elboujdaini, M. Lashgari	2008	Application of ligand field theory in corrosion inhibition (*Corrosion*, **64**, 283, 2008)
V.S. Sastri, J.R. Perumareddi, M. Elboujdaini	2008	Photochemical corrosion inhibition (*Corrosion*, **64**, 657, 2008)

Microbiological Corrosion

R.H. Gaines	1910	Sulfate-reducing bacteria in soils produce H_2S and cause corrosion

Role of thermodynamics

–	–	Corrosion of metals obeys the laws of thermodynamics; was recognized in the early development of corrosion science
M. Pourbaix	1940	Pourbaix diagrams involving pH and potential give regions of corrosion, immunity, and passivity

Kinetics

Evans, Hoar	1932	Quantitative correlation of corrosion rates with measured electrochemical reaction rates
F. Habashi	1965	Validity of single kinetic law irrespective of the metal, composition of the aqueous phase, and evolution of hydrogen when no insoluble products, scales, or films are formed

TABLE 1.3 Numbers of Corrosion Science Publications

Theme	1907	1950	2000	2007
Corrosion	35	922	10,985	15,903
Corrosion and protection	3	122	1,162	1,578
Corrosion inhibition	0	19	367	416

TABLE 1.4 Beginning Journal Years for Corrosion Developments

Title	Year
Corrosion	1945
Corrosion Science	1961
British Corrosion Journal	1965
Werkstoffe und Korrosion	1950
Corrosion Prevention and Control	1954
Anti-corrosion Methods and Materials	1962
Materials Performance	1962

TABLE 1.5   Organizations at Forefront of Corrosion Science Starting Year

American Society for Testing Materials (ASTM)	1898
American Society of Metals (ASM)	1913
Corrosion Division of the Electrochemical Society	1942
National Association of Corrosion Engineers	1943
Comité international de thermodynamique et cinétique	1949
électrochimique (CITCE)	
International Society of Electrochemistry (ISE)	1971
International Corrosion Council	1961
The Corrosion Group of the Society of Chemical Industry	1951
Belgium Centre for Corrosion Study (CEBELCOR)	1951
Commission of Electrochemistry	1952
National Corrosion Centre (Australia)	
Australian Corrosion Association	~1980
Chinese Society of Corrosion and Protection	~1980
National Association of Corrosion Engineers (in Canada)	–

Some of the research groups that became active in corrosion studies in the early stages are

Massachusetts Institute of Technology

National Bureau of Standards

Ohio State University

University of Texas

University of California, Los Angeles

National Research Council Canada

Cambridge University

Technical University, Vienna

Some industrial laboratories, such as U.S. Steel, International Nickel Company, Aluminum Company of America, and DuPont, initiated their own research in corrosion.

The advances made in the scientific approach and the degree of maturity attained will be obvious from the following two abstracts of papers. The abstract of a paper published by A.S. Cushman (*American Society for Testing Materials* **8**, 605, 1908) noted that, "the inhibitive power of some pigments on iron and steel were tested by agitating in water with a current of air and the loss in weight due to rusting was determined." It is instructive to compare this with a paper entitled "Selection of Corrosion Inhibitors." Its abstract is given below:

Data on the inhibition of corrosion of iron by methyl pyridines in HCl, H_2SO_4, and H_2S solutions, by *para*-substituted anilines in HCl solutions and by *ortho*-substituted benzimidazoles in HCl solutions, and on the inhibition of corrosion of aluminum by methyl pyridines have been analyzed in terms of the Hammett equation and in terms of a new equation relating the degree of inhibition with the fraction of the electronic charge due to the substituent in the inhibition molecule. The new relationship has been found to be useful in predicting new inhibitors offering a greater degree of inhibition than the currently known inhibitor systems. *Source:* (6).

The impact of corrosion is felt in three areas of concern—economics, safety, and environmental damage. Metallic corrosion, seemingly innocuous, affects many sectors of the national economy. The National Bureau of Standards (NBS) in collaboration with Battelle Columbus Laboratory (BCL) studied the costs of corrosion in the United States using the input/output model (7).

Elements of the costs of corrosion used in the model include those concerned with capital, design, and control, as well as associated costs. They are outlined next.

1.4 ECONOMICS OF CORROSION

Capital costs:

- Replacement of equipment and buildings
- Excess capacity
- Redundant equipment

Control costs:

- Maintenance and repair
- Corrosion control

Design costs:

- Materials of construction
- Corrosion allowance
- Special processing

Associated costs:

- Loss of product
- Technical support
- Insurance
- Parts and equipment inventory

The data resulting from the calculations using the I/O model are given in Table 1.6 Data for the year 2010 are estimates only.

The estimated costs of corrosion in Canada in 2010 along with the various sectors are given in Tables 1.7 and 1.8.

The cost of corrosion in other countries in the world is given in Table 1.9.

TABLE 1.6   Corrosion Costs in the United States (Billions of Dollars)

Industry		1975	1995	2010
All industries	Total	82.0	296	549
	Avoidable	33.0	104	194
Automotive	Total	31.4	94	167
	Avoidable	23.1	65	116
Aircraft	Total	3.0	13	25
	Avoidable	0.7	3	5.9
Others	Total	47.6	159	290
	Avoidable	9.3	36	70

TABLE 1.7   Corrosion Costs in Canada[8]

Sector	$ Billion
Utilities	10.5
Transportation	7.3
Infrastructure	5.5
Government	5.1
Production and manufacturing	4.4
Total	32.8

TABLE 1.8   Total Corrosion Costs in Canada

	$ Billion
Total direct cost of corrosion	32.8
Cost of corrosion (extrapolated to Canadian economy)	64.5
Estimated savings by corrosion control	19.5

TABLE 1.9   Corrosion Costs of Other Countries

Country	Year	Corrosion Costs ($)	Percent of GNP	Avoidable Cost ($)	Reference
United Kingdom	1969–1970	3.2 billion	3.5	0.73 billion	9
West Germany	1968–1969	1.5 billion	3.0	0.375 billion	10
Sweden	1964	58–77 million	–	15–19 million	11
Finland	1965	47–62 million	–	–	12
Russia	1969	6.7 billion	2.0	–	13
Australia	1973	470 million A$	1.5	–	14
India	1960–1961	320 million	–	–	15
Japan	1976–1977	9.2 billion	1.8	–	16

TABLE 1.10 Some Nations' Efforts to Combat Corrosion

United Kingdom		
National Corrosion Service	1975	Educating engineering undergraduate students in corrosion awareness. Published 15 guides on corrosion and 6 booklets on "controlling corrosion" for distribution. Published directory of personnel involved in corrosion prevention
National Corrosion Coordination Centre		Carrying out multiclient 50/50 cost-shared research on high-temperature corrosion, metal finishing, microbial corrosion, and expert systems in corrosion engineering
United States	~1980	National Association of Corrosion Engineers in collaboration with the National Bureau of Standards developed corrosion data program
Australia	1982	National Corrosion Centre. Established nationwide referral service for corrosion problems
Peoples Republic of China	1980	Educating and training of personnel in corrosion. Organized 15 corrosion courses. Established 11 institutes offering courses in corrosion
Canada	1994	A proposal to establish a national corrosion secretariat (19) to educate industrial personnel, operate a referral service, assemble a directory of corrosion experts, initiate site visits, show videos on corrosion, and establish 50/50 cost-shared projects in corrosion was submitted in 1994

The high costs of corrosion have a significant effect on the national economy, and therefore it is necessary that corrosion personnel adopt corrosion control measures in order to avoid corrosion losses. A useful report entitled "Economics of Corrosion" (17) has been produced by the National Association of Corrosion Engineers (NACE) Task Group T-3C-1. The report deals with (i) economic techniques that can be used by personnel as a decision-making tool; (ii) facilitating communications between corrosion scientists and management; and (iii) justifying the investments in corrosion prevention measures to achieve significant long-term benefits.

In general, corrosion costs amount to about 2–4% of GNP, and about 25% of the costs are avoidable when corrosion control measures are adopted. The measures taken to combat corrosion in the United Kingdom, United States, Australia, China, and Canada have been discussed (18). Some efforts to combat corrosion worldwide are given in Table 1.10.

1.5 SAFETY AND ENVIRONMENTAL CONSIDERATIONS

One of the most important impacts of corrosion is safety. While safety should be uppermost in the minds of industrial personnel, accidents do occur, in spite of great precautions. So, corrosion not only is expensive but also poses risks to human life and safety. An example, corrosion of iron hulls in ships and their resulting loss poses a threat to crew. Accidents are more likely to occur in chemical industries handling corrosive chemicals releasing cyclohexene (Flixborough, England) and hydrogen cyanide (Bhopal, India) than in those that do not. Fatal airline accidents, bridge collapse, bursting of gas pipelines, failure of steam pipes in nuclear power plants—all have caused loss of life.

Corrosion can also impact the environment. Corrosion-related failure of oil or gas pipelines or oil tanks can have severe detrimental effects on the environment in the form of water and air pollution, leading to the demise of aquatic life. Corrosion-related accidents can, in principle, destroy irreplaceable flora or fauna. Another aspect is the corrosion's effects on limited resources. Some decades ago, recycling was accorded scant attention. At present, it is widely practiced and recycling of metal products, paper, and plastics is commonplace, since recycling helps to conserve limited and finite resources. Corrosion prevention and protection arrests the degradation of metals and materials, and hence contributes in a significant way to the conservation of resources with minimum damage to the ecosystem.

1.6 FORMS OF CORROSION

Corrosion can be defined in general terms and of universal applicability or in specific terms depending upon the perspective from which it is defined. For instance, corrosion in aqueous media is defined as an electrochemical process. In more general terms, corrosion is defined as the degradation of material caused by an aggressive environment. The corrosive environment can be water, air, carbon dioxide, organic liquids, molten salts, or gaseous sulfur. Some less common corrosive environments are neutron beams, ultraviolet light, nuclear fission fragments, and gamma radiation.

Materials subject to corrosion include engineering materials, such as metals, plastics, rubber, and ionic and covalent solids; aggregates such as concrete, composite materials; and wood. The present discussion is concerned with metals, alloys, and aggregates. Corrosion can manifest in many forms, such as uniform or general corrosion, galvanic corrosion, crevice corrosion, pitting corrosion, intergranular corrosion, selective leaching, erosion corrosion, stress corrosion, corrosion fatigue, and fretting corrosion. The eight forms of corrosion defined by Fontana are general corrosion, pitting corrosion, intergranular corrosion, parting, galvanic corrosion, crevice corrosion, stress-corrosion cracking (SCC), and erosion corrosion.

Classification of the different forms of corrosion may be based on intrinsic and extrinsic modes. Intrinsic modes of corrosion independent of design are general corrosion, pitting, intergranular corrosion, parting, and stress-corrosion cracking. Extrinsic modes of corrosion affected by design are crevice or underdeposit corrosion, galvanic corrosion, erosion corrosion, fretting corrosion, and corrosion fatigue.

The forms of corrosion have been identified based on the apparent morphology of corrosion, the basic factor influencing the mechanism of corrosion in each form. Thus, the six forms of corrosion are as given in Table 1.11.

TABLE 1.11 Morphological Classification of Corrosion

1. General corrosion	Uniform, quasi-uniform, and nonuniform corrosion, galvanic corrosion
2. Localized corrosion	Pitting corrosion, crevice corrosion, filiform corrosion
3. Metallurgically influenced corrosion	Intergranular corrosion, sensitization, exfoliation, dealloying
4. Microbiological corrosion	
5. Mechanically assisted corrosion	Wear corrosion, erosion corrosion, corrosion fatigue
6. Environmentally induced cracking	Stress-corrosion cracking, hydrogen damage, embrittlement, hydrogen-induced cracking, high-temperature hydrogen attack, hot-cracking, hydride formation, liquid metal embrittlement, solid metal-induced embrittlement

1.6.1 General Corrosion

General corrosion can be even or uneven and is the most common form of corrosion. It is characterized by a chemical or electrochemical reaction that takes place on the exposed surface. The metal becomes thinner and eventually results in perforation and failure. General corrosion accounts for the greatest loss of metal on a tonnage basis. This mode of corrosion does not present a great threat from a technical standpoint since the life of the equipment can be estimated from the corrosion rates obtained from immersion of the sample material in the medium of interest. The corrosion rate data may then be used in the design of the equipment. General corrosion can be prevented or reduced by the proper choice of materials or by use of corrosion inhibitors or cathodic protection (Fig. 1.1).

1.6.2 Galvanic Corrosion

Galvanic corrosion occurs when a potential difference exists between two dissimilar metals immersed in a corrosive solution. The potential difference results in the flow of

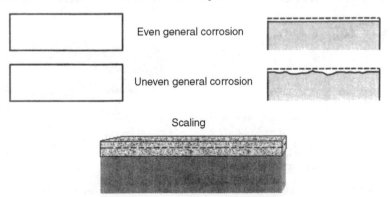

FIGURE 1.1 General corrosion and high-temperature scaling.

electrons between the metals. The less corrosion-resistant metal becomes the anode, and the more corrosion-resistant metal the cathode. Galvanic corrosion is generally more prominent at the junction of two dissimilar metals and the severity of attack decreases with increasing distance from the junction. The distance affected depends upon the conductivity of the solution. The cathode-to-anode ratio plays an important role in this form of corrosion. Severe galvanic corrosion occurs when a cathode of a large surface area and an anode of a small surface area are involved. Corrosion of leaded or nonleaded solders in copper pipes carrying drinking water is an example of this form of corrosion.

Some preventive measures to combat galvanic corrosion are (a) selection of metals that are close to each other in galvanic series; (b) maintenance of cathode/anode surface area ratio to the smallest possible minimum; (c) providing insulation between the two dissimilar metals; (d) use of coatings that are kept in good condition; (e) use of corrosion inhibitors to reduce the corrosivity of the medium; (f) avoiding threaded joints between the two dissimilar metals; (g) use of a suitable design such that replacement of anodic parts is easy; and (h) use of a third metal that is anodic to both the metals in galvanic contact. Galvanic corrosion of mild steel elbow fixed on copper pipe is illustrated in Fig. 1.2.

1.6.3   Crevice Corrosion

Crevice corrosion usually occurs within crevices and shielded areas on metal surfaces in contact with corrosive media. This type of corrosion is generally associated with small volumes of stagnant solution trapped in holes, gasket surfaces, lap joints, surface deposits, and crevices under bolt and rivet heads. Crevice corrosion is also known as deposit or gasket corrosion. Some of the deposits that cause crevice corrosion are sand, dirt, corrosion products, and other solids. Contact between a metal and nonmetallic surface such as a gasket can result in crevice corrosion. Stainless steels in particular are prone to crevice corrosion. The mechanism of crevice corrosion consists of the oxidation of the metal and the reduction of oxygen, yielding hydroxyl ion. After some time, the oxygen in the crevice is consumed and converted into hydroxyl ion. The metal continues to be attacked and the excess positive charge is

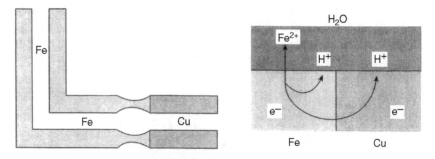

FIGURE 1.2   Galvanic corrosion of mild steel elbow connected to a copper pipe.

balanced by the migration of chloride anion from the bulk into the crevice. Thus, ferric or ferrous chloride builds up in the crevice. The ferric chloride can hydrolyze in the crevice and give rise to iron hydroxide and hydrochloric acid. The pH in the crevice can fall to 2 or 3 along with a high concentration of chloride.

Some preventive measures against crevice corrosion are (a) use of welded butt joints in place of riveted or bolted joints; (b) closure of crevices in lap joints by continuous welding; (c) caulking or soldering; (d) vessel design that allows complete drainage without stagnation; (e) removal of solid deposits; (f) use of nonabsorbent gaskets such as Teflon; and (g) flushing of the equipment with an inhibitor solution.

1.6.4 Pitting Corrosion

Pitting corrosion is a form of localized attack that results in localized penetration of the metal. This is one of the most destructive and insidious forms of corrosion. Pitting can cause equipment failure due to perforation, accompanied by a small percentage weight loss of the whole structure. Areas where a brass valve is incorporated into steel or galvanized pipeline are prone to pitting corrosion. The junction between the two areas is often pitted, and if the pipe is threaded, the thread in close contact with the brass valve pits rapidly, resulting in a leak. This occurs frequently in industry, homes, and farms. Pitting corrosion is difficult to measure in laboratory tests because of the varying number of pits and the depth under identical conditions. Pitting attack usually requires several months or a year to show up in service. Because of the localized and intense nature, pitting corrosion failures may occur suddenly. Pits usually grow in the gravitational direction. Most pits develop and grow downwards from horizontal surfaces. Pitting usually requires an extended initiation period on the order of months to years. Pitting can be considered a unique type of anodic reaction as well as an autocatalytic type of process. The metal in the pit dissolves along with the reduction of oxygen, as is the case with crevice corrosion. The rapid dissolution of metal in the pit results in a buildup of excessive positive charge in the pit followed by the migration of chloride ions into the pits to maintain the electroneutrality condition. Because of the high ionic potential of ferrous or ferric ion, hydrolysis of a ferric or ferrous ion results in lowering of the pH in the pit, which, together with the high chloride ion concentration in the pit, increases the corrosion rate.

The preventive measures cited for crevice corrosion also apply to pitting corrosion. The pitting corrosion resistance of some commonly used metals and alloys is in the order:

Titanium > Hastealloy C > Hastealloy F > Type 316 stainless steel > Type 304 stainless steel.

The addition of molybdenum to Type 304 stainless steel was found to improve the resistance of the steel to pitting corrosion. Localized corrosion morphologies are given in Fig. 1.3, which illustrates pitting and crevice corrosion.

Localized attack

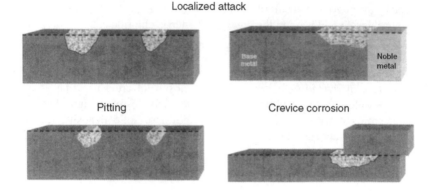

Pitting   Crevice corrosion

FIGURE 1.3   Forms of localized corrosion.

1.6.5 Dealloying or Selective Leaching

Selective leaching is the removal of an element from an alloy by corrosion. Selective removal of zinc from brass, known also as "dezincification," is a prime example of this form of corrosion. Selective leaching has also been observed with other alloys in which iron, aluminum, cobalt, and chromium are removed selectively. Selective leaching of zinc from yellow brass containing 30% zinc and 70% copper is a common example. The removal of zinc can be uniform or localized. As a general rule, the uniform type of dezincification is observed in brasses containing high zinc in an acidic medium. A localized type of dezincification is commonly observed in low zinc brasses exposed to a neutral, alkaline, or mildly acidic medium.

The addition of small amounts of arsenic, antimony, phosphorus, or tin to 70/30 brass resulted in minimizing dezincification. The added minor elements minimize dezincification of brass by forming a protective film on the brass. Brasses suffer severe corrosion in aggressive environments and hence cupronickel alloys (70–90% Cu, 30–10% Ni) are used in place of brasses. Gray cast iron is known to exhibit selective leaching, giving the appearance of graphite. This type of attack is known as graphitic corrosion. Corrosion inhibitors have been used in the inhibition of dezincification of brasses. Uniform dealloying in admiralty brass is illustrated in Fig. 1.4.

1.6.6 Intergranular Corrosion

This form of corrosion consists of localized attack at—and adjacent to—grain boundaries, causing relatively little corrosion of grains, but resulting in disintegration of the alloy and loss of strength. The impurities at the grain boundaries, enrichment of one of the alloying elements, or depletion of one of the elements in the grain boundary areas causes intergranular corrosion. This form of corrosion has been observed in the case of failures of 18-8 stainless steels. The 18-8 stainless steels on heating to temperatures of 950–1450°F fail due to intergranular corrosion. It is surmised that depletion of chromium in the grain-boundary location of the steels results in

FIGURE 1.4   Dezincification of a bolt in brass.

intergranular corrosion. When the carbon content of the steels is 0.02% or higher, the chromium carbide ($Cr_{23}C_6$), being insoluble, precipitates out of the solid solution and results in depletion of chromium in areas adjacent to grain boundaries. The chromium carbide remains unattacked while the chromium-depleted areas near the grain boundary corrode. Intergranular corrosion of austenitic stainless steels can be controlled or minimized by quench annealing or by the addition of small amounts of niobium or tantalum, which form carbides readily, or by lowering the carbon content of the steel to less than 0.02%. By heating the steel to 1950–2050°F followed by water quenching, the chromium carbide dissolves, resulting in a more homogeneous alloy resistant to corrosion.

Mechanically assisted corrosion consists of (i) erosion corrosion, (ii) cavitation damage, (iii) fretting corrosion, and (iv) corrosion fatigue. These four forms of mechanically assisted corrosion are illustrated in Fig. 1.5. Erosion corrosion consists of the increase in attack of a metal due to the relative movement between a corrosive medium and the metal surface. The rapid movement or flow of the medium results in mechanical wear. The metal is removed from the surface in the form of dissolved ions or in the form of solid corrosion products, which are mechanically swept from a surface. Erosion corrosion of a metal appears in the form of grooves. Erosion corrosion is observed in piping systems such as bends, elbows, tees, valves, pumps, blowers, centrifuges, propellers, impellers, agitators, heaters and condensers, turbine blades, nozzles, wear plates, grinders, mills, and baffles. All types of equipment exposed to moving fluids are prone to erosion corrosion.

Some of the factors involved in erosion corrosion are (a) the nature of the surface films formed on the metal surface, (b) velocity of the moving fluid, (c) amount of turbulence in the liquid, (d) impingement, (e) the galvanic effect, (f) chemical composition, (g) hardness, (h) corrosion resistance, and (i) the metallurgical history of the metals and alloys.

In general, increased velocity results in increased erosion corrosion, and this effect is more pronounced beyond a critical velocity. The increase or decrease in erosion

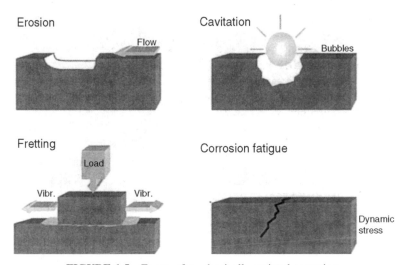

FIGURE 1.5   Forms of mechanically assisted corrosion.

corrosion with increase in flow of the medium depends upon the corrosion mechanisms involved. The increase in velocity increases the supply of oxygen, carbon dioxide, or hydrogen sulfide, resulting in greater attack of the metal. Unlike laminar flow, turbulence causes greater agitation of liquid at the metal surface, as well as more intimate contact between the metal and the environment. Turbulent flow occurs in the inlet ends of tubing in condensers and heat exchangers. Turbulence causes the erosion corrosion in impellers and propellers.

Failure due to impingement attack is a common failure mode where the aqueous medium is forced to change its direction, as it does in bends, tees, cyclones, and inlet pipes in tanks. Galvanic corrosion can accentuate erosion corrosion in a synergistic manner. At high velocities, the erosion-corrosion attack on steel coupled to copper or nickel becomes significant and can be attributed to the galvanic effect.

Stainless steels form passive oxide film and, as a result, offer resistance to corrosion. A hard, dense, and adherent film gives good protection from erosion corrosion, since the film cannot be displaced by mechanical forces. Titanium has probably the best resistance to erosion corrosion due to the titanium oxide film formed on the surface.

A noble metal has inherent corrosion resistance and, as a result, gives better performance when all other conditions are the same. An example of this is the better performance of 80/20 Ni/Cr alloy than 80/20 Fe/Cr alloy, because nickel has better resistance to corrosion than iron. The addition of a third metal such as iron to cupronickel improves the resistance to erosion corrosion in seawater. The addition of up to 13%chromium to steel and iron alloys results in increase in resistance to erosion corrosion in acid mine waters.

Some of the measures for prevention of erosion-corrosion damage are proper selection of materials with good resistance to erosion corrosion, proper design, coatings, cathodic protection, and reduction in the degree of the aggressive nature of

FIGURE 1.6 Erosion-corroded pump propeller.

the environment. The most economical solution is the selection of a suitable alloy with high resistance to erosion corrosion. Increasing pipe diameter reduces the flow velocity, giving rise to laminar flow, and streamlining bends reduces the impingement effects. Filtration of solids in the medium and lowering the temperature of the environment are useful in reducing erosion-corrosion damage. Hard facings and repair of damaged areas by welding as well as use of zinc plugs in pumps are some of the ways of combating erosion corrosion. The erosion-corrosion damage of a pump propeller in copper/tin/zinc alloy in the form of grooves, gullies, rounded holes, and valleys is shown in Fig. 1.6.

1.6.7 Cavitation Damage

Cavitation damage is a special form of erosion corrosion caused by the formation of vapor bubbles and their collapse in the liquid near a metal surface. This form of damage occurs in hydraulic turbines, pump impellers, and other surfaces that are in contact with high-velocity liquid flow along with high pressures. The formation and collapse of water vapor bubbles can produce shock waves with pressures as high as 60,000 lb/in^2, which can cause plastic deformation of metals.

Cavitation damage resembles pitting corrosion, except that the pitted areas are closely spaced, resulting in a very rough surface. Cavitation damage can be reduced by selection of more corrosion-resistant metals or alloys, choosing a design to minimize hydrodynamic pressure drops, or using pump impellers with a smooth finish and cathodic protection. Cavitation erosion damage of a cylinder liner of a diesel engine is shown in Fig. 1.7.

FIGURE 1.7   Cavitation damage in cylinder liner of a diesel engine.

1.6.8   Fretting Corrosion

Fretting corrosion is a combination of wear and corrosion in which material is removed from contacting surfaces when the motion between the surfaces is limited to very small amplitude oscillations. The oscillatory motion between the contacting surfaces in the tangential direction results in fretting corrosion. Oxidation is the primary factor in the fretting process. In oxidizing systems, fine metal particles removed by adhesive wear are oxidized and trapped between the fretting surfaces. The oxides act like an abrasive and increase the rate of removal of the material.

Some samples of vulnerable components are shrink-fits, bolted parts, and spines. The contacts between hubs, shrink- and press-fits and bearing houses on loaded rotating shafts or axles and many parts of vibrating machinery are prone to fretting corrosion. Fretting corrosion can destroy electrical contacts between gold and copper parts in electronic equipment. Fretting is also encountered in a stainless steel spacer used to mount a control rudder on the space shuttle. Fretting corrosion also occurs in bolted fish plates used in railway tracks and in loosening of wheels and turbines from the attached shafts.

Fretting corrosion may be controlled by using low viscosity, high tenacity oils and greases and phosphate coatings in conjunction with lubricants. Increasing the hardness of one or both the parts in contact, increasing the surface hardness by shot-peening or cold-working, use of lead coatings, use of gaskets, increasing the relative motion between the parts, use of laminated plastic on gold plate, and cast iron-on-cast iron with phosphate coating or rubber cement coating are some methods useful in combating fretting corrosion damage.

1.6.9 Corrosion Fatigue

Corrosion fatigue consists of cracking of materials under the combined action of fluctuating or cyclic stress and a corrosive environment. While purely mechanical fatigue failure occurs above a critical cyclic stress value known as the fatigue limit, corrosion fatigue can occur at stresses lower than the fatigue limit. The damage due to corrosion fatigue is usually greater than the sum of the damage due to corrosion and fatigue acting separately. The fracture tension as a function of the number of cycles in purely mechanical fatigue and in corrosion fatigue is illustrated in Fig. 1.8.

Corrosion fatigue differs from stress-corrosion cracking in the sense that it occurs in most of the aqueous media. The mechanism consists of exposure of oxide-free, cold-worked metal by extrusion of slip bands in grains of the metal surface. These parts of the metal become anodic and result in grooves, which develop into transcrystalline cracks.

Corrosion fatigue is affected by oxygen content, pH, temperature, and the composition of the solution. Iron, steel, stainless steels, and aluminum bronzes have good resistance to corrosion fatigue in water. On the other hand, aluminum bronzes and austenitic stainless steels have only 70–80% of normal fatigue resistance in seawater. High-chromium alloys have only 20–40% of fatigue resistance in seawater.

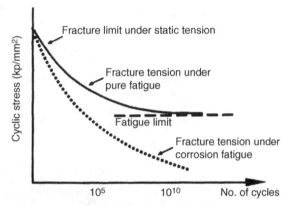

FIGURE 1.8 Fracture tension as a function of cycles in mechanical fatigue and corrosion fatigue.

Corrosion fatigue is generally defined in terms of the alloy and the environment. Corrosion fatigue may be eliminated or minimized by changes in design of the equipment, or by heat treatment to relieve stress or shot-peening of the metal surface to induce compressive stresses. Corrosion fatigue may also be minimized by the use of corrosion inhibitors and also by the use of electrodeposited zinc, chromium, nickel, copper, and nitride coatings. A typical corrosion fatigue crack through a sample of mild steel exposed to flue gas condensate is shown in Fig. 1.9.

1.6.10   Stress-Corrosion Cracking

Stress-corrosion cracking involves cracking of susceptible material caused by the simultaneous presence of tensile stress and a specific corrosive environment as shown in Fig. 1.10. During SCC the metal may be virtually unattacked over most of the surface, but fine cracks progress through the metal. Stress-corrosion cracking has serious consequences, because it can occur at stresses within the range of typical design stress. The important factors that influence this mode of damage are temperature, composition of the solution and the metal or alloy, stress, and the structure of the metal. Two typical examples of stress-corrosion cracking are the seasonal cracking of brass and the caustic embrittlement of steel. Seasonal cracking of brass cartridge cases

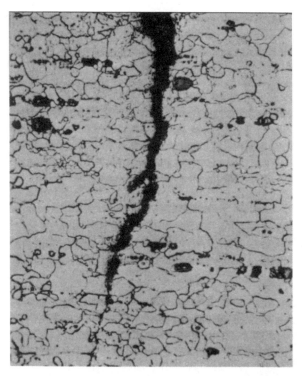

FIGURE 1.9   Corrosion fatigue crack in mild steel sheet exposed to flue gas condensate.

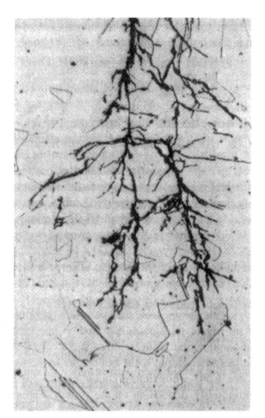

FIGURE 1.10 Transgranular stress-corrosion cracking in 304 steel.

develop cracks during heavy rainfall, coupled with hot temperatures in the tropics. Carbon steels subjected to stress close to the elastic limit and exposed to hot concentrated alkali solutions on nitrate solutions are susceptible to stress-corrosion cracking.

Stress-corrosion cracks appear like brittle mechanical fracture, although the cracks are due to corrosion. Intergranular SCC occurs along grain boundaries, and transgranular cracking occurs without preference to grain boundaries. Transgranular stress-corrosion crack propagation is generally discontinuous on the microscopic scale and occurs by periodic jumps on the order of micrometer (Fig. 1.10), while intergranular cracks propagate continuously or discontinuously, depending on the system (Fig. 1.11). Both types of cracking have been observed in high-nickel alloys, iron/chromium alloys, and brasses. In general, cracking occurs perpendicular to the applied stress. The type of crack, either single or branching, depends upon the structure and composition of the metal and the composition of the environment. For a given metal or alloy and a particular environment, a threshold stress can be defined, above which cracking occurs. The source of the stress can be applied, residual, thermal, welding, or corrosion products in constricted regions.

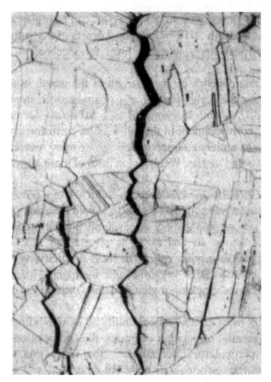

FIGURE 1.11   Intergranular stress-corrosion cracking in brass.

The time required for cracking is of importance. As the stress-corrosion cracks penetrate the metal, the cross-sectional area decreases, and the cracking failure occurs due to mechanical action. The cracking tendencies of the alloys can be defined only by carrying out long-term tests. Stress-corrosion cracking has been observed in the case of aluminum alloys in seawater or chloride solutions, copper alloys in ammoniac solutions, magnesium alloys in chloride dichromate, nickel in sodium hydroxide solutions, stainless steels in chloride solutions, and titanium alloys in seawater and fuming nitric acid solutions. Stress-corrosion cracking is influenced by metallurgical factors such as the chemical composition of the alloys, preferential orientation of the grains, composition and distribution of precipitates, degree of progress of phase transformation, and dislocations.

The precise mechanism of stress-corrosion cracking is not well known in spite of the large amount of work done on the subject. The main reason for this is the complex interplay of metal, tensile stress, and the corrosive environment. Corrosion plays an important role in the initiation of cracks. Any discontinuity like pits on the surface of the metal increases the stress. Stress concentration at the tip of the notch increases considerably as the radius of the notch decreases and cracks initiate at the base of the pit. The combined action of corrosion and stress in crack propagation was established by the observation that cathodic protection arrested crack propagation.

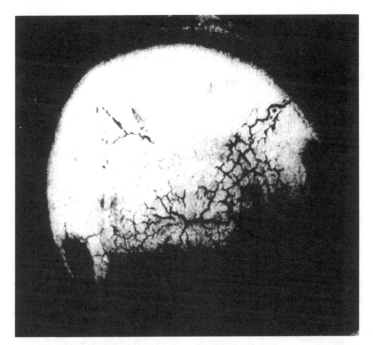

FIGURE 1.12   Stress corrosion of type 304 autoclave.

The methods used for the mitigation of SCC tend to be general because the exact mechanism of this form of degradation is not well understood. Annealing the alloy lowers the residual stress and hence lowers the propensity to stress-corrosion cracking. Carbon steels and austenitic steels can be annealed at 1100–1200°F and 1500–1700°F respectively, thereby lowering the tendency of the steels to suffer cracking. The corrosive environment can be made less aggressive by degasification or demineralization. Material selection such as substituting type 304 stainless steel by a more corrosion-resistant alloy such as Inconel can be done to mitigate the corrosive attack. Cathodic protection of the structure and addition of corrosion inhibitors to make the environment less corrosive are some methods to combat stress-corrosion cracking failures.

Stress-corrosion cracking damage of a type 304 stainless steel autoclave used under vapor condensation conditions in aqueous chloride solutions is depicted in Fig. 1.12.

1.7   CORROSION INHIBITION

In general, any corrosion retardation process can be considered corrosion inhibition. Corrosion inhibition has been achieved by the addition of a chemical compound that inhibits the oxidation of the metal. The chemical inhibitor added to the system may be in the form of a liquid or vapor or both.

TABLE 1.12   Number of Abstracts as a Function of Time

Year	Corrosion	Passivity	Corrosion Inhibitors
1908	22	2	1
1912	61	5	10
1957	>800	38	280
1994	>2,000	>100	>400
2007	15,903	–	416

The two steps involved in the action of the corrosion inhibitor are the transport of the inhibitor to the metal surface followed by the interaction of the inhibitor with the metal surface. The interaction of a corrosion inhibitor with a metal surface is similar to the interaction of a drug molecule with human physiology in that both involve transport of the moiety to the active site followed by the interaction with the site.

In early times, protection of iron by bitumen and tar was practiced by the Romans. At the turn of the nineteenth century, some of the fundamentals of the corrosion phenomena were understood (3, 4) soon after the discovery of the galvanic cell and the relationship between electricity and chemical changes (20). The basic electrochemical theory of corrosion accepted today is due to Wollaston (21). The use of corrosion inhibitors for the protection of metals can be traced to the last half of the nineteenth century. Marangoni and Stefanelli (22) used extracts of glue, gelatin, and bran to inhibit the corrosion of iron in acids. This and subsequent discoveries of effective inhibitors for metals were the result of empirical studies. The first patent given to Baldwin (British Patent 2327) consisted of the use of molasses and vegetable oils for pickling sheet steel in acids.

The progressive increase in the number of abstracts dealing with corrosion and corrosion inhibition is clear evidence of the increased activity in corrosion and corrosion inhibition studies (Table 1.12).

In early studies, the effect of inhibitors was tested by the determination of loss in weight of samples agitated in solutions containing inhibitors compared to solutions devoid of inhibitors. In earlier times, the selection of inhibitors was done on a trial and error basis. Intensive fundamental studies on corrosion inhibitors and the factors influencing their effectiveness have been in progress over the last 50 years. The advent of modern surface analytical techniques has made it possible to study the nature and composition of corrosion inhibitor films on metal surfaces. It is instructive to compare two abstracts of the work that appeared in 1908 and 1987. The abstract of the 1908 patent (22) reads

The Coslett Rust Preventing Process involving the immersion of clean steel in a boiling iron phosphate solution for 3 to 4 hours followed by drying at 100°C.

The abstract of the 1987 paper (23) states

Corrosion inhibition is a complex phenomenon and depends on the formation of protective layers on the metal surface. Of the many factors that affect the protective nature of the surface layers, the incorporation of the inhibitor in the surface layer is the most important. The manner in which the inhibitor is incorporated in the surface layer leads to three types of inhibition; namely, interface inhibition, interphase inhibition, and precipitation coating.

Determining the composition of the surface layers of metals is facilitated by modern surface analytical techniques, such as X-ray photoelectron spectroscopy (XPS), Auger electron spectroscopy (AES), secondary ion mass spectrometry (SIMS), reflectance spectroscopy (RS), and electron microprobe analysis (EMPA). The roles of these modern surface analytical techniques are discussed with respect to elucidating the interfacial inhibition of copper by mercaptobenzothiazole, interphase inhibition of mild steel AISI 1010 by oxyanions such as chromate, molybdate, and tungstate, and precipitation coating of AISI 1010 steel by oxalate in acid mine water.

Two recent papers (24, 25) with the titles "Selection of corrosion inhibitors based on structural features" and "Selection of inhibitors based on theoretical considerations" provide evidence for the advances that took place in the past 25 years.

The scientific basis of corrosion inhibition may well have begun with the recognition of the phenomenon of the adsorption of the inhibitor on a metal surface (26). Further *ortho*-substituted organic bases, such as *ortho*-toluidine, were found to be more effective corrosion inhibitors than the *meta*- and *para*-toluidines in oil field corrosion (27). The effect of the structure of organic bases such as polyethylene imines on corrosion inhibition was studied by varying the methylene rings and their influence on the CNC bond angles, which in turn affected the availability of the electron pair and hence the extent of corrosion inhibition.

The ready availability and capability of modern scientific instrumentation to study solid–liquid interfaces *in situ*, to analyze various species in solution, to monitor fast electrochemical reactions, and to determine inhibitor film composition and structure helped in the development of mechanistic models of corrosion inhibition such as the corrosion inhibition of copper by benzotriazole in a hydrochloric acid medium (28).

Hackerman proposed the type of bonding and the effect of the structure of the metal–inhibitor adsorbate on the strength and stability of the bond lead to the concept of a two-dimensional quasi-compound (29). The surface film retards either the anodic or the cathodic reaction, or both. The surface film may also impede the transfer of reactant on the product or influence the rate constant. The net effect is corrosion inhibition, although different principles are involved in the various processes.

Corrosion inhibitors that retard the cathodic reaction are known as cathodic inhibitors. Similarly, the inhibitors that retard the anodic reaction are known as anodic inhibitors. Some compounds that influence both the cathodic and the anodic reactions are known as mixed inhibitors. Corrosion inhibitors have also been classified as organic and inorganic inhibitors. Based upon the hard and soft acid and base principle (HSAB), the inhibitors have been termed as hard, soft, and borderline inhibitors (30). These concepts will be discussed in detail in the appropriate section.

In spite of the considerable progress made in the chemistry of inhibitors, the problem of inhibition in a corrosion system presents the same challenges today as it did in the past. The transport of the corrosion inhibitor from bulk solution to the surface of the metal and the active inhibitor species that is available to interact with the metal are the fundamental factors governing corrosion inhibition. Other equally important factors are the flow rate, temperature, solubility, inhibitor stability, surface adherence properties, and changes in corroding conditions that govern the effectiveness of corrosion inhibitors. Factors such as pit, temperature, flow rate of bulk fluid, and metal interface are of importance (31). Readily available modern computerized electrochemical and surface analytical instrumentation enables the understanding of the corrosion inhibition process to some extent. It is necessary to follow the corrosion inhibition process *in situ* by a combination of electrochemical and surface analytical probes to arrive at the actual mechanism of corrosion inhibition.

It is now useful to turn to the science of inorganic inhibitors and their action in the protection of metals from corrosion. The science of inorganic corrosion inhibitors has not reached the same level of advancement as that of organic inhibitors. The most common inhibitors, such as chromates used for corrosion inhibition of iron and its alloys in aqueous media of wide pH range and arsenates in neutral pH aqueous media, are considered unacceptable for use from an environmental point of view and are banned by governmental agencies. In most of the applications, environment-friendly, otherwise known as green inhibitors, such as molybdates and tungstates, have replaced chromates.

Other common inhibitors such as borates, carbonates, silicates, and phosphates continue to find applications. Some of the inorganic inhibitors, such as borates, offer corrosion protection by increasing the pH of the system while others such as phosphates form insoluble compounds on the metal surface to be protected. The mechanism of corrosion protection by oxyanions such as chromates, molybdates, and tungstates of mild steel AISI 1010 was studied by detailed electrochemical polarization along with a combination of surface analytical techniques such as X-ray photoelectron spectroscopy, auger electron spectroscopy, and electron microprobe analysis (32). These studies enabled the determination of the nature, composition, and thickness of the inhibitor films on the surface of the steel sample, which in turn threw light on the mechanism of corrosion inhibition.

The vast literature that is available on corrosion inhibitors and their use in different systems poses a daunting task in the selection and use of an inhibitor in a particular situation. The selection of inhibitors has been made on an empirical basis. The selection of corrosion inhibitors based on structural features (24) and theoretical considerations (25) are two significant approaches. With the advent of electronic computers, the selection of inhibitors can be done by using expert systems (33). In the normal course, the selection of an inhibitor from the literature is a time-consuming task, since the literature on the available inhibitors is vast and the choice depends on the metal to be protected, as well as the type of corrosive environment and the operating conditions.

With widespread access to computers, the expert's knowledge is stored and can be recalled easily and rapidly at will. The required information can be accessed in a number of ways, such as the database and the expert system. In the database system, data are stored in a structural manner and the user must have available all the important information about his or her system so as to be able to move logically through the database to extract the required data. When the expert system is used, the user is taken through the model by a series of questions from the model; the answers to the questions allow the appropriate advice to be given to the user.

Both the use of database and expert systems can result in the same conclusions, although the method of using each is different. The user must know important parameters of his system while using the database, while the expert system will elicit the required information by itself. Both the database and expert system have their own useful features. The scientist who wants to find an inhibitor for a particular application might not like the verbosity of the expert system, while an inexperienced user might find the database system unfriendly to use and also find the information possibly misleading.

In conclusion, it can be stated that the field of corrosion inhibitors is old and has developed into a mature science as a result of advances made in electrochemical and surface analytical aspects. It is interesting to note that inhibitors such as glue, gelatin, bran, and vegetable oils were used in corrosion inhibition of iron in acids during the nineteenth century. Incidentally, these inhibitors are nontoxic and can be considered environmentally acceptable green inhibitors. No consideration of environmental compatibility was shown with respect to the use of inhibitors until the period 1970–1980. The success achieved in the future will depend upon a balanced approach based on basic science, practice, and environmental considerations.

REFERENCES

1. J Cousteau, *Nat Geog Mag* **105**:1 (1954).

2. N Hackerman, In: RD Gundry (ed.), *Corrosion 93, Plenary and Keynote Lectures*, NACE International, Houston, Texas, 1993, pp. 1–5.

3. UR Evans, *Chem Ind* **986** (1952).

4. W Lynes, *J Electrochem Soc* **98C**:3–10 (1951).

5. H Davy, *Nicholson's J* **4**: 337 (1800).

6. VS Sastri, JR Perumareddi, M Elboujdaini, *Corros Eng Sci Technol* **40**:270 (2005).

7. LH Bennett, J Kruger, RI Parker, E Passaglia, C Reimann, AW Ruff, H Yakowitz, EB Berman, Economic Effects of Metallic Corrosion in the United States—A Three Part Study for Congress; (a) Part I NBS Special Publication 511-1, SD Stock No SN-003-003-01926-7; (b) Part II NBS Special Publication 511-2 Appendix B. A report to NBS by Battelle Columbus Laboratories, SD Stock No SN-003-003-01927-5, U.S. Govt. Printing Office, Washington, DC, 1978; (c) Part III, Appendix C, Battelle Columbus Input/Output Tables, NBS GCR 78-122, PB-279-430, National Technical Information Service, Springfield, VA, 1978.

8. VS Sastri, M Elboujdaini, JR Perumareddi, Economics of Corrosion and Wear in Canada. In: J Li, M Elboujdaini (eds.), *Proceedings of the International Conference on Environmental Degradation of Metals, Metallurgical Society of Canadian Institute of Mining*, Vancouver, BC, Canada, Aug 23–27, 2003.

9. TP Hoar, Report of the Committee on Corrosion and Protection, Department of Trade and Industry, H.M.S.O., London, UK, 1971.

10. D Behrens, *Br Corros J* **10**(3):122(1975).

11. KF Trädgäidh, *Tekn Tidskrift, Sweden* **95**(43):1191 (1965).

12. V Vläsaari, *Talouselämä Econ* **14/15**:351 (1965).

13. Y Kolotyrkin, *Sov Life* **9**:168 (1970).

14. RW Revie, HH Uhlig, *J Inst Eng Aust* **46**(3–4):3 (1976).

15. KS Rajagopalan, *The Hindu*, Madras, India, Nov 12, 1973.

16. Committee on Corrosion and Protection, *Corros Eng, Jpn* **26**(7):401 (1977).

17. NACE Technical Committee Report, Task Group T-3C-1 on Industrial Economic Calculational Techniques, Economics of Corrosion, NACE, Houston, Texas, 1972.

18. VS Sastri, *Corrosion Inhibitors*, John Wiley & Sons, Ltd., Chichester, UK, 1998, pp. 5–10.

19. VS Sastri, M Elboujdaini, JR Perumareddi, Economics of corrosion and wear in aerospace industry, *Proceedings of the International Conference of the Metallurgical Society of CIM on Wear and Corrosion of Metals*, Sudbury, Ontario, Canada, Aug 1997; *Proceedings of the International Conference on Aerospace Materials, Metallurgical Society of CIM*, Montreal, Aug 12–15, 2002. In: *CIM Bulletin* **90**(63), (1997).

20. H Davy, *Philos Trans R Soc* 114, 151, 242, 328 (1824).

21. WH Wollaston, *Philos Mag* **11**:206 (1801).

22. Anonymous, *Engineering* **85**:870 (1908).

23. VS Sastri, RH Packwood, JR Brown, Physical Metallurgy Laboratories, Canmet, Ottawa, Canada, PMRL 87-49 (OP-J), 1987.

24. JR Perumareddi, VS Sastri, PR Roberge, *Proceedings of the International Symposium on Materials Performance Maintenance*, Ottawa, Ontario, Canada, *30th Annual Conference of Metallurgists of CIM*, Pergamon Press, New York, p. 195.

25. VS Sastri, PR Roberge, JR Perumareddi, *Proceedings of the International Symposium on Materials Performance: Sulphur and Energy*, Edmonton, Alberta, Canada, Aug 23–27, 1992, *31st Annual Conference of Metallurgists of CIM*, Canadian Institute of Mining and Metallurgy, Montreal, p. 45.

26. CA Ulaamu, BE Lauer, CT Hultin, *Ind Eng Chem* **28**:159 (1936).

27. H Kaesche, N Hackerman, *J Electrochem Soc* **105**:191 (1958).

28. T Hashemi, CA Hogarth, *Electrochim Acta* **33**:1123 (1988).

29. N Hackerman, *Trans NY Acad Sci* **17**:7 (1954).

30. VS Sastri, PR Roberge, *Proceedings of the International Corrosion Congress*, Vol. 3, Florence, Italy, Apr 2–6, 1990, pp. 55–62.

31. N Hackerman, *Langmuir* **8**:922 (1987).

32. VS Sastri, RH Packwood, JR Brown, JS Bednar, LE Galbraith, VE Moore, *Br Corros J* **24**:30 (1989).

33. PA Brook, *Corros Prev Control* **36**:13 (1989).

2

ELECTROCHEMICAL PRINCIPLES AND CORROSION MONITORING

2.1 THERMODYNAMIC BASIS

It is useful to examine the corrosion process from the point of view of thermodynamics before we enter into a discussion of the electrochemical basis of corrosion. Thermodynamics deals with energy and its changes in reactions. According to thermodynamics, energy can be neither created nor destroyed, and free energy is released from the system to surroundings in all spontaneous reactions. Corrosion reactions are spontaneous and are governed by the laws of thermodynamics.

Consider the general reaction:

$$A + B \underset{k_b}{\overset{k_f}{\rightleftharpoons}} C + D,$$

where k_f and k_b are the forward and backward rate constants, respectively. According to transition state theory, A and B react to form an activated complex (A $\cdots$ B), which leads to products C and D under energetically favorable conditions. The energy profile of the reaction is depicted in Fig. 2.1.

The energy difference between the reactants and the products is ΔG, with a negative value since the reaction is spontaneous. The reactants have to surmount the energy barrier $\Delta G^{\ddagger}$ to give the products C and D. The energy barrier arises from the internal energy of A and B. The activated complex (A $\cdots$ B) can revert to A and B or

Green Corrosion Inhibitors: Theory and Practice, First Edition. V. S. Sastri.
© 2011 John Wiley & Sons, Inc. Published 2011 by John Wiley & Sons, Inc.

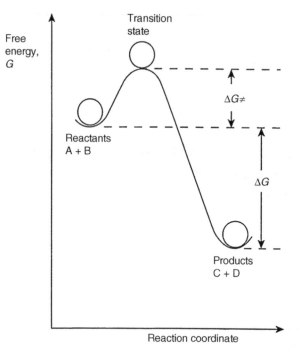

FIGURE 2.1   Energy profile of a reaction.

C and D, and the latter is favored energetically, since the reaction considered is spontaneous and the activated complex (A $\cdots$ B) attains lower energy by giving products C and D.

$$\text{Rate of forward reaction}\quad = k_f[A][B],$$
$$\text{Rate of backward reaction} = k_b[C][D].$$

At equilibrium $k_f[A][B] = k_b[C][D]$, and the equilibrium constant K is given as

$$K = \frac{k_f}{k_b} = \frac{[C]\,[D]}{[A]\,[B]}.$$

The value of K is large when $k_f \gg k_b$ and small when $k_f \ll k_b$.

The Arrhenius equation for the temperature dependence of the forward reaction may be written as

$$k_f = Ae^{-\Delta G^{\ddagger}/RT},$$

where A and R are constants and $\Delta G^{\ddagger}$ is the activation energy. The rate constant K increases with increase in temperature. For the reverse reaction, involving conversion of C and D into A and B, the activation energy required will be greater than $\Delta G^{\ddagger}$, and the reverse reaction is not favored.

From thermodynamics we have the relationship between free energy and equilibrium constant for a reaction:

$$\Delta G^{\circ} = -RT \ln K = -2.303RT \log K.$$

The value of ΔG° for the reaction can be calculated by the equation

$$\Delta G^{\circ}_{298} = [G^{\circ}_{298}(C) + G^{\circ}_{298}(D)] - [G^{\circ}_{298}(A) + G^{\circ}_{298}(B)].$$

It is possible to obtain the value of ΔG° for a reaction when the equilibrium constant K is known. It is also possible to calculate ΔG° using G° values given in thermochemical tables for reactants and products in a reaction. A negative value for ΔG° of a reaction indicates that the forward reaction is favored. Thus, the values of ΔG° should be negative for the corrosion reactions to occur. For copper and gold, the oxidation reactions are

$$Cu + H_2O + \tfrac{1}{2}O_2 \rightarrow Cu(OH)_2, \quad \Delta G^{\circ} = -119 \, kJ/mol,$$

$$Au + \frac{3}{2}H_2O + \frac{3}{4}O_2 \rightarrow Au(OH)_3, \quad \Delta G^{\circ} = +66 \, kJ/mol.$$

The ΔG° values of -119 and $+66$ kJ/mol predict that the corrosion of copper can occur, while gold will not corrode. These predictions are borne out by the familiar tarnishing of copper and nontarnishing of gold. The free energy values indicate whether the reactions as written will occur or not; they do not give any idea as to the rates of the reactions. The rates of the corrosion reactions are governed by the kinetics. The environment to which the metal is exposed has a decisive role in determining the occurrence of corrosion of the metal. This is exemplified by the resistance of gold to mineral acid attack and atmospheric oxidation, but it dissolves in aerated cyanide solutions. The role of the environment in corrosion phenomena is further exemplified by the rusting of iron by atmospheric corrosion and the preservation (free from corrosion) of iron, effected by keeping the metal in peat bogs.

2.2 NATURE OF CORROSION REACTIONS

Consider the system in which metallic iron is immersed in a copper sulfate solution. After some time, metallic copper is formed. This process is known as the cementation reaction. The various species present initially and after a lapse of time may be written as follows:

$$\text{Initial: } Fe^{\circ} \rightarrow Cu^{2+},$$
$$\text{Final: } Fe^{2+} \rightarrow Cu^{\circ}.$$

The reactions occurring may be written as

$$\text{Oxidation (anodic)}: Fe^\circ \rightarrow Fe^{2+} + 2e^-,$$
$$\text{Reduction (cathodic)}: Cu^{2+} + 2e^- \rightarrow Cu^\circ,$$
$$\text{Overall reaction}: Fe^\circ + Cu^{2+} \rightarrow Fe^{2+} + Cu^\circ.$$

The equilibrium constant K and the free energy change in the overall cementation reaction may be written as

$$\Delta G = \Delta G^\circ + RT \ln \frac{[Fe^{2+}][Cu^\circ]}{[Fe^\circ][Cu^{2+}]}.$$

Since the corrosion of iron in copper sulfate solution involves oxidation and reduction reactions with exchange of electrons, the reaction must involve electrochemical potential difference, related to the equilibrium constant, and may be written as (Faraday's law):

$$\Delta G^\circ = -nFE^\circ,$$

where F (Faraday) $= 96,494$ coulombs, E is the potential difference, and n is the number of electrons transferred.

Neglecting solids, one can write for the reaction of iron in copper sulfate solution

$$-nFE = -nFE^\circ + RT \ln \frac{[Fe^{2+}]}{[Cu^{2+}]}.$$

Division by nF leads to

$$E = E^\circ - RT \ln \frac{[Fe^{2+}]}{[Cu^{2+}]}.$$

In general terms, one can write

$$E = E^\circ - RT \ln \frac{[\text{products}]}{[\text{reactants}]} = E^\circ - \frac{0.059}{n} \log \frac{[\text{products}]}{[\text{reactants}]}.$$

This equation is known as the Nernst equation and is extensively used in electrochemical measurements. Under equilibrium conditions $E = E^\circ$, and the E° values obtained experimentally can be used as criteria that tell us whether the reaction will occur or not.

2.3 STANDARD ELECTRODE POTENTIALS

Having established the criterion for the oxidation of a metal in terms of oxidation potential and its sign and magnitude, it is useful to turn to experimental determination of the oxidation potentials. The oxidation potentials are obtained by measuring against a standard hydrogen electrode (SHE), consisting of a platinum electrode immersed in 1 M HCl with hydrogen gas passing through it at 1 atmosphere pressure. The reaction is

$$2H^+ (1\ M) + 2e^- \rightarrow H_2(1\ atm), \quad E^\circ = 0\ V.$$

The hydrogen electrode is known as standard hydrogen electrode, with a reduction potential of zero. Now we connect the zinc electrode system to a standard hydrogen electrode system, with a salt bridge as shown in Fig. 2.2.

The cell diagram for the system is as follows:

$$Zn_{(s)}||Zn^{2+}(1\ M)||H^+(1\ M)||H_2(1\ atm)|Pt_{(s)}.$$

Zinc is the anode

$$Zn_{(s)} \rightarrow Zn^{2+}_{(aq)} + 2e^-\ oxidation\ E^\circ_{Zn/Zn^{2+}} = 0.76\ V.$$

At the platinum electrode,

$$2H^+_{(aq)} + 2e^- \rightarrow H_{2(g)}\ reduction\ E^\circ_{H^+/H_2} = 0.0\ V.$$

When Zn^{2+} is 1 M, H^+ is 1 M, and $H_{2(g)}$ is at 1 atm (i.e., standard state conditions), the EMF of the cell is 0.76 V at 25°C.

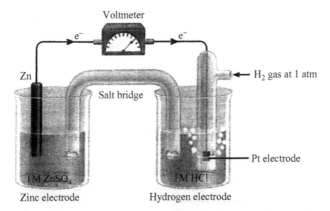

FIGURE 2.2 An electrochemical cell consisting of zinc and hydrogen electrodes.

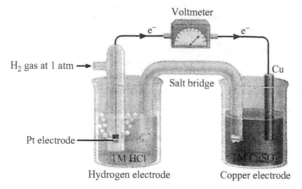

FIGURE 2.3   An electrochemical cell consisting of copper and hydrogen electrodes.

The standard oxidation potential of zinc is 0.76 V, and the overall reaction is the sum of oxidation and reduction potentials. The standard electrode potential of copper can be obtained by connecting a copper electrode to a hydrogen electrode through a salt bridge and a voltmeter as shown in Fig. 2.3.

In this system copper is the cathode, since reduction occurs.

$$Cu_{aq}^{2+} + 2e^- \rightarrow Cu_{(s)}.$$

The cell diagram for the system may be written as

$$Pt_{(s)}|H_2(1\ atm)|H^+(1\ M)||Cu^{2+}(1\ M)|Cu_{(s)}.$$

The half-cell reactions are

Anode (oxidation): $H_2(1\ atm) \rightarrow 2H^+(1\ M) + 2e^-$, $E^\circ_{H_2/H^+}$,
Cathode (reduction): $Cu^{2+}(1\ M) + 2e^- \rightarrow Cu_{(s)}$, $E^\circ_{Cu^{2+}/Cu}$,
Overall: $H_2(1\ atm) + Cu^{2+}(1\ M) \rightarrow 2H^+(1\ M) + Cu_{(s)}$, E°_{cell}.

The standard potential is 0.34 V at 25°C.

$$E^\circ_{cell} = E^\circ_{H^2/H^+} + E^\circ_{Cu^{2+}/Cu}$$

$$0.34\ V = 0 + E^\circ_{Cu^{2+}/Cu}.$$

The standard reduction potential of copper $E^\circ_{Cu^{2+}/Cu}$ is 0.34 V. Hence, the standard oxidation potential, $E_{Cu^{2+}/Cu}$ is -0.34 V.

Consider the familiar Daniel cell, shown in Fig. 2.4, in which zinc and copper are anodic and cathodic sites.

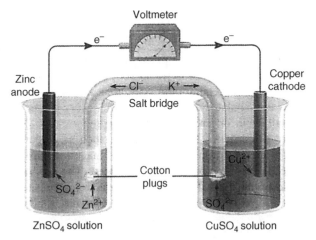

FIGURE 2.4 The Daniel electrochemical cell.

The anodic and cathodic reactions are

$$\text{Anode (oxidation): } Zn_{(s)} \rightarrow Zn^{2+} (1 \text{ M}) + 2e^-, \ E^{\circ}_{Zn/Zn^{2+}},$$

$$\text{Cathode (reduction): } Cu^{2+} (1 \text{ M}) + 2e^- \rightarrow Cu_{(s)}, \ E^{\circ}_{Cu^{2+}/Cu},$$

$$\text{Overall: } Zn_{(s)} + Cu^{2+} (1 \text{ M}) \rightarrow Zn^{2+} (1 \text{ M}) + Cu_{(s)}, \ E^{\circ}_{cell}.$$

The EMF of the cell

$$E^{\circ}_{cell} = E^{\circ}_{Zn/Zn^{2+}} + E^{\circ}_{Cu^{2+}/Cu} = 0.76 \text{ V} + 0.34 \text{ V} = 1.10 \text{ V}.$$

The standard potential series obtained is given in Table 2.1. In the table, the potential values given are reduction potentials and the oxidation potentials are obtained by reversing the sign. The sign and magnitude of the potentials give a rough idea as to the ease of oxidation or reduction, as the case may be.

The potentials for reduction for copper and gold are

$$Cu^{2+} + 2e^- \rightarrow Cu, \quad 0.344 \text{ V},$$

$$Au^{3+} + 3e^- \rightarrow Au, \quad 1.42 \text{ V}.$$

The reduction potential for gold is greater than that for copper, and this indicates the ease of reduction of gold compared to copper. Increasing positive values of the potential indicate the ease of reduction, and increasing negative values indicate the ease of oxidation. The potentials given in the table are with respect to a hydrogen electrode, which is difficult to use in practice and hence a calomel electrode is commonly used as a standard electrode.

Let us recall the two examples of copper and gold that were shown to be feasible, and not feasible, respectively, with respect to corrosion, based on the calculated free

TABLE 2.1   Standard Potential Series

Electrode	Reaction	$E^\circ_{red}(V)$
Li^+, Li	$Li^+ + e^- \rightarrow Li$	-3.024
K^+, K	$K^+ + e^- \rightarrow K$	-2.924
Ca^{2+}, Ca	$Ca^{2+} + 2e^- \rightarrow Ca$	-2.87
Na^+, Na	$Na^+ + e^- \rightarrow Na$	-2.714
Mg^{2+}, Mg	$Mg^{2+} + 2e^- \rightarrow Mg$	-2.34
Ti^{2+}, Ti	$Ti^{2+} + 2e^- \rightarrow Ti$	-1.75
Al^{3+}, Al	$Al^{3+} + 3e^- \rightarrow Al$	-1.67
Mn^{2+}, Mn	$Mn^{2+} + 2e^- \rightarrow Mn$	-1.05
Zn^{2+}, Zn	$Zn^{2+} + 2e^- \rightarrow Zn$	-0.761
Cr^{3+}, Cr	$Cr^{3+} + 3e^- \rightarrow Cr$	-0.71
Fe^{2+}, Fe	$Fe^{2+} + 2e^- \rightarrow Fe$	-0.441
Co^{2+}, Co	$Co^{2+} + 2e^- \rightarrow Co$	-0.277
Ni^{2+}, Ni	$Ni^{2+} + 2e^- \rightarrow Ni$	-0.250
Sn^{2+}, Sn	$Sn^{2+} + 2e^- \rightarrow Sn$	-0.140
Pb^{2+}, Pb	$Pb^{2+} + 2e^- \rightarrow Pb$	-0.126
Fe^{3+}, Fe	$Fe^{3+} + 3e^- \rightarrow Fe$	-0.036
H^+, H_2	$2H^+ + 2e^- \rightarrow H_2$	0.000
Saturated calomel	$Hg_2Cl_2 + 2e^- \rightarrow 2Hg + 2Cl^-$ (sat. KCl)	0.244
Cu^{2+}, Cu	$Cu^{2+} + 2e^- \rightarrow Cu$	0.344
Cu^+, Cu	$Cu^+ + e^- \rightarrow Cu$	0.522
Hg_2^{2+}, Hg	$Hg_2^{2+} + 2e^- \rightarrow 2Hg$	0.798
Ag^+, Hg	$Ag^{2+} + 2e^- \rightarrow 2Ag$	0.799
Pd^+, Pd	$Pd^+ + e^- \rightarrow Pd$	0.83
Hg^+, Hg	$Hg^+ + e^- \rightarrow Hg$	0.854
Pt^{2+}, Pt	$Pt^{2+} + 2e^- \rightarrow Pt$	ca. 1.2
Au^{3+}, Au	$Au^{3+} + 3e^- \rightarrow Au$	1.42
Au^+, Au	$Au^+ + e^- \rightarrow Au$	1.68

energy values for the respective reactions. The same conclusions can be reached, based on the electrochemical potentials for the reactions of copper and gold.

$$2Cu_{(s)} + 2H_2O_{(l)} + O_{2(g)} \rightarrow 2Cu(OH)_{2(s)}, \quad E = 0.63 \text{ V},$$
$$4Au_{(s)} + 3O_{2(g)} + 6H_2O_{(l)} \rightarrow 4Au(OH)_3, \quad E = 0.23 \text{ V}.$$

Using the above values for the potentials and the relationships between the potential and free energy,

$$\Delta G^\circ = -nFE.$$

We get -120 and 66.5 kJ/mol for the reactions of copper and gold, which are in good agreement with the values obtained earlier.

It is well known that gold is extracted and assayed by the familiar cyanide leaching process:

$$8NaCN + 4Au + O_2 + 2H_2O \rightarrow 4NaAu(CN)_2 + 4NaOH.$$

The potential for the reaction is 1.0 V, corresponding to -96.5 kJ/mol for the free energy of the reaction, showing the feasibility of this reaction as well as the important role played by the environment in the reaction.

Having established the electrochemical nature of the corrosion phenomenon, it is useful to examine the familiar rusting of iron due to corrosion. The mechanism of corrosion is complex and a simplistic mechanism is as follows. Some sites on the iron metal surface act as anodic facilitating oxidation at these sites:

$$\text{Oxidation: } Fe_{(s)} \rightarrow Fe^{2+}_{(aq)} + 2e^-.$$

The oxygen present in the atmosphere is reduced at the cathodic sites on the metal surface:

$$\text{Reduction: } O_{2(g)} + 4H^+ + 4e^- \rightarrow 2H_2O_{(l)}.$$

The source of H^+ can be dissolved CO_2 or SO_2 in industrial atmospheres. The overall corrosion process is

$$\text{Overall: } 2Fe_{(s)} + O_{2(g)} + 4H^+_{(aq)} \rightarrow 2Fe^{2+}_{(aq)} + 2H_2O_{(l)}.$$

The hydrated iron oxide, $Fe_2O_3 \cdot nH_2O$ appears as a brown deposit on the surface (Fig. 2.5).

The standard potential series can be used as only a rough guide with respect to the ability of a metal to resist corrosion. In most of the corrosion reactions, the potential values given in the table are not applicable, because of the presence of a film on the surface of the metal, and the change in potential, because the activity of metal ions is less than unity.

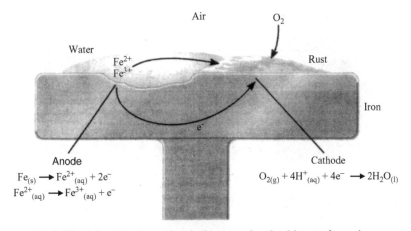

FIGURE 2.5   The electrochemical process involved in rust formation.

Galvanic series provides an alternate guide to the corrosion resistance of metals and alloys. When two metals are coupled together and immersed in an electrolyte, an electrode potential difference is observed due to exchange of electrons and ions. Using this procedure, galvanic series in seawater medium has been developed as shown in Fig. 2.6. From this galvanic series, it is possible to assess the

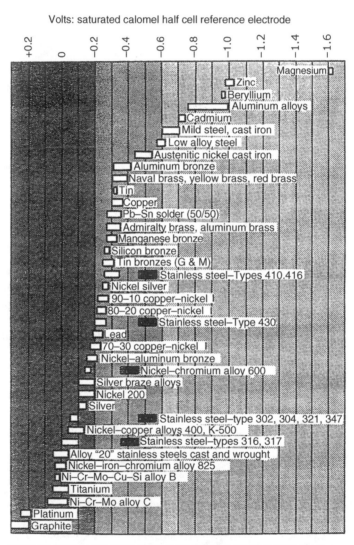

Alloys are listed in the order of the potential they exhibit in flowing sea water. Certain alloys indicated by the symbol: ▦ in low velocity or poorly aerated water, and at shielded areas, may become active and exhibit a potential near −0.5 V.

FIGURE 2.6   The galvanic series in seawater.

TABLE 2.2   Reference Electrodes

System at 25°C	Electrolyte	Potential (V Versus SHE)
$2Hg + 2Cl^- = Hg_2Cl_2 + 2e^-$	0.01 M KCl	0.389
Calomel	0.1 M KCl	0.333
	1.0 M KCl	0.280
	Sat. KCl	0.241
$Cu = Cu^{2+} + 2e^-$	0.1 M CuSO$_4$	0.284
Copper–copper sulfate	0.5 M CuSO$_4$	0.294
	Sat. CuSO$_4$	0.298
$Ag + Cl^- = AgCl + e^-$	0.001 M KCl	0.400
	0.01 M KCl	0.343
	0.1 M KCl	0.288
	1.0 M KCl	0.234
Platinized platinum	0.1 M NaCl	Approx. −0.12
Gold	0.1 M NaCl	Approx. −0.25
Zinc	Seawater	Approx. −0.79

behavior of metals and alloys with respect to corrosion in flowing seawater. It is clear from the galvanic series that magnesium and zinc with negative potentials occupying the top portion of the figure corrode easily, while platinum and graphite at the bottom of the figure with positive potentials exhibit greater resistance to corrosion.

The potentials of various metals given in Table 2.1 are with respect to the standard hydrogen electrode. The hydrogen electrode is not very convenient to use in practice as a reference electrode in the context of measuring electrode potentials. Some of the reference electrode systems used in practice are given in Table 2.2.

The measured electrode potential with respect to the calomel electrode is added to the value of the calomel electrode to obtain the value of the potential with respect to the standard hydrogen electrode.

$$E_{SHE} = E_{meas} + E_{cal.}$$

2.4   POURBAIX DIAGRAMS

When an electrochemical reaction is perturbed from its equilibrium state, the relative stabilities of the species in the reaction are changed. The change due to the perturbation is reflected in the measured electrode potential, which is different from the equilibrium electrode potential of the reaction. When the measured electrode potential is positive with respect to the equilibrium electrode potential, the reaction proceeds irreversibly from left to right.

$$\text{Reduced species} = \text{oxidized species} + ne^-.$$

When the measured potential is negative with respect to the equilibrium value, the reaction favors the reduced species.

Consider water, which is an electrochemically active species. The electrochemical reactions involving water are

$$2H_2O = 4H^+ + O_2 + 4e^-,$$
$$H_2 + 2OH^- = 2H_2O + 2e^-.$$

For solutions in which the activity of water and the fugacities of oxygen and hydrogen gases are unity, the equilibrium electrode potentials for the above reactions at 25°C are

$$E_{0,H_2O/O_2} = +1.23 - 0.059\, pH,$$
$$E_{0,H_2/H_2O} = \pm 0.059\, pH.$$

The electrode potentials given are with respect to the standard hydrogen electrode.

Pourbaix diagrams consist of plots of electrode potentials of electrochemical reactions against the pH of the solutions and these diagrams are useful in identifying regions of stability of various chemical species in solution. The Pourbaix diagram for the system H_2O–H_2–O_2–H^+–OH^- is shown in Fig. 2.7.

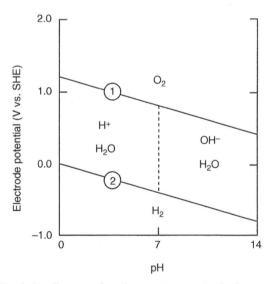

FIGURE 2.7 Pourbaix diagram for the system: water–hydrogen–oxygen–hydrogen ion–hydroxyl ion. Activity of water is unity; fugacities of hydrogen and oxygen are unity. Temperature 25°C.

Lines ① and ② represent plots of $E_{0,H_2O/O_2}$ and $E_{0,H_2/H_2O}$ versus pH, respectively. Inspection of line ① and the reaction

$$2H_2O = 4H^+ + O_2 + 4e^-$$

shows that water is a stable entity at potentials below this line and, conversely, unstable at potentials above this line. Similarly, analysis of line ② and the reaction

$$H_2 + 2OH^- = 2H_2O + 2e^-$$

shows that water is a stable entity at electrode potentials above line ② and unstable below this line. Pourbaix diagrams may be considered a form of electrochemical phase diagram. The region where a species is stable is identified with the chemical symbol of the species.

The Pourbaix diagram is extremely useful in determining stable chemical species of metals in contact with water. When the Pourbaix diagram for water is superimposed on the metal in question, two types of behavior are observed. This behavior classification of metals is based on the relationship between the metal–metal ion equilibrium reaction and the region of stability of water. In one class of metals, the metal–metal ion equilibrium potential falls within the region of stability of water, and copper is an example of this type of behavior. In the second class of metals, the metal–metal ion equilibrium electrode potentials fall below (outside) the region of stability of water, and iron is an example of this class of metal.

The Pourbaix diagram for copper and some of its ionic species, as well as compounds in aqueous solutions at 25°C is given in Fig. 2.8. The equilibrium electrode potential for the copper–cupric ion reaction is located within the region of stability of water represented by dashed lines.

Thus, the measurement of the equilibrium electrode potential is possible and the kinetics of the copper–cupric ion system can be studied without the interference from reactions involving solvent decomposition. All the reactions involved in this system are electrochemical except the reaction

$$Cu^{2+} + H_2O \equiv CuO + 2H^+$$

and can be studied by standard electrochemical techniques. If the measured electrode potential and pH are known, the Pourbaix diagram (Fig. 2.7) can be used to determine the stable form of copper and its compounds under those conditions.

The use of the Pourbaix diagram for copper can be illustrated by considering copper metal at +0.150 V versus SCE in an aqueous solution of pH 2.5 and a cupric ion activity of 0.01 at 25°C. In order to use the Pourbaix diagram for copper, the potential is converted to SHE scale.

$$E = +0.150 + 0.241 = 0.391 \text{ V versus SHE.}$$

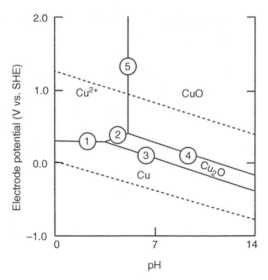

FIGURE 2.8 Pourbaix diagram for the system copper–cupric ion–cuprous oxide–water. Activity of cupric ion is 0.01. The dashed lines denote the stability range of water. The temperature is 25°C. The reactions considered ① copper going to cupric ion and two electrons; ② cuprous oxide reacting with hydrogen ion to give cupric ion, water, and two electrons; ③ cuprous oxide reacting with water to give cupric oxide, hydrogen ion, and two electrons; ④ copper metal reacting with water to give cupric oxide and hydrogen ion.

Now reference to the Pourbaix diagram given in Fig. 2.7 for copper and water shows the stable species to be Cu^{2+}, H^+, and H_2O, leading to the conclusion that corrosion of copper is possible under these conditions.

The Pourbaix diagram for iron and its compounds in an aqueous medium at 25°C is given in Fig. 2.9. The equilibrium potential of the reaction $Fe° = Fe^{2+} + 2e^-$ falls outside the stability region of water represented by dashed lines. Hence, measurement of the equilibrium electrode potential is complicated by the solvent undergoing a reduction reaction, while the iron is undergoing electrochemical oxidation. This is the basis of the mixed potential model of corrosion.

Let us consider the case of iron metal at -0.750 V versus SCE in a solution of pH 5.0 and a ferrous ion activity of 10^{-6} at 25°C. The electrode potential with respect to SHE is -0.509, and reference to the Pourbaix diagram in Fig. 2.8 for water and iron shows the stable species to be Fe^{2+} and H_2 and that corrosion is possible under the specified conditions.

In the potential–pH region where iron metal is the stable species, corrosion cannot occur since the reactions are not thermodynamically favored. This region is termed as "immune" to corrosion. With reference to the Pourbaix diagram for iron (Fig. 2.10), the region of stability of iron oxide shows that the film of iron oxide on the metal surface forms a barrier between the metal and environment. This condition is known as passivity and is characterized by the measured electrode potentials in the regions

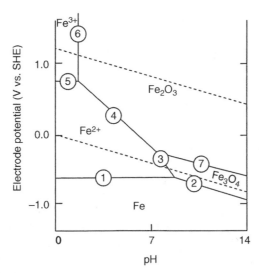

FIGURE 2.9 Pourbaix diagram for the system iron–ferrous ion–ferric ion–hematite–ferric oxide. Activities of ferrous and ferric ions are 10^{-6}. The temperature is 25°C. The reactions considered (1) iron metal giving ferrous ion and two electrons; (2) iron reacting with water to give hematite, hydrogen ion, and electrons; (3) ferrous ion reacting with water to give hematite, hydrogen ion, and electrons; (4) ferrous ion reacting with water to give ferric oxide, hydrogen ion, and electrons; (5) ferrous ion giving rise to ferric ion and electron; (6) ferric ion reacting with water to give ferric oxide and hydrogen ion; and (7) hematite reacting with water to give ferric oxide, hydrogen ion, and electrons.

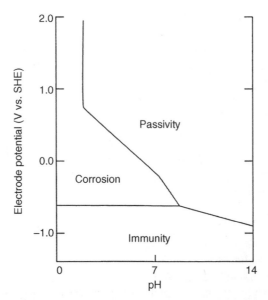

FIGURE 2.10 Pourbaix diagram for iron in terms of corrosion, passivity, and immunity.

where the passive oxide film is stable. Figure 2.9 is a simplified Pourbaix diagram for iron, which clearly distinguishes the regions of immunity, corrosion, and passivity.

In conclusion, Pourbaix diagrams reveal information on the stability of chemical species and whether corrosion is likely to occur under a given set of conditions.

2.5   DYNAMIC ELECTROCHEMICAL PROCESSES

The potential series and the Pourbaix diagrams shed light on the feasibility of the corrosion process based on thermodynamics under equilibrium conditions. These concepts do not shed light on the rates of the corrosion reactions. In order to ascertain the corrosion rates, it is necessary to understand the intimate dynamical processes occurring at the metal exposed to the electrolyte solution.

Let us consider an electrode immersed in an electrolyte solution. A potential difference arises at the interface between the electrode and the surrounding electrolyte solution due to charge separation. When electrons leave the electrode and reduce the cations in solution, the electrode becomes positively charged, and anions will move toward the positively charged electrode. This scenario is depicted in Fig. 2.11. Thus, we have a pair of positive and negative sheets known as the electrical double layer.

The process of contact adsorption of an ion on the electrode consists of desolvation of the anion, removal of the solvent molecules from the electrode surface, followed by the adsorption of the anion on the electrode surface. The steps of anion adsorption on the electrode in terms of energy are

$$E(\text{electrode} - \text{anion}) > E(\text{water on electrode}) + E(\text{desolvation of anion}).$$

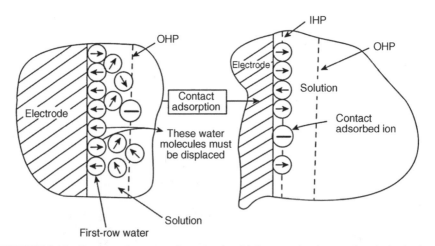

FIGURE 2.11   Process of contact adsorption in which a negative ion can lose its hydration and displace first-row water. The locus of the centers of the ions defines the inner Helmholtz plane.

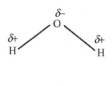

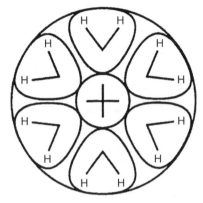

FIGURE 2.12 Schematic of a positive ion surrounded by a sheet of apex-inward water dipoles.

The disposition of water molecules attached to positive ions and negative ions with metal–oxygen and anion–hydrogen are shown in Figs. 2.12 and 2.13, respectively.

The first-row water molecules near the electrode surface play an important role in the sense that they have to allow either an anion or a solvated cation to come close to

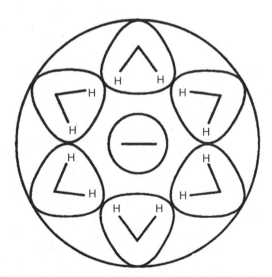

FIGURE 2.13 Schematic of a negative ion surrounded by a sheet of apex-outward water dipoles.

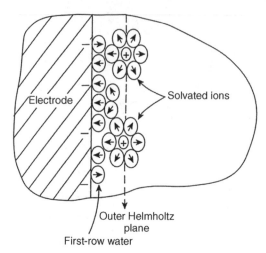

FIGURE 2.14 A layer of hydrated positive ions whose hydration sheath cannot be stripped, on the first layer of water molecules. The locus of the centers of these ions define the outer Helmholtz plane.

the electrode surface. The locus of the center of the first-row molecules on the electrode surface defines the inner Helmholtz plane, and the distance between the inner Helmholtz plane and the electrode is ~ 3 Å. Similarly, the locus of the center of the water molecules adjacent to the first row is the outer Helmholtz plane and is about 5–6 Å from the electrode surface.

The solvation of cations and anions by water molecules is possible due to the dipolar nature of water in which oxygen and hydrogen have partial negative and positive charges, respectively. It is useful to note that the bonding between a cation and a water dipole is stronger than an anion and a water dipole.

Let us now consider a negatively charged electrode, shown in Fig. 2.14. It is important to note that solvated cations lie at the outer Helmholtz plane, unlike the case in which anions are adsorbed on the positive electrode surface (Fig. 2.11). In terms of interaction forces, the cation–water interactions are stronger than negatively charged electrode–water interactions. In other words, the water molecules attached to the cation do not exchange with water molecules adsorbed on the negative electrode surface. The solvated cation is located at the outer Helmholtz plane.

The double layer model assumes a fixed layer of charges on the electrode and the outer Helmholtz plane. This model has been modified by Guoy–Chapman analysis, which assumes that ions of charge opposite the charge on the electrode distribute themselves in a diffuse manner as depicted in Fig. 2.15. According to this model, the variation of potential with distance is exponential, while it is linear according to the Helmholtz model. Stern proposed a synthesis of Helmholtz and Guoy–Chapman models by taking into account both the fixed layer of charges and diffuse layer of charges.

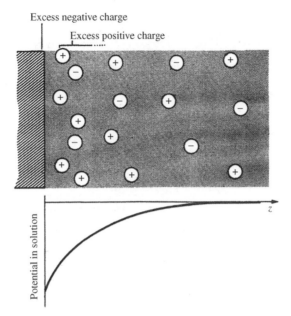

FIGURE 2.15   Guoy–Chapman model of the double layer.

The charge distribution and the variation of potential with distance according to the Stern model is shown in Fig. 2.16. By assuming the layer of charges as a parallel plot capacitor, we determine that the total differential capacitance is related to the capacitances due to Helmholtz (C_H) and Guoy–Chapman (C_G):

$$\frac{1}{C} = \frac{1}{C_H} + \frac{1}{C_G}.$$

The variation of potential is both linear and nonlinear in accordance with the fact that Stern's model embraces both Helmholtz and Guoy–Chapman models.

The variation of potential with distance when a test charge located at a distance approaches the electrode surface passing through the solution is shown in Fig. 2.17. The trend in the variation of potential is a predicted trend. The potential difference between the metal and the solution is $\Delta\Phi$ and is given by

$$\Delta\Phi = \Phi_m - \Phi_s,$$

where Φ_m and Φ_s are the potentials of metal and solution, respectively. The potential difference $\Delta\Phi$ (m,s) is the measured parameter when a metallic electrode is immersed in an electrolyte.

In the corrosion dynamic, electrochemical processes are important and hence the consequences of perturbation of a system at equilibrium are considered. Let us consider the Daniel cell consisting of metallic copper in cupric sulfate, and metallic zinc immersed in zinc sulfate solution, depicted in Fig. 2.18.

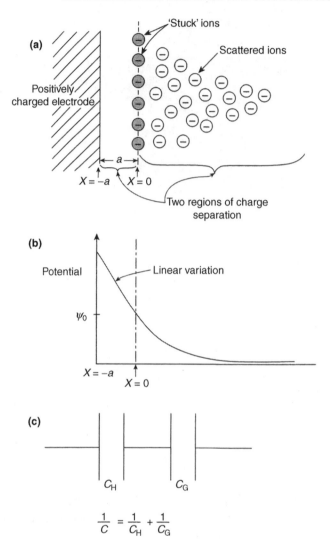

FIGURE 2.16 The Stern model. (a) A layer of ions stuck to the electrode and the remainder scattered in cloud fashion; (b) the potential variation according to this model; (c) the corresponding total differential capacitance C is given by the Helmholtz and Gouy capacitances in series.

This cell has an electromotive force of 1.1 V when there is no flow of current. The potential will decrease below 1.1 V when a small current flows through the resistance R. On continued flow of current, the potential difference between the electrodes approaches nearly a zero value. The potential difference decreases as the current increases since the electrodes polarize. The effect of net current flow on the voltage of the cell is represented by plotting the individual potentials of copper and zinc electrodes against current, as depicted in Fig. 2.19.

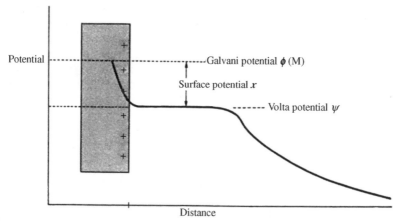

FIGURE 2.17   Variation of potential with distance from an electrode.

This representation is known as a polarization diagram. The potentials Φ_{Cu} and Φ_{Zn} when the current is zero are known as open-circuit potentials. The zinc electrode polarizes along a–b–c while the copper electrode polarizes along d–e–f. At a current value of I we have the polarization of zinc equal to the potential at b minus Φ_{Zn}, the open-circuit potential of zinc. Similarly, we have for copper the polarization potential equal to $\Phi_e - \Phi_d$ or $\Phi_e - \Phi_{Cu}$. We also have

$$\Phi_b - \Phi_e = I_1(R_e + R_m),$$

where R_e and R_m are electrolytic and metal resistance in series, respectively.

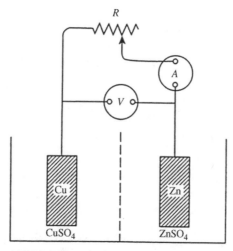

FIGURE 2.18   Polarized copper–zinc cell.

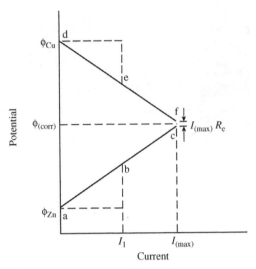

FIGURE 2.19 Polarization diagram for copper–zinc cell.

At point C we have maximum current and the potential difference at a minimum and equal to $I_{max} \cdot R_e$. The potential corresponding to I_{max} is known as the corrosion potential. Using the equivalent weight of zinc (65.32/2), the Faraday value of 96.500 C/equiv, and the value of I_{max}, the amount of zinc corroding in unit time can be calculated. An equivalent amount of copper will be deposited due to the cathodic reaction. The lines (a–b–c) and (d–e–f) are known as the anodic and cathodic branches of the polarization diagram.

According to Faraday's law, the charge Q is related to the ionization of M moles of metal.

$$Q = ZFM,$$

where Z is the valence of the metal, F the Faraday value, and M the number of moles. Thus,

$$\frac{dQ}{dt} = ZF\frac{dM}{dt}.$$

Noting that $dM/dt = J$, the flux of the substance, I the current flowing through the unit area of the cross section, and i the current density, we have

$$i = ZFJ.$$

The flux of the material is the corrosion rate and the corrosion rate is equated to the current density.

Polarization involves perturbation that can disturb an equilibrium resulting in a dynamic situation. The three types of polarization are concentration polarization, activation polarization, and IR drop.

Consider the system of copper metal immersed in water and let us assume that the system produces cupric ions:

$$Cu \rightarrow Cu^{2+} + 2e^-.$$

The environment must provide sufficient energy to cause corrosion of copper to give cupric ions by surmounting the energy barrier $\Delta G^{\ddagger}$. For some time, corrosion will occur, resulting in the production of cupric ions in solution. The corrosion rate of copper will decrease with increase in current from zero and the value of ΔG will decrease along with potential in accordance with Faraday's law. The thermodynamic energies of the cupric ions and copper atoms approach each other. After some time, an equilibrium state is reached when the rate of production of cupric ions equals the rate of reduction of cupric ions to produce metallic copper. The equilibrium may be written as

$$Cu \underset{i_c}{\overset{i_a}{\rightleftharpoons}} Cu^{2+} + 2e^-$$

and for a general case,

$$M \underset{i_c}{\overset{i_a}{\rightleftharpoons}} M^{2+} + 2e^-.$$

At equilibrium $i_a = i_c$, and the measured current density $i_{meas} = i_a - i_c$, and no net current flows. There will be flow of current, but it is equal and opposite, which cannot be measured. This quantity is known as exchange current I_o, or as i_o when I_o is divided by the area.

It is useful to note that current density is a measure of corrosion rate. Corrosion rates are generally expressed as mpy (mils per year), ipy (inches per year), ipm (inches per month), and mdd (loss of weight) in milligrams per square decimeter per day.

Perturbation of a system at equilibrium in terms of potential is known as polarization. The polarization of the equilibrium is also known as overpotential or overvoltage, denoted by η. Consider the system

$$M \underset{i_c}{\overset{i_a}{\rightleftharpoons}} M^{2+} + 2e^-$$

for which the corrosion rate r can be written as (1)

$$r = k_{corr}[\text{reactants}],$$

where $k_{corr} = Ae^{-\Delta G\ddagger/RT}$ and A is constant. We then may write

$$r = Ae^{-\Delta G\ddagger/RT}[\text{reactants}].$$

At equilibrium $K_f = K_r$ or $i_a = i_e$,

$$i_a = i_o = A_o e^{-\Delta G/RT}.$$

By noting anodic polarization as $\alpha'\eta$ and cathodic polarization as $(1 - \alpha'\eta)$, we have

$$i_a = A_o e^{(-\Delta G\ddagger + \alpha\eta F)/RT} = A_o e^{-\Delta G\ddagger/RT} e^{\alpha\eta F/RT},$$

$$i_a = i_o e^{\alpha\eta ZF/RT},$$

$$i_e = i_o e^{(1-\alpha)\eta ZF/RT}.$$

Since bulk current flow, $i_{meas} = i_a - i_c$

$$i_{meas} = i_o(i_o e^{\alpha\eta ZF/RT} - i_o e^{(1-\alpha)\eta ZF/RT}).$$

The above equation is known as the Butler–Volmer equation. By substituting $\alpha ZF/RT = A^1$,

$$i_a = i_o e^{A^1\eta}.$$

By taking logarithms, we have

$$\ln i_a = \ln i_o + A^1\eta,$$

$$\ln\left(\frac{i_a}{i_o}\right) = A^1\eta,$$

$$\eta = \frac{2.303}{A^1}\log\left(\frac{i_a}{i_o}\right).$$

If

$$\beta = \frac{2.303}{A^1} = \frac{2.303RT}{\alpha ZF},$$

$$\eta_a = \beta_a \log\left(\frac{i_a}{i_o}\right).$$

The equation is known as the Tafel equation. For the anodic and cathodic processes, we have

$$\eta_a = \beta_a \log i_a - \beta_a \log i_o,$$
$$\eta_c = \beta_c \log i_c - \beta_c \log i_o,$$

where

$$\beta_a = \frac{2.303RT}{\alpha ZF}, \quad \beta_c = \frac{2.303RT}{1 - \alpha ZF}.$$

The Tafel equations consist of β_a and β_c, the anodic and cathodic Tafel constants, respectively. Tafel plots are useful in obtaining corrosion rates. Let us consider a sample of a metal polarized 300 mV anodically and 300 mV cathodically from the measured corrosion potential, E_{corr} at a potential scan rate of 0.1–1.0 mV/s. The resulting current is plotted on a logarithmic scale as shown in Fig. 2.20. The corrosion current i_{corr} is obtained from the plot by extrapolation of the linear portions of the anodic and the cathodic branches of the curve to the corrosion potential, E_{corr}.

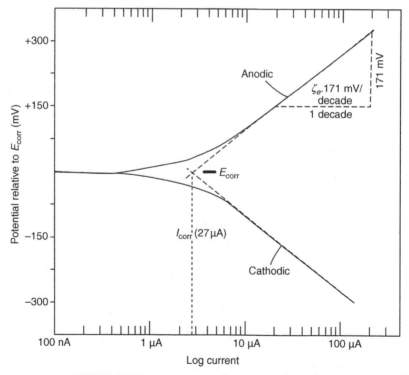

FIGURE 2.20 Experimentally measured Tafel plot.

The corrosion current is then used to calculate the corrosion rate using the equation:

$$\text{corrosion rate (mpy)} = \frac{0.13 i_{corr} (\text{Eq wt})}{d},$$

where i_{corr} is the corrosion current density (in A/cm^2), d the density of the corroding metal sample (in g/cm^3), and Eq wt is the equivalent weight (mass) of the corroding metal (in g).

The linear polarization technique is rapid and gives corrosion rate data, which correlate well with weight loss method-based corrosion rates. The technique involves scanning through 25 mV above and below the corrosion potential and plotting the resultant current against potential as shown in Fig. 2.21. The corrosion current i_{corr} is related to the slope of the line:

$$\frac{\Delta E}{\Delta i} = \frac{\beta_A + \beta_c}{2.3(i_{corr})(\beta_A + \beta_C)},$$

$$i_{corr} = \frac{\beta_A \beta_c}{2.3(\beta_A + \beta_c)} = \left(\frac{\Delta i}{\Delta E}\right),$$

$$\text{Corrosion rate (mpy)} = \frac{0.13 i_{corr}(\text{Eq wt})}{d}.$$

Polarization studies on a corrosion system are carried out by means of a potentiostat. The experimental arrangement of the cell consists of a working electrode, a reference electrode, and a counterelectrode. The counterelectrode is

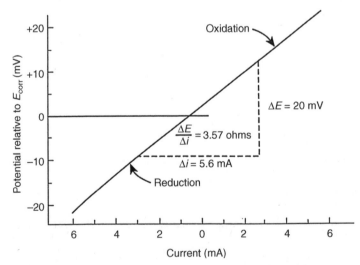

FIGURE 2.21 Experimentally measured polarization resistance.

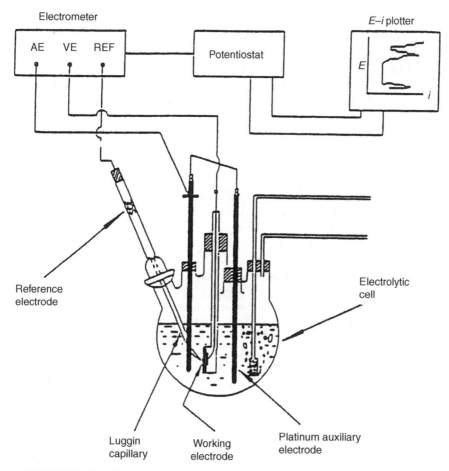

Electrometer

E–i plotter

AE   VE   REF

Potentiostat

E

i

Reference
electrode

Electrolytic
cell

Luggin
capillary

Working
electrode

Platinum auxiliary
electrode

FIGURE 2.22   Electrochemical cell used for potentiodynamic polarization studies.

used to apply a potential on the working electrode both in the anodic and the cathodic direction and to measure the resulting currents. The electrochemical cell assembly depicted in Fig. 2.22 is used in polarization studies.

2.6 MONITORING CORROSION AND EFFECTIVENESS OF CORROSION INHIBITORS

Monitoring corrosion involves obtaining corrosion rates in a particular industrial plant in operation, which can be converted into useful information for a corrosion management program. The useful techniques are real-time data gathering systems, process control techniques, knowledge-based systems, and smart structures. The complex nature of corrosion monitoring is evident when it is recognized that there exist many forms of corrosion; corrosion can be either uniform or localized, variation of corrosion rates over distances; and there is a lack of one corrosion monitoring technique capable of measuring corrosion rates under variable conditions.

Before adopting a corrosion monitoring protocol, it is advisable to consider the data on hand and the types of corrosion problems that need to be considered. It is prudent to use several monitoring techniques that either complement or supplement one another. Monitoring corrosion continuously can help detect corrosion damage over time and provide remedial action before the damage is out of control.

It is surprising that some industries do not pay attention to corrosion control until a corrosion-related failure occurs. Some industries use corrosion monitoring for inspection, repairs, and maintenance periodically. Other uses for corrosion monitoring are to alert operators of the need for injection of inhibitors or use of coatings or for use of corrosion-resistant materials. Another use of corrosion monitoring is to instruct operators to apply corrosion control when needed.

Some benefits of corrosion monitoring to industries are (i) improved safety, (ii) reduced downtime, (iii) early warnings of impending damage or failure, (iv) reduced maintenance costs, (v) reduced environmental damage, (vi) longer intervals between scheduled maintenance, (vii) reduction in operating costs, and (viii) plant life extension. Corrosion monitoring systems consist of the familiar simple coupon exposure or handheld data loggers or completely integrated plant process surveillance systems with remote data access and data management capabilities.

Corrosion sensors or probes have to be highly sensitive, since they form the heart of corrosion monitoring systems. The retrievable probe can be inserted or removed through a high-pressure access fitting. A block diagram of the monitoring system is given below:

Sensors are now available that have an inherent capability of processing the signal and displaying the degree of corrosivity. For on-line and real-time monitoring of corrosion, computers located in mobile laboratories are used. Corrosion monitoring data are later converted into meaningful process-relevant information. Data from other sources such as process parameter logging and inspection reports are combined with corrosion sensor data and used as input to a management information system. Corrosion monitoring in many industries has shown that corrosion damage is not uniform with time and that corrosion damage can be correlated with the upsets in operational conditions. Thus, operational upsets can be identified only with real-time monitoring methods.

In general, it is useful to note that one technique alone is not sufficient to monitor corrosion in complex industrial activity. It is recommended that many (two or three) monitoring techniques be used so that the weakness of a single technique is offset by another technique. It is also recommended that long-term coupon exposure should be carried out in addition to the chosen corrosion monitoring technique. One corrosion monitoring technique that is extensively used is linear polarization resistance (LPR) in water treatment plants. The corrosion rate data given by the LPR technique is valid only when (i) there is one cathodic reaction and one anodic reaction, (ii) the anodic and cathodic Tafel constants are known and do not vary with time, (iii) the corroding surface is clean without corrosion products, scale on the surface, (iv) corrosion is uniform, (v) solution resistance is negligible, and (vi) the corrosion potential has reached a steady-state value. These conditions are rarely met in actual industrial operational conditions.

2.6.1 Objectives of Corrosion Monitoring

The most important step is defining the objective of corrosion monitoring in the industrial scenario. When the objective is corrosion control it is necessary to limit the severity of corrosion so that the corrosion allowance is not used before the end of design life of the equipment. It is also necessary to know the corrosion rates and the corrosion mechanism when corrosion control is the objective of corrosion monitoring. Knowledge of the corrosion mechanism can be used in the choice of corrosion inhibitors and their use for corrosion control. Corrosion monitoring may also be needed in the context of maintaining the integrity of a plant, which enables the scheduling of periodic inspections.

2.6.2 Corrosion Monitoring Probe Location

Location or position of the corrosion monitoring probe or sensor is crucial. The obvious location is a point where corrosion is most severe. The point where corrosion is most severe can be determined based on principles of corrosion, in-service failure records, and in consultation with service personnel. In the case of water tanks, corrosion is severe at the water–air interface, and this can be monitored by attaching sensors to a floating platform.

In many cases, the most favorable positions for the corrosion probe may not be accessible. The flush or protruding electrode probe should be placed at the 6 o'clock position in a pipeline or at the bottom of a vessel or at a solution accumulation point in a separator tower. Sometimes the design of the sensor has to be changed to be placed in the location where corrosion occurs. An example of this is a protruding electrode at the 12 o'clock position with a long body that is in contact with the aqueous phase (1).

The corrosion sensor should be positioned to reflect the state of the system being monitored. If, for example, turbulence occurs around a protruding probe in a pipeline, the sensor will not be able to detect localized corrosion in the pipeline wall. In this case, a flush-mounted sensor should be used to monitor localized corrosion (Fig. 2.23).

The choice of corrosion monitoring points or locations is governed by available access points, as in the case of pressurized systems. It is common to use the access points for the placement of sensors. An alternative to this is bypass lines with customized sensors and access fittings. A significant advantage of a bypass is that

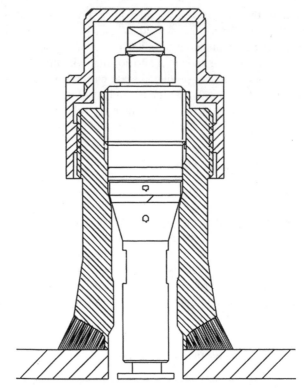

FIGURE 2.23   Flush-mounted corrosion sensor in an access fitting.

testing can be done under prevailing conditions of highly corrosive atmospheres without disturbing plant operations.

Consider a distillation column (2), shown in Fig. 2.24. The unit has seven corrosion monitoring points. The feed point, overhead product receiver, and the bottom product line are points of extreme temperature as well as points where products of various degrees of volatility settle. In most instances, a high degree of corrosion occurs at an intermediate height of the column due to the presence of high amounts of corrosive species. Initially, there are seven corrosion monitoring points, which are reduced to a lower number after analyzing the initial corrosion data.

Corrosion monitoring in oil and gas gathering lines and process piping is usually done by metal loss coupons or electrical resistance (ER) probes inserted into the process fluids through access fittings (3). A corrosion coupon or probe is placed at the 6 o'clock position on horizontal sections of pipeline, because produced water is heavier than crude oil or gas condensate. The probe installation at the 6 o'clock position is shown in Fig. 2.25b. This probe is in the best position to monitor corrosion events. Figure 2.25a illustrates a gas production line with the probe at the 9 o'clock position, and the probe in this case cannot accurately measure the corrosion rates associated with the aqueous phase at the bottom of the pipeline.

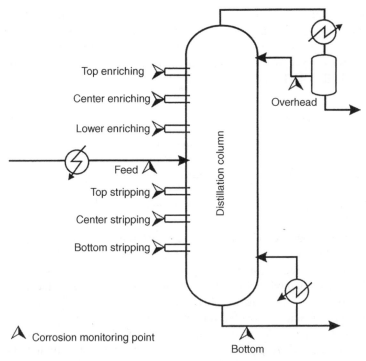

FIGURE 2.24   Corrosion monitoring points in a distillation column.

Some pipeline and facility designers have installed corrosion monitoring locations on the sides of the pipeline instead of the bottom. The side locations provide easier access for coupon crews. The coupons or probes on the side cannot give accurate data unless the pipe is full of aqueous phase. It is advisable to design and install the coupon or probe access fittings during the initial fabrication and installation of pipeline. An intrusive probe in a pipeline containing a three-phase production with large water

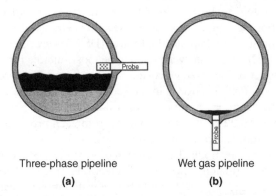

FIGURE 2.25   (a) Probe at 9 o'clock position; (b) probe at 6 o'clock position.

content gives a satisfactory performance in monitoring corrosion. In general, a flush-mounted probe is preferred, since it is less dependent on the amount of water in the pipeline. Linear polarization probes may also be used in pipelines carrying water. Electric resistance probes are best suited for monitoring corrosion when hydrocarbons are present in the pipeline.

2.6.3   Probe Type and its Selection

The design of the corrosion probe is important, since it has to interface with the process environment and also be suitable for installation. The surface roughness, residual stresses, corrosion products, surface deposits, preexisting corrosion damage, and temperature are some of the factors to be considered in procuring the monitoring probe. Corrosion sensors may also be made from precorroded material that has been exposed to plant conditions. Sensor designs such as spool pieces in pipes and heat exchanger tubes, flanged parts of candidate materials, or test paddles bolted to agitators may be used to make sensors of representative material. Probes of high sensitivity are required for short-term monitoring and thicker probe elements are needed for long-term monitoring. A velocity shield is required when the medium has suspended solids. An electric resistance probe may be used to measure the erosion rate in the presence of sand. A noncorroding metal may be used (3).

Flush-mounted electrode design is most suitable for monitoring corrosion in oil and gas flow lines where pigging operations are common. The limitation of this electrode is that the exposure is to a limited area and is not suited for low-conductivity atmospheres (1). Care must be taken in the fabrication of the probe so that a crevice is not created between the outer circumference of the electrode and the surrounding insulating material, which might lead to crevice corrosion.

Protruding electrode design has enjoyed wider application than has flush-mounted electrode. This design is suitable for replacing electrodes, and the probability of crevice corrosion is minimal. The protruding electrode design is such that since the whole electrode surface area is exposed to the corrosive medium, difficulties might arise when the flow regime becomes turbulent (3).

Corrosion probes have been developed to enable the working electrode of a three-electrode system to be prestressed to yield stress in order to match the operating condition of a pipe or a vessel. Several methods of on-line localized corrosion monitoring of the double-shell tanks (DSTs) were studied and electrochemical noise (ECN) technique was found to be most suitable for monitoring and identifying the onset of localized corrosion (4). The monitoring probe consists of three nominally identical electrodes immersed in the tank waste. Each system is composed of in-tank probe and ex-tank data collector hardware. Electrodes are made from UNS K02400 steel. Four channels from each probe are formed from bullet-shaped electrodes and four channels are formed from thick-walled C-rings of the dimension $44 \, cm^2$/electrode. The unstressed bullet-shaped electrodes monitor pitting and general corrosion. The working electrode on each C-ring channel is notched, precracked, and stressed-to-yield before installation to enable monitoring stress-corrosion cracking.

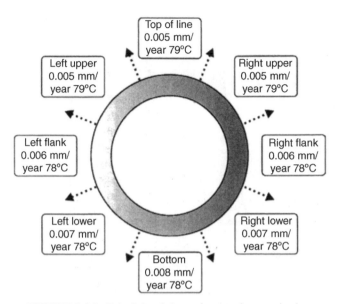

FIGURE 2.26   Principle of ring pair corrosion monitoring.

A circumferential spool probe has been developed for monitoring corrosion in hydrocarbon environments. The elements in the form of thin rings are obtained by slicing a pipe and reassembling pairs of the rings, separated by insulation, to remake a pipe section that can retain line pressure. Each electrically isolated ring gives a measurement. When wires are attached at equally spaced intervals around the ring, the overall metal loss and the loss in each segment between each pick-up point can be obtained (Fig. 2.26). The ring pair corrosion monitoring system is held in compression by a clamp arrangement producing an inner-pressure tight cylinder. The cylinder is placed in a housing. This is then placed in an outer housing that is a pressure-tight assembly. This system can be used in a pipeline submerged in the sea.

The coupled multielectrode array system consists of a resistor positioned between each electrode and the common coupling point, as shown in Fig. 2.27. Electrons from a corroding electrode flow through the resistor connected to the electrode and produce a potential drop of the order of a few microvolts. This potential drop is measured by a highly sensitive voltage sensor. The principle of the technique is illustrated in Fig. 2.27. The coupled multiple electrode assembly system gives (i) a greater sampling of current fluctuations, (ii) a greater ratio of cathode-to-anode areas that enhance localized corrosion, and (iii) pit penetration rate and macroscopic spatial distribution of localized corrosion (5).

The probe converts all the measured values of current into a single parameter and hence the probe can be used for real-time and on-line monitoring of corrosion. Since the anodic electrodes in the probe simulate the anodic sites on the surface of the metal, the most anodic current can be considered as the corrosion current from the most corroding site on the metal. The corrosion current from the most corroding site on the

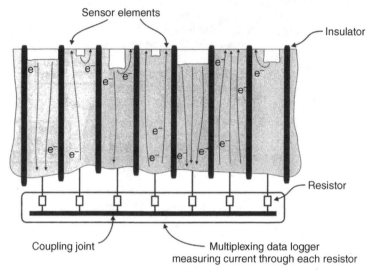

FIGURE 2.27   Multielectrode array multiplexed current and/or potential measurement.

metal may be represented by a value based on three times the standard deviation of the current. The standard deviation value may be from either the anodic currents or both the anodic and cathodic currents.

In the case of less corrosive media or in the case of a more corrosion-resistant metal, the most anodic electrode may not be fully covered with anodic sites until the electrode is completely corroded. The electrode in such a case may have available cathodic sites and the electrons from the anodic sites may flow internally to the cathodic sites within the same electrode. In this case, the total anodic corrosion current, I_{corr}, and the measured anodic current, I_a^{ex}, is related by the equation

$$I_a^{ex} = \varepsilon I_{corr},$$

where ε is a current distribution factor, which represents the fraction of electrons that flows through the external circuit due to corrosion. The value of ε can vary between 0 and 1, depending on factors such as the heterogeneous nature of the metal surface, the corrosive environment, and the size and number of electrodes. When the electrode is severely corroded and more anodic than other electrodes in the array, the value of ε will be nearly 1 and the measured external current will equal the localized corrosion current. A coupled multielectrode array system has been used in monitoring localized corrosion of a variety of metals and alloys in a variety of conditions and environments: deposits of sulfate-reducing bacteria (SRB); deposits of salt in air; high-pressure simulated natural gas systems; hydrogen sulfide media; oil–water mixtures; cathodically protected systems; cooling water systems; simulated crevices in seawater; salt-saturated aqueous solutions; concentrated chloride solutions; concrete; soils; low-conductivity drinking water; process streams of chemical plants at high temperatures; and coatings.

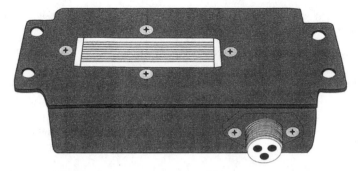

FIGURE 2.28   Multielectrode array multiplexed current and/or potential measurement.

Atmospheric corrosion monitoring presents some difficulties, such as the formation of fine mist or the discontinuous film by condensing humid aid on the surface of the object or metal, which give rise to high resistivity. The high resistivity of the electrolyte requires that the corrosion monitoring probe electrodes be close to each other and be electrically insulated. The success of the probe in monitoring atmospheric corrosion requires (i) high sensitivity, (ii) minimal IR drop, and (iii) relatively large electrodes to enable recording corrosion signals.

An electrochemical probe that has been successfully used in low-conductivity condensing environments for monitoring aircraft corrosion is given in Fig. 2.28. The probe consists of closely spaced probe elements made from uncoated aluminum alloy.

Another type of probe used in atmospheric corrosion monitoring is shown in Fig. 2.29. This probe uses microcircuit technology consisting of a polyamide film electroplated with gold and cadmium in a pattern to maximize the galvanic current produced in a moderately corrosive environment. The galvanic current is then integrated with a coulometer as a function of time. The data are stored in a memory chip for future use with the help of a computer.

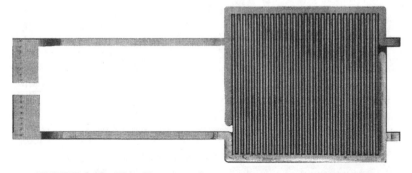

FIGURE 2.29   Thin film sensor for atmospheric corrosion monitoring.

TABLE 2.3 Sensitivity and Response Time of Typical Corrosion Monitoring Applications

Application	Sensitivity Range (mm/year)	Response Time	System Characteristics
Corrosion tests	0.1–100	1 h–5 d	Continuous
Inhibition control	0.1–20	0.5–2 d	Continuous optimization
Corrosion control (upsets)	1–100	1 h–2 d	Continuous monitoring (upsets)
Corrosion control performance demonstration	1–10	1 wk–1 mo	Continuous/interval measurement
Inspection planning	0.2–10	1 mo–0.5 y	Interval
Inspection	1–20	3 mo–10 y	Interval

The two most important properties of corrosion monitoring systems are (i) sensitivity and (ii) response time. Both the sensitivity and response time have an inverse relation to each other (6). The sensitivity and response time of typical corrosion monitoring applications are given in Table 2.3.

The sensitivity–response windows for various corrosion and corrosion applications are depicted in Fig. 2.30.

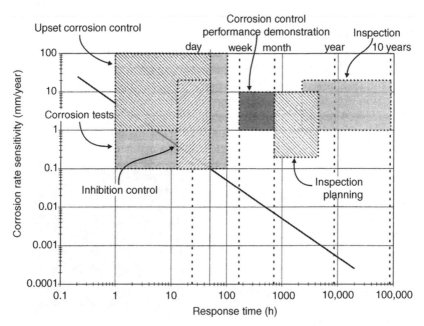

FIGURE 2.30 Application windows in sensitivity–response time plot.

TABLE 2.4   Direct Corrosion Measurement Techniques

Intrusive techniques
  Physical techniques
    Mass-loss techniques
    Electrical resistance
    Visual inspection
  Electrochemical DC techniques
    Linear polarization resistance
    Zero-resistance ammeter (ZRA) between dissimilar alloy electrodes: galvanic
    Zero-resistance ammeter between the same alloy electrodes
    Potentiodynamic–galvanodynamic polarization
    Electrochemical noise
  Electrochemical AC techniques
    Electrochemical impedance spectroscopy (EIS)
    Harmonic distortion analysis
Nonintrusive techniques
  Physical techniques for metal loss
    Ultrasonics
    Magnetic flux leakage (MFL)
    Electromagnetic: eddy current
    Electromagnetic: remote field technique (RFT)
    Radiography
    Surface activation and gamma radiometry
    Electrical field mapping
  Physical techniques for crack detection and propagation
    Acoustic emission
    Ultrasonics (flaw detection)
    Ultrasonics (flaw sizing)

Most corrosion monitoring and inspection techniques have been reviewed and strengths and weaknesses of the techniques have been discussed (7). Direct corrosion monitoring techniques are techniques that measure corrosion-affected parameters, and indirect corrosion monitoring techniques provide data on parameters that either affect or are affected by the corrosivity of the environment or by the products of the corrosion processes. Further, a monitoring technique can be intrusive if it requires access through a pipe or a vessel to enable measurement of corrosion rates. Indirect techniques can be on-line or off-line. Tables 2.4 and 2.5 list the various monitoring techniques.

2.6.4   Direct Intrusive Corrosion Monitoring Techniques

The three major groups of techniques are physical techniques, electrochemical DC techniques, and electrochemical AC techniques.

2.6.4.1   Physical Techniques   In this group of techniques, corrosion damage is determined by measuring the changes in the geometry of exposed test coupon

TABLE 2.5 Indirect Corrosion Measurement Techniques

On-line techniques
  Corrosion products
    Hydrogen monitoring
  Electrochemical techniques
    Corrosion potential (E_{corr})
  Water chemistry parameters
    pH
    Conductivity
    Dissolved oxygen
    Oxidation reduction (redox) potential
  Fluid detection
    Flow regime
    Flow velocity
  Process parameters
    Pressure
    Temperature
    Dew point
  Deposition monitoring
    Fouling
  External monitoring
    Thermography
Off-line techniques
  Water chemistry parameters
    Alkalinity
    Metal ion analysis (iron, copper, nickel, zinc, manganese)
    Concentration of dissolved solids
    Gas analysis (hydrogen, H_2S, other dissolved gases)
    Residual oxidant (halogen, halides, and redox potential)
    Microbiological analysis (sulfide ion analysis)
  Residual inhibitor
    Filming corrosion inhibitors
    Reactant corrosion inhibitors
  Chemical analysis of process samples
    Total acid number (TAN)
    Sulfur content
    Nitrogen content
    Salt content in crude oil

samples. As a result of corrosion, many properties of the exposed test sample may be altered. Some of the properties that change as a result of corrosion are its mass, electrical resistance, magnetic flux, reflectivity, stiffness, or any other mechanical properties.

The mass loss coupon method of determining corrosion damage can be used to monitor corrosion on existing equipment and to evaluate alternate materials of construction, the effect of process conditions that cannot be simulated in the laboratory, and the efficiency of corrosion inhibitors with respect to their performance. This simple, low-cost method consists of exposing small specimens in the environment of interest for a specific duration of time and then removing them, followed by cleaning and weighing them. It is recommended that a comprehensive guide such as ASTM G-4 be consulted (8).

Special devices are available (Metal Samples Company) commercially for inserting coupons into the operating equipment. Special tools are available on the market for insertion and withdrawal of sensors under high-pressure conditions.

In the course of mass loss coupon testing, a large variety of metals or alloys can be exposed simultaneously to the process environmental conditions and can be ranked with respect to their integrity. A slip-in rack may be inserted into the process stream and removed from the equipment with a retractable coupon holder (9) (Fig. 2.31). The main advantage of the coupon holder is that many samples in duplicate or triplicate can be tested as well as simulate conditions such as welding, residual stresses, or crevices, which in turn can be helpful in material selection or maintenance or repair. Coupons can be designed to assess localized corrosion in process flow conditions or in a side stream branching off from the main process stream. The coupon may be designed according to the objectives of the tests, as, for example, flat coupons for general corrosion or pitting corrosion, welded coupons for corrosion in weldment, or precracked samples for stress-corrosion cracking tests (7).

Duration of the corrosion test should be such that a reasonable amount of corrosion is apparent. A minimum exposure of 3 months has been used for evaluating pitting and crevice corrosion. According to ASTM G-31 in the case of general corrosion, the duration of the test is obtained as (10):

$$\text{Duration of test (h)} = \frac{\sim 50}{\text{corrosion rate (mm/year)}}.$$

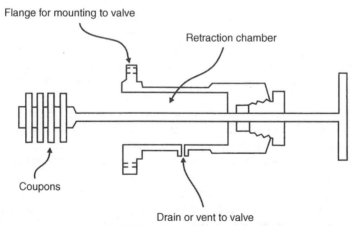

Flange for mounting to valve

Retraction chamber

Coupons

Drain or vent to valve

FIGURE 2.31 Retractable coupon holder (9).

The duration of a test will be ∼500 h when the corrosion rate is 0.1 mm/year. The duration of corrosion testing can vary depending upon whether or not the test is done in the laboratory or upon plant conditions. It is useful to note the limitations of coupon testing, as gleaned from the literature (7, 9):

(i) Mass loss coupon technique gives the average metal loss over the time of exposure.

(ii) Incapable of detecting rapid changes in corrosivity of the process.

(iii) Localized corrosion is not guaranteed to initiate before the coupons are removed.

(iv) Reinsertion of used coupons is not recommended.

(v) Short exposure periods give unrepresentative average metal loss.

(vi) Corrosion rate data based on the coupon weight loss may not represent the corrosion rate of plant due to (a) multiphase flow, (b) turbulence from mixers, elbows, valves, pumps, which accelerate in areas removed from the sites where coupons were placed.

(vii) Procedure for mass loss coupon analysis may be labor intensive.

(viii) Simulation of erosion-corrosion and heat-transfer effects requires well-planned placement of the coupons.

(ix) Data from coupons may not be valid when corrosion rates vary markedly with time, due to unforeseen process factors.

Rectangular coupons with stencil stamping are generally used for evaluating uniform corrosion. Surface finish of the coupons is not critical in evaluating the performance of corrosion inhibitors in water treatment programs. Surface finish of the coupons is critical in ranking candidate alloys, and, ideally, the surface finish should match the process equipment.

Pairs of test coupons with relative exposed areas from 1:1 to 10:1 can be coupled electrically to monitor galvanic corrosion. The method described in ASTM G-71 should be followed in testing for galvanic corrosion (9, 11).

Equipment crevices are common in complex systems. The two most common crevice geometries in field coupon testing use insulating flat washers or multiple crevice washers made of ceramics or soft thermoplastic resins. ASTM G-78 and G-48 may be consulted when one is designing a crevice corrosion test (12, 13).

The residual stresses resulting from forming and welding operations and the assembly stresses in interference-fitted parts cause stress-corrosion cracking. The suitable coupons for plant tests are self-stressed bending and residual stress samples. The useful coupons are the cup impression, U-bend (14), C-ring (15), tuning fork, and welded panel (16).

The test coupons are removed after the test, cleaned, and weighed according to the procedures described in ASTM standard (17) G-1. Examination of coupons after cleaning should reveal the forms of corrosion that appear in the equipment.

Visual examination followed by microscopy, both binocular and scanning electron microscope, may prove useful in detecting the corrosion damage effects (9).

The test coupons may be sectioned and examined metallographically in order to observe the types of corrosion damage. The coupons are cleaned repeatedly in a thorough manner and weighed. The corrosion rate is estimated by using the relationship

$$R(\text{corrosion rate}) = \frac{K(W_1 - W_2)}{A(t_1 - t_2)\rho},$$

where R is the corrosion rate, W_1 and W_2 are the initial and final masses (g), ρ is the density, t_1 and t_2 are the starting and ending times, respectively, and K is a constant. The incremental weight loss is plotted against the number of times of cleaning the coupons in Fig. 2.32 and the weight W_2 corresponds to the intercept made by extrapolation of the line to y-axis. The reproducibility of the corrosion rate depends upon the degree of uncertainty that each variable, such as time, mass loss, and area, contributes to the total uncertainty. Minimum uncertainty in the corrosion rate is possible when (i) localized corrosion does not occur, (ii) there is uniform penetration of the coupon, (iii) the projected and actual surface areas are nearly identical, (iv) corrosion product removal results in the same weight, (v) areas do not change during exposure, and (vi) penetration rate is independent of time.

As long as the environment remains unchanged, the longer the duration of exposure of the coupons, and the smaller will be the error in the corrosion rate. An electronic balance capable of weighing to 0.1 mg is preferable. Each coupon may be weighed three times to obtain an average value. It is recommended that it will

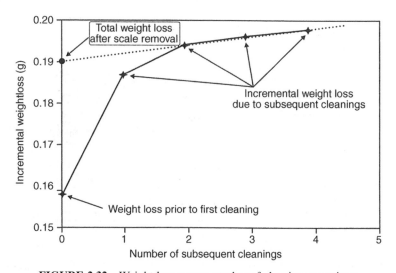

FIGURE 2.32   Weight loss versus number of cleaning operations.

be useful to adhere to ASTM G-31. In practice, long exposure times may not be possible because of one's inability to control the environment for long periods of time.

2.6.4.2 Electrical Resistance   The electrical resistance method of monitoring corrosion was introduced by Dravnieks (19, 20). The principle of the method involves the increase in the electrical resistance of a sensing element as its cross-sectional area decreases due to corrosion. The electrical resistance of a metal is given by the relationship

$$R = r\frac{L}{A},$$

where L is the probe element length in centimeters, A is the cross-sectional area in square centimeters, and r is the specific resistance of the probe metal in Ω-cm.

Reduction or metal loss in the cross section, A, as a result of corrosion results in an increase in the resistance, R. Because temperature affects the electric resistance of the probe element, it is prudent that one set the electric resistance sensors usually to measure the resistance of a corroding sensor element relative to an identical shielded probe element (Fig. 2.33). Commercial sensor elements of different forms such as plates, tubes, or wires are available.

The sensitivity of these sensors can be increased by reduced thickness at the expense of reduced sensor lifetimes. Electric resistance probes usually have useful lifetimes up until they reach half of the original thickness. In the case of wire-shaped electric resistance sensors, the lifetime corresponds to a quarter original thickness loss. These sensors cannot be used when conductive corrosion products or surface deposits form on the sensor element. Formation of iron sulfide in sour gas media or microbiological corrosion and carbonaceous deposits in atmospheric corrosion prevent the use of electric resistance sensors in these systems.

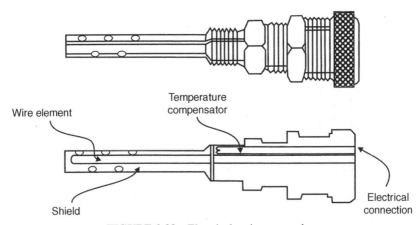

FIGURE 2.33   Electrical resistance probe.

The corrosion rates are determined by plotting a series of measurements at different times of exposure and obtaining the slope of the plot (21). Since the probes are small, the installation is easy and the system can be connected to a portable resistance bridge or to a control room where a computer may be used. The results obtained by this technique give a good picture of general corrosion but not localized corrosion. Special probes have been prepared to monitor crevice corrosion; multiple crevices on the probe element, such as beads on wire loop probes, accomplish this task.

2.6.4.3 Inductive Resistance Probes (22) Probes have been developed by Cormon Ltd. These probes have very high resolution with long life and the capability of intrinsically safe operation in hydrocarbon process plant environments (22). The reduction of thickness of a sensing element is measured by the changes in the inductive resistance of a coil embedded in the sensor. The sensitivity is claimed to be several orders of magnitude higher than other comparable electrical resistance probes.

The operation of the probe is not affected by process variables such as temperature, hydrostatic pressure, slugging, or flow regimes. The probe is also unaffected by industrial noise such as electromagnetic induction and thermally induced electromotive force voltages.

2.6.4.4 Electrochemical Techniques Corrosion is an electrochemical process and hence the electrochemical properties of metal–solution interface, such as corrosion potential, corrosion current density, and electrochemical impedance, play a decisive role in monitoring corrosion. The metal of the actual plant equipment is not normally part of the measurement circuit. Corrosion potentials on corrosion current density or polarization resistance are measured. Electrochemical data such as corrosion current density are converted into corrosion rate using equations or algorithms that are specific to the technique that is used. The measured corrosion current density is converted into corrosion rate via Faraday's law and empirically determined Tafel slopes and other factors in the equation. Some techniques use analysis of the interface with direct current methods. Other techniques use alternating current (AC) methods to study corrosion interface and conductivity of the process fluids.

The application of electrochemical techniques to field situations is fraught with problems such as probe geometry. For example, the Luggin capillary used in laboratories to reduce the interference of solution resistance cannot be used with ease in field applications (7).

The seemingly popular electrochemical polarization techniques are fraught with some difficulties. The rate of scanning of potential or current may have a significant effect on polarization (23). The scan rate must be slow enough to minimize surface capacitance charging. Otherwise, part of the generated current may charge the surface capacitance, with the result that the measured current will be greater than the corrosion current.

The distance between the reference electrode and the working electrode should be minimized so that the solution resistance effect is minimized. In solutions of high resistance such as concrete, soils, and organic media, this effect can be very significant. The corrosion reactions occur on the metal surface exposed to the process environment. The surface of the metal or alloy may be modified by changing process conditions, which may reflect in the polarization curves (23). Corrosion rates are obtained based on the assumption that corrosion is uniform. Sometimes a hysteresis loop is observed in the polarization curves, indicating localized corrosion.

In multiphase systems, nonconductive phases can deposit on the electrodes and give erroneous results until the electrodes are cleaned.

2.6.4.5 Linear Polarization Resistance

In this technique, a potential perturbation of 10 to 20 or 30 mV is applied to the sensor electrode and the resulting current measured. The ratio of the potential to current perturbation is the polarization resistance, which is inversely proportional to the uniform corrosion rate. The polarization resistance of a metal is the slope of the potential–current density ($\Delta E/\Delta i$) plot at the free-corrosion potential. The polarization resistance is related to the corrosion current, i_{corr}, through the Stern–Geary equation.

$$\frac{\Delta E}{\Delta i_{app}} = \frac{\beta_a \beta_c}{2.3 i_{corr}(\beta_a + \beta_e)}$$

when β_a and β_c are the estimated or the measured Tafel slopes of the anodic and cathodic reactions, respectively, ΔE is the applied potential change, Δi_{app} is the resultant current density change, and i_{corr} is the corrosion current density at the free-corroding potential. Tafel slopes can also be determined by curve-fitting of polarization resistance curves or by potentiodynamic polarization curves or by harmonic distortion analysis.

In a plant situation, commercially available probes (Metal Samples Company, www.metalsamples.com) are placed in the area of the container where the corrosion rate is to be determined (Fig. 2.34). A power supply polarizes the sample from the corrosion potential and the resulting current is recorded as a measure of the corrosion rate. The LPR probes can be interfaced with a computer data acquisition module. An alarm system may be used to alert plant operators when high corrosion rates occur (9, 24).

The linear polarization resistance probe can be of either a two- or a three-electrode configuration with flush or projecting electrodes. In the case of the three-electrode system, the corrosion measurements are made on the test electrode and a stable reference electrode is not necessary because of the short duration of measurement. The reference electrode for field monitoring is made of stainless steel or of the same alloy as that being monitored on the test electrode. The auxiliary electrode is also of the alloy that is being monitored. The compensation for the solution resistance depends upon the close positioning of test and reference electrodes. In the case of a two-electrode system, the corrosion measured is an average of the rate for the two electrodes, and both electrodes are made of the alloy being monitored (7).

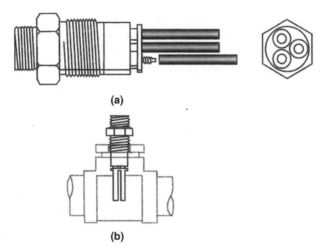

FIGURE 2.34   (a) Linear polarization resistance probe; (b) LPR probe in a pipe tee.

A combination of linear polarization resistance and zero-resistance ammeter was used to determine the rate of localized corrosion in a flowing environment with a large-area electrode in a fast-flow condition and a small electrode in slow-flow conditions, as shown in Fig. 2.35. The large electrode acted as the cathode and the small electrode as the anode due to differential aeration. In the course of time, the small electrode was covered with deposits and undergoing underdeposit or localized corrosion.

Some limitations of the LPR technique are (i) the environment must be an electrolyte with low resistivity and this requirement prevents use of this technique

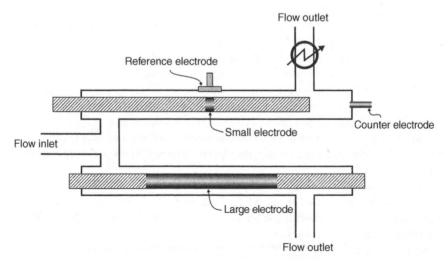

FIGURE 2.35   Schematic of differential flow cell.

in low-conductivity media such as oil and gas, refinery, and chemical systems; (ii) the vessel or pipe wall needs to be penetrated, which may cause leaks and unsafe conditions; (iii) the corrosion rates obtained are approximate and the method is best suited for situations where significant changes in corrosion rates occur (9, 21).

2.6.4.6   Zero-Resistance Ammetry   Galvanic currents between two dissimilar metals or two electrodes of the same alloy, but in a different metallurgical or electrochemical state, are measured with a zero-resistance ammeter. The differences in the electrochemical behavior of the two electrodes exposed to a process stream give rise to differences in the redox potential at these electrodes. The more noble electrode acts as cathode and the other electrode acts as anode. When the anodic reaction is relatively stable, the galvanic current monitors the response of the cathodic reaction to the process stream conditions and vice versa (7). The technique has been used to monitor depolarization effects of the cathode of a galvanic pair of electrodes to low levels of oxygen or the presence of bacteria, which depolarize the cathode of the galvanic pair and increase the coupling current.

The technique can throw light on effectiveness of inhibitors, alloy-specific inhibitors, for example, passivators or copper inhibitors, presence of depassivators (O_2, NH_3, HCN), film formation, passivation/depassivation, velocity-induced corrosion, and the presence of alloy-specific chelants. Dissolved oxygen at 0–50 ppb can be detected.

Some pitfalls of the technique are (i) the data obtained from the galvanic probes do not always reflect the actual galvanic corrosion rates, since galvanic corrosion depends upon relative areas and specific geometries of the components, which can vary between the probe design and the actual plant components being monitored. (ii) The technique cannot distinguish between either cathodic activation or anodic activation. An example would be either cathodic activation by dissolved oxygen or anodic activation through increased bacterial activity.

2.6.4.7   Potentiodynamic–Galvanodynamic Polarization   In this technique, a three-electrode corrosion probe is used to polarize the working electrode, which serves as the sensing element. The current or the potential response is measured as the potential (potentiodynamic) or current (galvanodynamic) is varied from the free-corrosion potential. Unlike the LPR technique, the applied polarization can be a few hundred millivolts. A typical plot of potential versus log of current is shown in Fig. 2.36. This technique is commonly used in the laboratory and very rarely in field studies. It is useful to estimate the anodic and cathodic Tafel slopes (Fig. 2.37). The onset of localized corrosion such as pitting can be detected, which may help one assess the corrosion risk.

Potentiodynamic polarization is used in aqueous media, while galvanodynamic polarization has been used in an oil medium to control the current density. The technique has been used to estimate the susceptibility of various materials to localized corrosion in process streams (7).

In an ideal situation, each point in the current–potential curve should be established by allowing a polarization time of tens of seconds to several minutes to enable

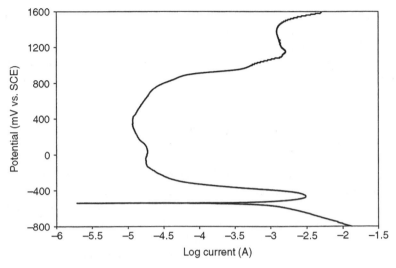

FIGURE 2.36 Potentiodynamic polarization plot for S43000 steel in 0.05 M sulfuric acid solution.

complete charging of the double-layer capacitance at the metal–liquid interface. Such long time intervals allow the electrode surface to oxidize or reduce all of the surface deposits and to polarize to the applied potential. In practice, the applied potential or current is varied continuously in an analog form at some preset rate (potentiodynamic or galvanodynamic) or in digital form of small discrete steps at some preset rate

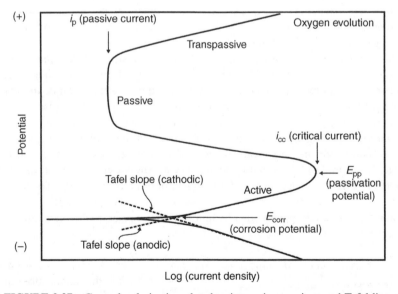

FIGURE 2.37 General polarization plot showing various regions and Tafel lines.

(potential or galvano staircase). In either case, the faster the rate of variation of the applied signal, the greater the lag of the measured signal behind the true steady-state values. This compromise between a practical and reasonable time to complete a scan and the extent of lag from measured-to-ideal response with this technique poses a serious question as to use of this technique in the field where the environmental conditions are continually changing.

The technique of potentiodynamic or galvanodynamic polarization is not considered an on-line or real-time method, because the electrodes are normally replaced after one or two test runs, and running the test changes the electrode surfaces.

2.6.4.8 *Electrochemical Noise*   Electrochemical noise is a technique that measures naturally occurring fluctuations of potential or current. It also can measure current noise under an applied potential or measure potential noise under an applied current. Current noise is a measurement of current fluctuations between two nominally identical electrodes. Potential noise is a measurement of the potential fluctuations between one electrode and a reference electrode or between two nominally identical electrodes. The frequency of the fluctuations is typically below 1 Hz.

It is common to couple nominally identical electrodes through a zero-resistance ammeter and measure the potential of this pair with respect to a reference electrode or a third nominally identical electrode. The ZRA forces the two electrodes to the same potential. The potential of the pair is then measured against the third reference electrode.

The relationships between potential and current noise are more complex to analyze quantitatively because the naturally occurring fluctuations do not have controlled frequencies, like the applied frequencies are in electrochemical impedance spectroscopy. The technique measures current and potential noise on the corroding electrodes, which are consequences of the uniformity or nonuniformity of the adjacent metal electrodes exposed to a process stream.

This technique has been used in the determination of the performance of two corrosion inhibitors both in the laboratory and in field conditions. The electrochemical cell and experimental arrangement for the determination of the performance of two corrosion inhibitors by the electrochemical potential noise technique are shown in Figs. 2.38 and 2.39, respectively. This technique is simple and well suited for monitoring the comparative performance of the inhibitors. It is useful for on-line monitoring of corrosion processes for long periods of time. The noise amplitude in process water with and without inhibitors as a function of time is depicted in Fig. 2.40. The data show the two inhibitors to be equally effective with inhibition efficiency in the range of 79–84% (26). It is preferable to use this technique in conjunction with other techniques, such as electrochemical impedance spectroscopy.

2.6.4.9 *Electrochemical Impedance Spectroscopy*   In electrochemical impedance spectroscopy, the sensing element is polarized by the application of an alternating potential, which in turn produces an alternating current response. For

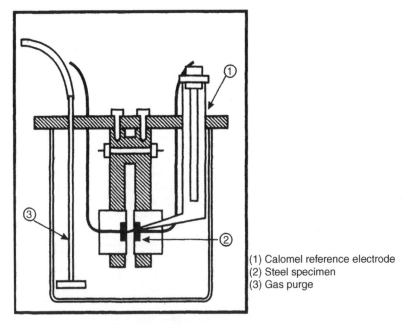

(1) Calomel reference electrode
(2) Steel specimen
(3) Gas purge

FIGURE 2.38   Electrochemical cell.

monitoring corrosion, the frequency range of the applied AC polarization is typically in the range of 0.1 Hz to 100 kHz, with a polarization level within 10 mV of the corrosion potential.

Alternating current techniques have some advantages over direct current techniques (i) AC techniques use very small excitation amplitudes in the range of 5–10 mV peak-to-peak; (ii) data on electrode capacitance and charge transfer kinetics provide

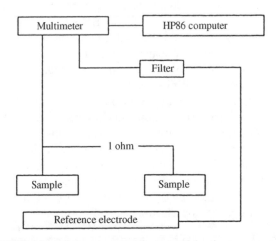

FIGURE 2.39   Apparatus used for potential noise measurements.

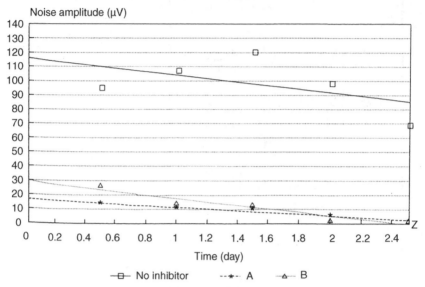

FIGURE 2.40   Potential noise amplitude versus time.

mechanistic information; and (iii) AC techniques can be applied to low-conductivity solutions, while DC techniques are subject to serious potential errors in these media.

Consider the application of a small sinusoidal potential ($\Delta E \sin \omega t$) on a corroding sample, which results in a signal along with current flow of harmonics 2ω, 3ω, and so on. Then the impedance $\Delta I \sin (\omega t + \Phi)$ is the relation between $\Delta E/\Delta I$ and phase Φ. In the case of corrosion studies, the sample is made part of a system known as equivalent circuit (26), which consists of the solution resistance R_s, charge transfer resistance R_{CT}, and the capacitance of the double layer C_{dl}. The measured impedance plot appears in the form of a semicircle (Nyquist plot). Both the equivalent circuit and the impedance plot are shown in Fig. 2.41. The electrochemical experimental arrangement consisting of an AC impedance analyzer, electrochemical cell, and the computer to acquire the data over a period of time is depicted in Fig. 2.42. The polarization resistance data obtained are used to calculate the corrosion rate of the corroding sample. The polarization resistance of the corroding sample may be monitored over an extended period of time. The polarization resistance can be converted into corrosion rate by an empirical measurement of Tafel slopes by potentiodynamic polarization and harmonic distortion analysis or from the literature (27).

Both laboratory and field monitoring of the performance of two corrosion inhibitors A and B were done by potential noise and electrochemical polarization measurement. The polarization resistance as a function of time is depicted in Fig. 2.43a. The data obtained by electrochemical noise gave 79–84% inhibition, figures that are in reasonable agreement with 84%obtained by AC impedance technique.

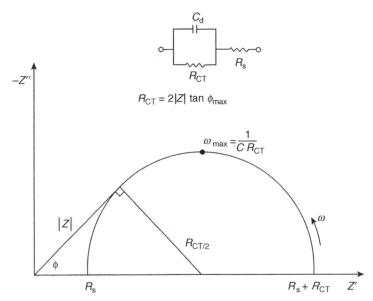

FIGURE 2.41   Representation of equivalent circuit.

In order to simplify the analysis of field EIS data, a method was developed consisting of finding the geometric center of an arc formed by three successive data points on a complex impedance diagram (Fig. 2.44). The three-point analysis technique consists of permuting the data points involved in the projection of centers

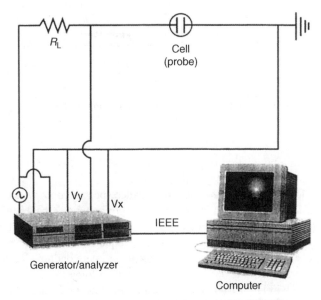

FIGURE 2.42   Experimental setup for AC impedance.

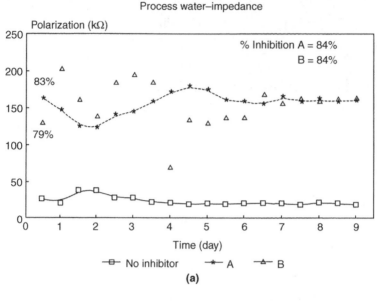

Process water–impedance

(a)

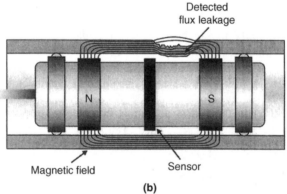

Detected
flux leakage

(b)

FIGURE 2.43 Pipe inspection using MFL.

in order to obtain a population of projected centers (28). This technique was tested extensively both in the laboratory and field trials.

Recently, a low-cost electrochemical impedance spectroscopic wireless monitoring system was developed (5 cm in diameter; 1.2 cm in height) that requires 10 mW power during the 200 s measurement period. The sensor in the wireless probe determines the impedance at 15–20 independent frequencies by measuring amplitude and phase at each frequency. It computes corrosion rate, conductivity, and coating impedance and transmits the data wirelessly to a data logger.

The miniature wireless system can be located in concrete or in other hidden and inaccessible locations. Since the sensor consumes minimal power it is useful to

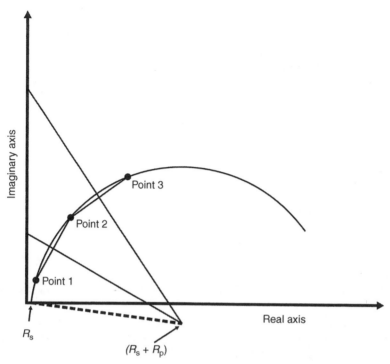

FIGURE 2.44   Schematic of the extrapolation method to obtain polarization resistance from EIS data.

monitor long-term integrity of coatings (29). The wireless system has been tested in a variety of environments, such as concrete, water, and coatings.

2.6.4.10   Harmonic Distortion Analysis   In this technique, a low-frequency sinusoidal potential is applied to a three-electrode measurement system and the resulting current is measured. Since the corrosion process is nonlinear in nature, a potential perturbation by one or more sine waves generates responses at more frequencies than the frequencies of the applied signal. The current responses can be measured at zero, harmonic, and intermodular frequencies.

The measurement of the DC frequency "zero" is known as the Faraday rectification (FR) technique. The FR technique can be used to measure the corrosion rate provided one of the Tafel parameters is known. The corrosion rate and both Tafel parameters can be obtained with one measurement by analysis of the harmonic frequencies. Tafel slopes can be determined in under 1 min (30).

In harmonic distortion analysis, a single, low-frequency and low-distortion, sinusoidal voltage is applied to the corrosion interface. As a quality check, three different frequencies are used to verify the repeatability of the technique. The amplitude is in the range of 10–30 mV, peak to peak. Typical frequency used is 0.1–10 Hz. Other frequencies may be used, depending on the corrosion process.

The theoretical analysis for the computation of Tafel slopes and corrosion current makes no assumptions about solution resistance effects, and the measurements cannot be performed unless the system is free of significant electrical noise in the frequency range of applied potential or one of its harmonic frequencies (7).

In principle, analyzing the primary frequency and the harmonics gives all the information one needs to calculate Tafel slopes and corrosion rate. This technique has been used in a very limited number of applications. The nonlinear techniques have been applied to activation-controlled processes, such as corrosion rate measurements in acid media with and without inhibitors.

2.6.5   Direct Nonintrusive Techniques

These techniques do not require access to the inside of the system being monitored and hence are free from the usual risks associated with the on-line insertion and retrieval operations of intrusive devices. In these techniques, the actual plant material is monitored, unlike intrusive monitoring techniques, in which test coupons are tested. The area of the plant material monitored by nonintrusive techniques is generally longer than inserted coupons in intrusive techniques. In general, the measurement techniques that make measurements on large areas of plant material do not have as high a resolution as do intrusive techniques. The lower sensitivity of nonintrusive techniques produces a longer response time to corrosion changes and the probability of missing a short upset altogether.

2.6.5.1   Ultrasonics   Ultrasonic testing to measure thickness has been used for a long time. A transducer (piezoelectric crystal) is made to oscillate at high frequencies, coupled directly or indirectly to one surface of the sample whose thickness is to be measured, and the time a wave of known velocity takes to travel through the material is used to determine its thickness. Ultrasonic equipment is vastly improved by combining electronics with computers. Rugged systems based on microcomputers with motor-driven robotic devices are capable of measuring wall thickness of corroded components at tens of thousands of points over $0.1\,\mathrm{m}^2$ ($1\,\mathrm{ft}^2$). This capability coupled with increased precision of field measurements possible with computerized systems has made the automated system suitable for on-line corrosion monitoring.

With more advanced systems, a larger number of thickness measurements over small areas are made possible, and hence statistical comparisons of the areas scanned can allow rapid comparison of selected spots used in a corrosion monitoring program. The volume of the material in the area scanned can be calculated, data that can be used to develop volumetric changes over time (or mass loss). Knowledge of the change in area of corrosion and the remaining wall thickness and pit depth may be used to obtain pitting rates.

Ultrasonic technique is generally used for inspection such as the integrity of a vessel, but in cases of severe metal loss, the technique may be used as an ongoing corrosion monitor. Continuous monitoring is possible with some new developments in transducers. The technique can be used on-line, but the sensitivity does not permit its use for real-time measurements. The accuracy of the technique can approach ± 0.025 mm (± 0.001 in. or ± 1 mil) in the laboratory. The accuracy in field testing is ± 0.1 mm (± 0.004 in. or ± 4 mil). With some new transducers, the accuracy is ± 0.005 mm (± 0.0002 in. or ± 0.2 mil).

Since corroded areas of the actual component are being measured directly, analytical techniques can be used to determine the integrity of the damaged item. The method published in ANSI/ASME B 31G can be used (31) to calculate maximum allowable operating pressure of piping systems and integrity of corroded storage tanks (32) and pressure vessels.

The accuracy of corrosion-monitoring data and sensitivity to small changes of metal loss is not as high as with physical techniques for metal loss or with electrochemical techniques. This technique is normally used for corrosion prediction over longer periods of time and is not suited for monitoring real-time changes in corrosion rate.

2.6.5.2 *Magnetic Flux Leakage*

Magnetic flux leakage technique uses a combination of permanent magnets and sensor coils to identify corrosion in steel tube pipe and plate. Magnetic flux is channeled into the sample component being inspected as the flux makes a circuit between the opposite poles of two magnets. When there is an anomaly in the metal, some flux leaks out and is detected by the sensors placed near the metal's surface and between the magnets. The sensors measure the amplitude of the leaking flux. The amplitude of the leaking flux is compared with results from known anomalies in similar materials after the inspection, and these lead to wall-loss predictions. MFL has been applied in examining tanks, heat exchangers, pipelines, and other tubular products.

Different MFL systems are used for different applications. Some systems such as in-line pipeline inspection tools are highly sophisticated self-contained systems that gather and store data remotely. Signal analysis can be real-time or done after the fact, depending on the application. Software for signal analysis is available.

Magnetic flux leakage is used to monitor corrosion over long periods of time. Information from ongoing inspections is compiled and corrosion rates are determined. Some inspections can be done on-line while others such as tank floor inspections are done when the tank is out of service. The MFL technique is useful in locating defects where volumetric metal loss such as pitting occurs. The technique requires calibration against the actual part or component.

Magnetic flux leakage can work through coatings of up to 10 mm thickness and is generally limited to pipe thicknesses of 12–15 mm. Above this thickness it is difficult to obtain magnetic saturation. Typical wall thickness sizing accuracy is of the order of $\pm 10\%$ and length of accuracy of ± 1 cm. The typical arrangement of an MFL device on a pipe is depicted in Fig. 2.43b.

2.6.5.3 *Eddy Current Technique*

An eddy current is defined as a circulating electrical current induced in a conductive material by an alternating magnetic field.

The equipment consists of an electronic signal generator/processor and a probe containing at least one coil. The signal generator uses an alternating current to induce eddy currents in an electrically conductive sample material. The eddy currents then induce an alternating current in the sensing coils. The fields in the generator and the sensor coils are balanced by adjustments of frequency, amplitude, and distance. A change in the balance of the two fields indicates various flaws in the sample or material thinning. The system may be calibrated with respect to cracks, pitting, wall thinning, and other material changes. Calibration standards are given in ASME Boiler and Pressure Vessel Code, Section VII (33), and calibration may be made for other flaws as well.

Eddy current technique is commonly used off-line to identify defects in nonmagnetic metals, such as heat exchanger tubing. Very modern computerized instruments are used extensively in industry, such as manufacturing and nuclear power generation plants. Signal analysis may be performed real-time, or, afterwards, software for automated signal analysis is readily available.

The technique is generally an inspection technique. With repeated testing of the same area, it is possible to obtain data on metal loss over a known period and the long-term corrosion rates calculated. The technique can be used on-line, but is not common, and the sensitivity is too low to provide real-time measurements. Eddy current technique is sensitive to pitting, general corrosion/erosion, wear from parts, dents, bulges, cracks, plating, and dealloying. Because it is fast it can be used in tube bundles and in the aerospace industry.

2.6.5.4 *Remote Field Eddy Current Technique*   Remote field eddy current technique used for inspection of metallic pipes and tubing consists of exciting a relatively large, low-frequency AC coil inside the pipe and a pick-up coil, offset approximately two pipe diameters, is used to detect changes in the flux field due to tube wall conditions such as thickness, permeability, and conductivity. The probe is shown in Fig. 2.44. The wall thickness affects the phase between the excitation and detector signals, and the extent of the defect affects the amplitude. Although magnetic flux can be used to measure changes in thickness of both magnetic and nonmagnetic materials, it is generally used to identify defects in magnetic materials for which eddy currents are not effective. Advanced multichannel digital instruments are available to measure the magnetic flux. The system has the capability to generate, manipulate, and display the magnetic flux signals and a probe to introduce the magnetic field into the sample component being tested. This technique has been used in inspection of oil field tubulars for corrosion. It is now used in small-bore tubes and in plate. Other applications include inspection of heat exchanger and boiler tubes for corrosion or erosion.

Remote field eddy current technique is an inspection technique, but with repeated surveys of the same area at a frequency determined by the severity of corrosion, metal loss over time can be determined and long-term corrosion rates calculated. The method can be used on-line, but this is not common and the sensitivity is too low to permit real-time measurements (Fig. 2.45).

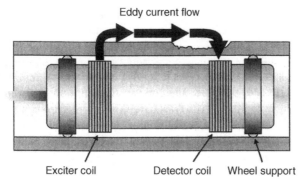

FIGURE 2.45   Risers remote field ET of a pipe (34).

2.6.5.5 Radiography   The thickness of corroded piping and other equipment can be obtained from radiographic images. The difference in optical density of the film in a noncorroded area of the image compared with the optical density of the film in the pitted area can be correlated with the difference in thickness of the two areas, and thereby the pit depth is determined. With repeated surveys of specific areas on a frequency determined from the severity of corrosion, the changing depth and area of corrosion can be resolved and corrosion rates calculated. The technique has been used in harsh oil field environments. The method can be used on-line, but is too insensitive to provide real-time data.

The radiographic technique can be used inexpensively to determine the integrity of piping and equipment over large areas, using either manual or automated real-time radiographic inspection systems. Since the technique does not require access to the component being inspected, insulated, clad, bundled, or otherwise inaccessible piping can be inspected and the degree of corrosion determined.

2.6.5.6 Thin-Layer Activation and Gamma Radiography   This technique involves exposing a small section of material of a component to a high-energy beam of charged particles to produce a radioactive surface layer. A proton beam may be used to produce radioactive Co^{56} within a steel surface, and this isotope decays to Fe^{56} with the emission of gamma radiation. The metallurgical properties of the component are not affected by the radiation. The changes in gamma radiation emitted from the surface layer are measured with a detector to study the rate of material removed from the surface. The measured gamma radiation depends on the quantity of the original radioactive isotope present and its half-life. This technique has been used to measure wear, erosion, and corrosion when the corrosion product is removed from the surface. Different components of the system can be irradiated at the same time with different isotopes and studied simultaneously. Areas less than $1.0\,mm^2$ can be monitored, and, specifically, welds and weld heat-affected zones serve as examples.

Gamma radiation can penetrate up to 5 cm of steel. The measured gamma radiation depends on the quantity of the original radioactive tracer present and the natural decay of the isotope with time. The technique can measure wear, erosion, and corrosion when the corrosion product is removed from the surface. The generation of radio-active tracer in the thin surface layer may be used on coupons, sensor, or on plant components. Separate components of a system can be irradiated simultaneously and measurements can be made. The technique may be used on-line to make real-time measurements for high metal loss rates. Actual plant components such as impellers and valve seats can be evaluated.

Thin-layer activation and gamma radiography measures material loss and cannot distinguish between erosion and corrosion. The corrosion products adhering to the metal surface are not measured as material loss.

2.6.5.7   Electrical Field Mapping   The technique of electrical field mapping, also known as field signature method, can monitor corrosion of a pipe wall directly. This technique was developed for use in the oil and gas production industry. Electrical field mapping is accomplished by measuring local variations in voltage from an induced current applied directly to a pipe to monitor changes due to the effect of internal or external cracking, pitting, corrosion, or erosion in the measurement location. The induced current is fed into the pipe, typically 3–10 m (about 9–30 ft) apart for a large pipe, or a few centimeters or inches for small pipes, to provide uniform current distribution. Welded, glued, or spring-loaded contact pins make electrical connections externally. The voltage readings are monitored and com-pared with one another to detect a nonuniformity that could be due to cracking or pitting in the monitored section. The average voltage drop is compared to a reference voltage and used to monitor general metal loss. The area monitored depends on the pin spacing. Increasing the pin spacing reduces the resolution for localized corrosion and increases the resolution for general corrosion. The reso-lution of general corrosion is inversely proportional to wall thickness. The technique is used on-line to provide data on metal loss. The technique is useful in inaccessible locations, since no access is required after initial installation. There are no consumable parts to the technique except the pipe spool itself when used in high corrosion rate conditions.

2.6.6   Indirect On-Line Measurement Techniques

Many techniques have been used to measure the indirect changes occurring either in the environment or in the metallic component during the corrosion processes. Corrosion processes encompass many branches of science such as metallurgy, physics, biology, chemistry, chemical engineering, and mechanical engineering, and as a result many tools have been developed to monitor and assess the effects and factors involved in corrosion damage.

2.6.6.1   Hydrogen Monitoring   In a nonoxidizing acidic medium hydrogen is formed as a cathodic reaction product, which in its atomic form can diffuse through the vessel wall or pipe wall and recombine to form molecular hydrogen at the exterior surface. This "uptake" of hydrogen occurs when the hydrogen recombination poisons, such as arsenic, antimony, cyanides, selenium, or sulfur compounds are present. The generation of atomic hydrogen can be used for corrosion monitoring. Hydrogen monitoring devices can be attached to the outside walls of vessels and pipes. The diffusion of atomic hydrogen through the vessel or pipe results in hydrogen-induced cracking (HIC).

Hydrogen monitoring is applied in oil refining and petrochemical industries with hydrocarbon process streams. Hydrogen sulfide present in oil operations results in the absorption of hydrogen into plant components. The three types of hydrogen probes function as follows:

(i) Hydrogen pressure (or vacuum) probe can be intrusive and involves the insertion of a steel tube or cylinder that has an inner cavity. The pressure in the inner cavity of the tube or cylinder is measured with a pressure gauge. Another nonintrusive probe consists of a patch or foil welded or sealed outside of the pipe or vessel to create a cavity. Hydrogen passing through the wall of the tube or cylinder is detected as an increase in pressure in the cavity as a function of time. The higher the flux, the greater the rate of pressure increase with time (7). Various hydrogen probes attached to a pipe are shown in Fig. 2.46.

(ii) The electrochemical hydrogen patch probe is a nonintrusive probe and is attached to the outer surface of a pipe or vessel that is being monitored. The probe consists of an electrochemical cell containing an electrolyte in contact with a pipe or vessel wall (Fig. 2.47). The cell has a nickel electrode in contact with an electrolyte put directly on the pipe. The electrolyte is often isolated from the pipe or vessel by a thin palladium foil. The palladium foil is also used as an electrode in the detection circuit. The cell operates at a potential that oxidizes hydrogen and the current used to maintain the potential is proportional to the hydrogen flux in the cell (7).

(iii) The hydrogen fuel cell probe shown in Fig. 2.48 is another nonintrusive device, containing a solid electrolyte membrane, a cathode, and an anode surface catalyzed with palladium. The hydrogen entering the cell is reacted on

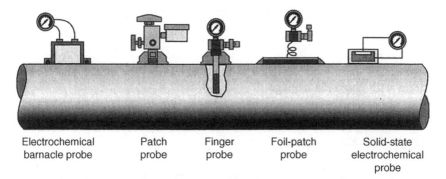

| Electrochemical barnacle probe | Patch probe | Finger probe | Foil-patch probe | Solid-state electrochemical probe |

FIGURE 2.46 Various hydrogen probes attached to a pipe.

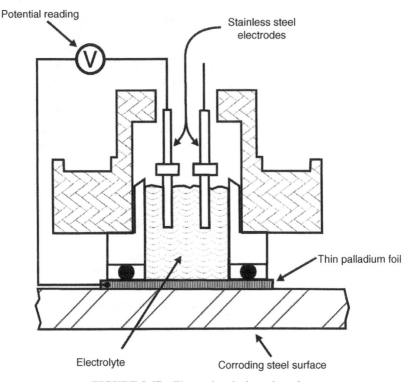

FIGURE 2.47   Electrochemical patch probe.

the activated palladium surface, while the cathode reacts with the oxygen in air, resulting in a current flow in the external circuit between the cathode and the anode. The current is proportional to the hydrogen flux through the palladium membrane (35).

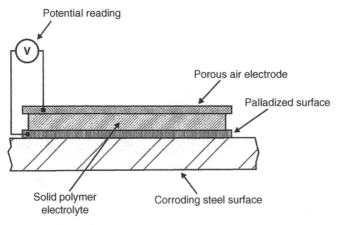

FIGURE 2.48   Schematic of hydrogen detecting fuel cell.

Only when the distribution of atomic hydrogen in a sample material is constant can the permeating fraction of hydrogen give the relative corrosion rate. The technique is used to detect high flow rates of hydrogen passing through steels, causing blisters and hydrogen-induced cracking. Hydrogen evolution is one of the many possible cathodic reactions that can occur with the actual anodic corrosion reaction. The technique is incapable of giving information on corrosion rates when other cathodic reactions such as reduction of oxygen is occurring.

2.6.6.2 Corrosion Potential   Corrosion potential (also known as rest potential, open-circuit potential, or freely corroding potential) is the potential of a corroding surface in an electrolyte relative to a reference electrode under open-circuit conditions. The potential is generally measured with respect to a reference electrode such as a calomel electrode, silver/silver chloride, or copper/copper sulfate.

Corrosion potential is useful when it is related to other measurements of the same phenomenon. The value of corrosion potential may be used to help predict the corrosion behavior by comparison with polarization data obtained from laboratory or site polarization scans. The corrosion potential is also useful in obtaining Pourbaix diagrams (E versus pH). The corrosion potential may be used to ascertain whether the stainless steel is in an active or passive region. The technique is useful in predicting corrosion behavior such as passivity, uniform corrosion, onset of pitting, and sensitivity to stress-corrosion cracking. The technique is useful in controlling cathodic and anodic protection systems, which hold the material at passive potentials. The technique can only indicate corrosion risk and is incapable of measuring corrosion rate.

The reference electrode can be unstable in process stream. The reference electrode can also cause contamination. Special reference electrodes are required for use at temperatures above 100°C.

2.6.6.3 On-Line Water Chemistry Parameters   A variety of chemical analyses can give useful information to corrosion monitoring. Parameters such as pH, conductivity, dissolved oxygen, metal ion concentrations, water alkalinity, concentration of suspended solids, inhibitors, and scaling indices play an important role in corrosion. Many of these parameters can be measured on-line and used to generate indices such as the Langelier index, which in turn determines the corrosivity of the aqueous medium.

2.6.6.3.1 pH   pH is defined as the negative logarithm of the hydrogen ion activity.

$$pH = -\log_{10}[a_{H^+}].$$

pH denotes whether the solution is acidic or basic. A value of 7.0 denotes that the solution is neutral, a value less than 7.0 means the solution is acidic, and a value greater than 7.0 means the solution is basic or alkaline. A difference of 1.0 pH unit indicates a 10-fold difference in hydrogen ion activity. In general, for steel, low pH (high acidity) produces a corrosive environment. This value can vary for other alloys.

pH is the most important measure of the corrosiveness of aqueous process streams toward plant materials. The technique of looking at pH is used on-line for real-time measurement. The hydrogen ion is a reactant for cathodic reaction. The pH also affects the solubilities of corrosion reactions, such as passivation involving oxides, sulfides, or carbonates.

pH is an input variable in indices such as the Langelier index used in aqueous systems. pH is an important factor in Pourbaix diagrams, in which potential is plotted against pH (E versus pH). pH can also be measured directly and continuously in-line or in side streams.

pH can be measured up to 99°C (210°F) and up to a pressure of 2 MPa (300 psig) and some special probes are available for up to 70 MPa (10,000 psi). pH measurements are affected by ions such as lithium, sodium, and potassium, which interact with the sensing glass membrane of a pH electrode. Lithium ions are not common, and real serious interference is due to sodium ions.

Fouling of the pH measuring probe by hydrocarbons can lead to serious problems. Low-conductivity solutions can also present problems. Even low-ionic strength pH probes can have problems at conductivities less than 20 micro ohms/cm. Some low-conductivity probes inject an electrolyte to correct for high solution resistance.

2.6.6.3.2 Conductivity   Conductivity is the current-carrying capacity of a liquid in terms of the current that flows when a constant potential is applied. Conductivity of a solution depends upon the concentration of dissolved ionic solids. Since corrosion is electrochemical in nature, an increase in conductivity increases the degree of corrosion. Measurement of conductivity can be done on-line and for real-time. Conductivity is a function of the dissolved ionic solids in the process stream, which in turn contributes to the corrosivity of the process streams.

This technique gives a general indication of the concentration of ionic species in liquid phase processes. The technique also measures the salinity in oil. The technique is an indirect measure of total dissolved electrolytes. The technique can be used directly and continuously in-line or in side streams. Conductivity measurements are made down to 1 μs/cm. Although the technique is simple and rapid, routine cleaning of the conductivity cell, effect of electrochemical reactions at the electrode interfaces, and temperature sensitivity are some of the limitations.

2.6.6.3.3 Dissolved Oxygen   Dissolved oxygen is the amount of oxygen dissolved in a solution (liquid), usually expressed as parts per million (ppm) or parts per billion (ppb). The solubility of oxygen depends on temperature, pressure, and molarity of the solution. Increased pressure increases oxygen solubility, while an increase in temperature decreases its solubility.

Dissolved oxygen can be measured by ion-selective electrodes or by a galvanic probe shown in Fig. 2.49. The oxygen ion-selective electrode consists of a thin organic

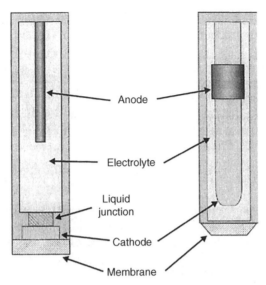

FIGURE 2.49 Galvanic probes for measuring dissolved oxygen.

membrane covering a layer of electrolyte and two metallic electrodes. Oxygen diffusing through the membrane is reduced electrochemically at the cathode. The fixed voltage between the cathode and the anode allows only the reduction of oxygen. The oxygen diffusing through the membrane is reduced and the resulting current is proportional to the amount of oxygen (36).

The corrections for sample and membrane temperatures are made by temperature sensors built in the probe. The cathode current, sample and membrane temperatures, barometric pressure, and salinity data are used in the calculation of the dissolved oxygen content of the sample (ppm or ppb or mg/L) or percent saturation.

Oxygen is probably the most well-known corrosive agent for metals and alloys used in structures. Oxygen takes part in corrosive attack as well as in passivation. Controlling corrosion of carbon steels used in oil fields and in boilers is achieved by removing all or most of the oxygen from the system. Lowering the oxygen levels to less than 20 ppb or 20 μg/L lowers the corrosion rate significantly. The oxygen level of <1 ppb or <1 μg/L can be achieved by the use of oxygen scavengers.

In some boilers using high-purity deionized water, a small amount of oxygen or hydrogen peroxide is added to the feed water, which act as a passivating agent under some carefully controlled boiler water chemistry conditions.

The measurement of dissolved oxygen concentration is limited to a maximum temperature of 65°C and a maximum pressure of 200–300 kPa. The solution should be stirred to enable correct measurement of oxygen by the probe. Due to the depletion of the electrolyte over time, the electrolyte must be replenished at regular intervals of a few weeks or months. Another problem is the poisoning of the electrode, which may preclude its use in environments such as heavy hydrocarbons and water.

2.6.6.3.4 Oxidation–Reduction Potential   Oxidation–reduction (redox) potential is the potential of a reversible oxidation–reduction electrode measured with respect to a reference electrode, corrected to the hydrogen electrode, in a given electrolyte. The measurement system consists of a potentiometer (high-sensitivity and high-imped-ance voltmeter) connected to a reference electrode and a noble metal sensing electrode. The change in potential of the sensing electrode produced by the oxidizing or reducing species is measured with respect to the reference electrode. This voltage difference is then converted electronically to display on a meter or recorder for real-time on-line measurement.

This technique has been used to detect the endpoint of oxidation or reduction reactions during water treatment so that one may control more closely the addition of oxidizing biocides, such as chlorine and bromine, which may have a significant effect on increasing the corrosion rates.

Redox potential is an important parameter in the evaluation of soil corrosivity, wherein the redox potential is essentially related to the degree of aeration. A high value of redox potential indicates a high level of oxygen in the system. Low redox potential values may indicate that the conditions are conducive to anaerobic micro-biological activity. Sampling of soil will lead to exposure of the soil to oxygen, which in turn results in unstable redox potential measured in the soil sample. This technique is useful in measurement of microbiological activity.

2.6.7   Fluid Detection

2.6.7.1   Flow Regime   Corrosion usually occurs on the metal surface wetted by the aqueous phase. The wetting and transport of the corrosive agent to the metal surface is controlled by the flow regime. In a flow regime consisting of slug flow, high surface shear stresses and extreme turbulence can make corrosion inhibition difficult and increase corrosion rates to high values of the order of several centimeters per year. At high flow rates, impingement, cavitation, and erosion conditions created by two-phase flows can result in an extremely corrosive scenario.

The flow regime in single-phase flows can be laminar, transition, or turbulent flow as defined by the value of the Reynolds number, which depends on the flow velocity, pipe diameter, fluid viscosity, and fluid density. In multiphase flows, the relative pattern of multiple phases such as mist flow, annular flow, or slug flow is important. In single-phase environments, the flow regime affects the mass transfer to the metal surface and the shear stresses on the metal surface. These factors directly impact the corrosion rates at the metal surface.

In multiphase flows, the flow regime varies with varying flow rates of each phase, elevation changes, or a specific geometry of a line or conduit. Acoustic monitoring and on-line γ-radiography have been used to detect slug flow. Flow characteristics can be used to predict corrosion rates based on modeling and other empirical data.

Flow regime is not a measurement technique but only an indication of problem area or frequency of a problem. This problem is identified based on a thorough understanding of the precise corrosion mechanisms as they occur in process streams. This can at times enable monitoring of the vital characteristics of the corrosivity of the flow.

2.6.7.2 Flow Velocity   In single-phase flow, the velocity can be determined from the mass flow rate using the necessary corrections for pressure and temperature. In multiphase flow, the velocity is more difficult to determine due to the slip between the phases, and the velocity is often not constant. Correlations may be used to estimate the slip and determine flow velocities for each phase. This technique is used on-line for real-time measurement.

Flow velocity has strong influence on corrosion. Corrosion is usually severe at extremes of velocity such as stagnant conditions or velocities two or three times higher than API RP14E limits (37). Flow/no flow detection can be used, particularly when oxygen content is important to corrosion. In other situations, orifices may be used to create a minimum or maximum flow.

2.6.7.3 Process Parameters   These parameters are related to fluid hydrodynamics and fluid mass transfer properties, which have a bearing on the corrosivity of the fluid.

2.6.7.4 Pressure   Pressure has an effect on the proportion of phases present in a vessel or pipe, or the composition of the process fluids. Different phases and constituents can produce quite different corrosive environments. For example, the partial pressure of CO_2 affects the amount of dissolved CO_2, which in turn affects the corrosivity of the fluid due to the presence of carbonic acid. Similarly, the partial pressure of H_2S is a major determining factor in the susceptibility of various alloys to sulfide stress cracking.

Total pressure can be measured on-line and it is a real-time measurement. Determination of partial pressure utilizes knowledge of the composition of the process fluid, the temperature, and the total pressure. This involves sampling of the process fluid. In some gas–liquid systems, pressure affects gas solubility, but the relationships are complex. Hence, the pressure is usually used only to analyze and predict which phases may be present, rather than as an on-line measurement.

2.6.7.5 Temperature   The temperature of a process fluid can have a direct effect on its corrosivity. These temperature effects can be nonlinear. Low temperatures can produce condensation of water or other corrosive liquids. High temperatures increase chemical reaction rates and can change the composition of the process. The temperature can lead to either vaporization (a dry condition) or condensation (from a dry to wet condition). Both vaporization and condensation of liquids can affect corrosion. As a process measurement, this technique is used on-line for real-time measurement.

2.6.7.6 Dew Point   Dew point is the temperature at which a liquid begins to condense. This phenomenon can be complex in some cases where multiple gases and condensable combinations are present. Dew point monitoring is important because the region of water and corrosive fluids in otherwise dry environments has a major impact on corrosion rates. In aqueous environments in which the condensing liquid is water, the dew point temperature is measured with a wet-bulb thermometer. This type of measurement is not suitable in a process system; cooling of the flow to the point of condensation on an optical mirror is used to measure the dew point.

The direct effect of dew point conditions can be measured with mass-loss coupon or electrical resistance techniques. Electrochemical methods such as electrochemical noise and multielectrode array systems have been used either as a time-of-wetness indicator or for corrosion rate measurements, since these techniques function even when the surfaces are partially wet. Forced cooling of the corrosion measurement surfaces can be used to generate the appropriate dew point on the measurement surfaces. This technique can be used on-line for real-time measurement.

2.6.7.7 Fouling   Fouling consists of the accumulation of both organic and inorganic substances from fluid streams onto the surfaces of equipment through which fluid circulates as well as corrosion products and inorganic salts such as hardness salts that deposit on the metal surface. Fouling causes underdeposit corrosion, since it produces highly anodic areas and gives rise to hot spots in boilers. Fouling also can restrict the flow by an increase in pressure drop through the equipment or retard heat transfer by forming an insulating deposit. Side-stream and on-line intrusive measurement techniques, together with visual inspection, have been used for real-time determination of fouling. Either heat-transfer monitoring or pressure-drop monitoring has been used.

2.6.8   Indirect Off-Line Measurement Techniques

A variety of chemical analytical techniques have been used to assess the corrosivity of the environments, effect of added chemical agents, such as corrosion inhibitors, in order to provide useful information to operators and providers of corrosion control services (7).

2.6.8.1 Off-Line Water Chemistry Parameters

2.6.8.1.1 Alkalinity   Alkalinity is an important parameter in water, since carbonate and hydroxide scales are common problems. Hydroxides, bicarbonates, and carbonates impart alkalinity to water. Alkalinity is a measure of the acid-neutralizing capacity of water.

Alkalinity can be determined by titration with standard acid to methyl orange endpoint (pH = $\sim$4.5). Alkalinity affects fluid chemistry and can be integrated in a chemical balance for total metal loss in low corrosion rates (used in the nuclear industry). It is useful in treatment of water to prevent corrosion where acid additions may occur and to assure correct neutralization occurs and that the desired pH is attained.

2.6.8.1.2 Metal Ion Analysis Metal ion analysis of a process stream is used as a technique for the determination of the amount of metal lost that has dissolved in the process stream or that has been carried along as corrosion product (e.g., iron, copper, nickel, zinc, manganese). Some operators have routinely used analysis. The technique is not used on-line or for real-time measurement. The assumption that the metal loss occurred over the total surface area or that the concentration of metal ions in solution is proportional to the corrosion rate is not correct, and it is only an indication of the corrosion trend. An increase in metal ion concentration indicates an increase in corrosion rate qualitatively. A low concentration of metal ion does not translate into a low corrosion rate, since localized corrosion, deposition of metal ion due to changes in pH and temperature, or a considerable time delay before analysis might account for the low concentration of metal ion.

Metal ion analysis is useful in closed systems or when the corrosion products are soluble or can be related to particular concentrations of soluble species. Relative changes in metal ion concentrations between locations in open systems can give useful information. This technique is not suitable in environments such as hydrogen sulfide and alkalis due to formation of insoluble metal sulfides and hydroxides.

Care should be exercised in the choice of sampling point since some points may accumulate corrosion products. It should also be noted that factors such as fluid velocity, temperature, and pressure can vary greatly in the process stream and may contribute to the difficulty in obtaining a suitable sample.

2.6.8.1.3 Concentration of Dissolved Solids Total dissolved solids (TDS) is the sum of all the minerals dissolved in water. This is an important parameter in the determination of scaling index. Total dissolved solids is determined by evaporation of the solution of a known weight. A calculated TDS value is obtained by adding the various cations and anions from an analysis. This parameter is related to conductivity; dissolved ionic solids affect fluid chemistry.

2.6.8.1.4 Gas Analysis Gas analysis is generally done in the laboratory but can also be done for certain gases in the field. The analysis is commonly done for hydrogen, hydrogen sulfide, or other dissolved gases. This technique is not used on-line or for real-time measurement.

The technique is used to determine potentially corrosive constituent gases, such as acid gases, which become corrosive when hydrated or the temperature falls below the dew point. The technique is used in the analysis of gaseous corrosion products such as hydrogen. The equipment for analysis is expensive, especially for plant equipment.

2.6.8.1.5 Residual Oxidant Ozone, chlorine, chlorine dioxide, and bromine are powerful oxidizing agents and are extensively used to control microbiological fouling in aqueous media. Residual halogens can directly oxidize the inhibitors used to protect against corrosion or fouling. Dissolved halogens can be measured using redox potential or any colorimetric techniques.

Halides are salts derived from the acids HCl, HBr, HI, and HF. Halides have been implicated in stress-corrosion cracking and are indirectly linked to galvanic corrosion because they increase the conductivity of the medium. Halides can be determined using ion-selective electrodes or by colorimetric methods. This technique is not used on-line or for real-time measurement.

2.6.8.1.6 Microbiological Analysis (38)–(48)   There are several techniques for microbiological analysis. The most common and well-developed technique of analysis is bacterial culturing from the sample to determine the concentration of viable bacteria. Successive dilutions of the sample are put into the culture media and left to grow. The samples that develop give an indication of the original concentration of bacteria. The usual incubation period for the test is 14–28 days, but some results may be seen in a few days. Some of the general bacteria techniques are as follows:

(i) Measurement of adenosine triphosphate (ATP) present in all living cells, but which disappears rapidly on death, can be used as a measure of living material. ATP is measured using an enzymic reaction, which generates flashes of light that are detected by a photomultiplier.

(ii) The total number of viable cells may be determined by using specific stains that fluoresce when irradiated with ultraviolet light. This technique is done in the laboratory.

(iii) The test consists of analysis for the hydrogenase enzyme that is produced by bacteria that use hydrogen as an energy source. Since cathodic hydrogen is important in microbiologically influenced corrosion (MIC), this test indicates the potential for this type of corrosion. The sample is treated with enzyme-extracting solution and then the extent of hydrogen oxidation in an inert atmosphere, as detected by a dye, is noted.

(iv) The three techniques to quantify sulfate-reducing bacteria are as follows:

(a) Radiorespirometry is specific to SRBs. The test is completed in 2 days. The sample is incubated with a known trace amount of radioactive-labeled sulfate. After incubation, the reaction is terminated with acid to kill the cells, and the radioactive sulfide (the SRBs reduce sulfate to sulfide) is fixed in zinc acetate followed by quantification in the laboratory. This technique is expensive and involves use of radioactive substances.

(b) Antibody fluorescence microscopy uses fluorescent dye to bind to antibodies' specific SRBs. Only the bacteria recognized by the antibodies fluoresce. Test results can be obtained in 2 h. The technique detects both viable and nonviable bacteria. The technique detects only the type of SRB used in the manufacture of these antibodies.

(c) Adenosine phosphosulfate (APS)-reductase measurement consists of reduction of sulfate to sulfide by SRBs in the presence of the enzyme APS-reductase. Measurement of the amount of APS-reductase in a sample gives an estimate of the total numbers of SRBs present. The test does not require bacterial growth and takes 15–20 min.

Bacteria in the system can present either in suspension in the fluid (planktonic) or attached to the surface (sessile). It is the sessile bacteria that cause corrosion. The presence of bacteria does not necessarily cause a problem; bacteria that cause a problem in one system may be harmless in another.

The monitoring techniques of bacteria give an indication of the effectiveness of biocide treatments to kill bacteria. The measurement of sessile bacteria can give an indication of the potential for microbiologically induced corrosion. Rapid techniques have limitations, but rapidly obtained results permit quick remedial action to be taken in a timely manner.

The techniques used for the bacterial analysis are slow and involve manual sampling and are not conducive to automation or remote operation.

2.6.8.1.7 Residual Inhibitor   Measurement of corrosion inhibitor residuals in a system gives an indication of the concentration of inhibitor at various locations in the system. When the reliability of an analytical test method is established and the minimum acceptable inhibitor concentration has been established, measurement of inhibitor concentrations throughout a system can indicate whether necessary corrosion protection is likely to be achieved at each sampling location. Inhibitor loss due to reaction with the system hardware or reaction with the environment such as absorption, neutralization, precipitation, adsorption on solids, or corrosion products can be detected and compensated for by suitable dosage selection.

Inhibitor injection points can be evaluated and the number of injection points determined for lengthy or complex systems by measurement of carry-through of inhibition. Inhibitor selection can be carefully refined after selection of potential inhibitors by determination of the products that are transported through the system at optimum usable concentrations with minimum loss due to oxidation, reduction, adsorption byproducts, or other solids or reaction with system physical equipment.

2.6.8.1.8 Filming Corrosion Inhibitor Residual   Filming corrosion inhibitors are chemical substances that are used in low concentrations to adsorb on system surfaces. The adsorbed inhibitor film shields the system equipment from corrodents in the environment. Inhibitor residual measurements are generally made to ensure that an adequate supply of inhibitor has been introduced into the system to compensate for reduction of concentration by adsorption and maintenance of inhibitor film.

Procedures can be used to measure and ensure that adequate transport of preselected inhibitor exists throughout the system. Sudden or unexpected loss of inhibitor due to system changes can be detected and corrected before significant damage to equipment occurs. Inhibitor residual measurements can be used to ensure adequate treatment of the system when it is not possible to use direct corrosion-measuring devices, as in the case of buried pipelines.

2.6.8.1.9 Reactant Corrosion Inhibitor Residual   Reactant corrosion inhibitors are chemical reagents that react with potential corrosive agents in a system to neutralize the harmful effect of the corrodents and combine with constituents in a system to produce *in situ* corrosion inhibitors. Reactant corrosion inhibitor residual measurements are made to ensure that sufficient amounts of the reactant corrosion inhibitor have been injected in the system to provide an excess of the inhibitor. Measurement of residual reactant inhibitor can be done after the point of introduction of oxidant, such as oxygen or air bubbling, to ensure that sufficient inhibitor has been injected to complete the reaction while leaving behind some residual inhibitor.

Reactant corrosion inhibitors can be alkalis to neutralize acidic systems to maintain a safe operating pH. For instance, it may be necessary to adjust the pH at the point of condensation of acid gases in a distillation system. The important parameter in determining the effectiveness of such an inhibitor is the ability to measure pH at appropriate points in a system. Reactant corrosion inhibitors may also be used to react with or reduce oxidants in a system; examples of these inhibitors, also known as oxygen scavengers, are sulfites, bisulfites, and hydrazine.

Reactant corrosion inhibitors such as chromates, ferrocyanides, sulfur or sulfide compounds, phosphates, or acetylenic alcohols must be present in sufficient amounts to provide a protective film and leave a residual to repair breaks in the protective barrier.

2.6.8.1.10 Chemical Analysis of Process Samples   Chemical analysis of process samples taken at times of high and low corrosion rates can be useful in identifying the constituents that are responsible for high corrosion rates. Using this information, the source of aggressive species can be identified and corrected. In petroleum production, crude oil and gas condensate samples are usually analyzed for organic nitrogen and acid content. Sulfur, organic acid, nitrogen, and salt content are usually analyzed for refining purposes. These parameters are used in combination to predict the degree of corrosivity of the oil.

In gas handling and gas processing operations, samples of produced and processed natural gas and gas–liquids are generally analyzed to determine hydrogen sulfide (H_2S), carbon dioxide (CO_2), water (H_2O), carbonyl sulfide (COS), carbon disulfide (CS_2), mercaptans (RSH), and/or oxygen (O_2) to predict and assess corrosion potential in producing wells, gas gathering systems, and gas processing operations.

In chemical analysis of process samples, a representative sample of a process stream is taken and kept in a vessel that maintains the original condition of the sample for subsequent analysis. The sampling methods can be quite complex since the samples need to be kept under the pressure conditions of the process stream. This can be particularly difficult when one is dealing with high-pressure process systems.

2.6.8.1.11 Sulfur Content   Sulfur is the most abundant element in petroleum other than carbon and hydrogen. Corrosion of carbon steel may become high when the sulfur content is greater than 0.2%. Sulfur can be present as elemental sulfur, dissolved hydrogen sulfide, mercaptans, sulfides, and polysulfides. The total sulfur content is analyzed according to ASTM D4294. Halides and heavy metals interfere with this method. The sulfur compound forms H_2S during heating in the refinery process, which correlates with the corrosion in plants.

2.6.8.1.12 Total Acid Number   The acid content of a system is generally expressed in terms of TAN, or neutralization (neut) number. For the purpose of predicting corrosion in crude distillation units in refineries, the TAN threshold is believed to be $\sim$0.5 for whole crude and 2.0 for the cuts. In petroleum production it has been found that the corrosion rate is usually inversely proportional to the algebraic product of the nitrogen concentration and TAN.

In this monitoring technique, an oil sample is dissolved in a mixture of toluene and isopropyl alcohol containing a small amount of water. The solution is then titrated with an alcoholic potassium hydroxide solution. Both ASTM D664 and ASTM D974 can be used to determine TAN. It should be noted that the ASTM D664 method gives data 30–80% higher than the ASTM D974 method.

Inorganic acids, esters, phenolic compounds, sulfur compounds, lactones, resins, salts, and additives such as inhibitors and detergents sometimes interfere with the measurement. Universal Oil Products (UOP) methods 565 and 587 give procedures for an operator to remove most of the interfering substances before performing the analysis for organic acids.

2.6.8.1.13 Nitrogen Content   The total nitrogen is generally determined using ASTM D3228 method in order to assess the corrosivity of process feed stocks. High nitrogen content indicates corrosion-inhibitive properties of the crude oil or condensate in petroleum production. As noted before, the corrosion rate is inversely proportional to the algebraic product of the organic nitrogen concentration in weight percent and the TAN. However, in refining, when the organic nitrogen concentration exceeds 0.05% cyanide, ammonia can form, collect in the aqueous phase, and corrode certain materials.

2.6.8.1.14 Salt Content of Crude Oil   Salt (primarily sodium chloride with lesser amounts of calcium and magnesium chloride) is present in produced water, and the produced water can be dispersed, entrained, and/or emulsified in crude oil. Salt can cause corrosion of refinery equipment and piping. Salt precipitates can form scale in heaters and heat exchangers and this can cause accelerated corrosion of equipment. Refineries generally limit salt content of crude oil for processing to 2.5–12 mg/L.

The salt content of crude oil can be determined by the method given in ASTM D3230. The analytical procedure assumes that calcium and magnesium are present as chlorides and all the chloride is calculated as sodium chloride.

REFERENCES

1. DC Eden, MS Cayard, JD Kintz, RA Schrecengost, BP Breen, E Kramer, *Corrosion 2003*, Paper No. 376, Houston, TX, 2003.
2. ASTM STP908, *Corrosion Monitoring in Industrial Plants Using Nondestructive Testing and Electrochemical Methods*, Philadelphia, PA, 1986.
3. DE Powell, DI Maruf, IY Rahman, *Mater Perf* **4**:50–54 (2001).
4. GL Edgemon, Electrochemical noise based corrosion monitoring, Hanford Site Program. *Corrosion 2005*, National Association of Corrosion Engineers, Houston, TX, Paper No. 584, 2005.
5. L Yang, IV Sridhar, O Pensado, DS Dunn, *Corrosion* **58**:1004–1014 (2002).
6. MJJS Thomas, S Terpsta, *Corrosion 2003*, NACE International, Houston, TX, Paper No. 431, 2003.
7. *Techniques for Monitoring Corrosion and Related Parameters in Field Applications*, NACE, 3T199, NACE International, Houston, TX, 1999.
8. ASTM G4-01, *Standard Guide for Conducting Corrosion Tests in Field Applications*, ASTM, West Conshohocken, PA, 2001.
9. SW Dean, Corrosion monitoring for industrial processes. In: DS Cramer, BS Covino (eds.), *Corrosion: Fundamentals, Testing and Protection*, Vol. 13A, Metals Park, OH, ASM International, pp. 533–541, 2003.
10. ASTM G31-72, 2004, *Standard Practice for Laboratory Immersion Corrosion Testing of Metals*, ASTM, West Conshohocken, PA.
11. ASTM G71-81, *Standard Guide for Conducting and Evaluating Galvanic Corrosion Tests in Electrolytes, Annual Book of ASTM Standards*, Vol. 03.02, ASTM, Philadelphia, PA, 2003.
12. ASTM G78-01, *Standard Guide for Crevice Corrosion Testing of Iron-Base and Nickel-Base Stainless Alloys in Seawater and Other Chloride-Containing Aqueous Environ-ments*, Vol. 03.02, ASTM, West Conshohocken, PA, 2001.
13. ASTM G48-03, *Standard Test Methods for Pitting and Crevice Corrosion Resistance of Stainless Steels and Related Alloys by Use of Ferric Chloride Solution, Annual Book of ASTM Standards*, Vol. 03.02, ASTM, Philadelphia, PA, 2003.
14. ASTM G30-97, *Standard Practice for Making and Using U-bend Stress-Corrosion Test Specimens*, Vol. 03.02, ASTM, West Conshohocken, PA, 2003.
15. ASTM G38-01, *Standard Practice for Making and Using C-Ring Stress-Corrosion Test Specimens*, Vol. 03.02, ASTM, West Conshohocken, PA, 2001.
16. ASTM G58-85, *Standard Practice for Preparation of Stress-Corrosion Test Specimens for Weldments*, Vol. 03.02, ASTM, West Conshohocken, PA, 1999.
17. ASTM G1-03, *Standard Practice for Preparing, Cleaning, and Evaluation of Corrosion Test Specimens*, Vol. 03.02, ASTM, West Conshohocken, PA, 2003.
18. RH Hausler, Corrosion inhibitors. In: R Baboian (ed.), *Corrosion Tests and Standards*, 2nd edition, ASTM, West Conshohocken, PA, 2005, pp. 480–499.
19. A Dravnieks, HA Cataldi, *Corrosion* **10**:224–230 (1954).
20. AJ Freedman, ES Troscinski, A Dravnieks, *Corrosion* **14**:175–178 (1958).
21. ASTM G96-90, *Standard Guide for On-Line Monitoring of Corrosion in Plant Equipment (Electrical and Electrochemical Methods), Annual Book of ASTM Standards*, Vol. 03.02, ASTM, Philadelphia, PA, 2001.

22. AF Denzine, MS Reading, *Mater Perf* **37**:35–41 (1998).

23. AC Van Orden, Applications and problem solving using the polarization technique, *Corrosion 98*, Paper No. 301, NACE International, Houston, TX, 1998.

24. ASTM G96-90, *Standard Guide for On-Line Monitoring of Corrosion in Plant Equipment (Electrical and Electrochemical Methods), Annual Book of ASTM Standards*, Vol. 03.02, ASTM, Philadelphia, PA, 2001.

25. B Yang, *Corrosion* **51**:153–165 (1995).

26. PR Roberge, VS Sastri, *Corrosion 93*, Paper No. 396, 1993.

27. R Grauer, PJ Moreland, GA Pini, *Literature Review of Polarization Resistance Constant (B) Values for the Measurement of Corrosion Rate*, NACE International, Houston, TX, 1982.

28. PR Roberge, VS Sastri, *Corrosion* **50**:744–754 (1994).

29. GD Davis, S Raghu, BG Carkhuff, F Garra, R Srinivasan, TE Phillips, Corrosion health monitor for ground vehicles, *Tri-Service Corrosion Conference*, Paper No. 103, Orlando, FL, Nov 14–18, 2005.

30. RW Bosch, J Hubrecht, WF Bogaerts, BC Syrett, *Corrosion* **57**:60–70 (2001).

31. American National Standards Institute (ANSI)/ASME B 31G, New York, 1971.

32. American Petroleum Institute (API), Washington, DC 20005, 1972.

33. ASME Boiler & Pressure Vessel Code, Section VIII, Division 1 (latest revision), Rules for Construction of Pressure Vessels, ASME, New York, NY, 1970.

34. M Lozev, B Grimmelt, E Shell, R Spencer, Evaluation of Methods for Detecting and Monitoring of Corrosion and Fatigue Damage in Risers, Project No. 45891GTH, 11-4-2003, Minerals Management Service, U.S. Department of the Interior, Washington, DC, 2003.

35. JR Vera, C Mendez, S Hernandez, S Cerpa, Field results of the hydrogen permeation sensor based on fuel cell technology, *Corrosion 2002*, Paper No. 346, NACE International, Houston, TX, 2002.

36. ASTM D888-03, *Standard Test Methods for Dissolved Oxygen in Water, Annual Book of ASTM Standards*, ASTM, West Conshohocken, PA, 2003.

37. API RP 14E (latest revision), *Recommended Practice for Design and Installation of Offshore Production Platform Piping Systems*, API, Washington, DC, 1965.

38. API RP 38 (latest revision), *Recommended Practice for Biological Analysis of Subsurface Injection Waters*, API, Washington, DC, 1966.

39. JA Hardy, KR Syreh, *Eur J Appl Microbiol Biotechnol* **17**:49–51 (1983).

40. GL Horacek, LJ Gawel, New test kit for rapid detection of SRB in oilfield, *63rd Annual Technical Conference of Petroleum Engineers*, Paper No. SPE 18199, SPE, Richardson, TX, 1988.

41. ES Lihmann, Oilfield bactericide parameters as measured by ATP analysis, *International Symposium on Oilfield Chemistry*, Paper No. 5312, SPE, Dallas, TX, 1975.

42. *Microbiologically Influenced Corrosion and Biofouling in Oilfield Equipment*, TPC #3, NACE, Houston, TX, 1990.

43. *Microbiological Methods for Monitoring the Environment Water and Wastes*, U.S. Environmental Protection Agency, Cincinnati, OH, 1978.

44. NACE Standard TM0173 (latest revision), *Methods for Determining Water Quality for Subsurface Injection Using Membrane Filters*, NACE, Houston, TX, 1973.

45. DH Pope, TP Zintel, Methods for the investigation of under-deposit microbiologically induced corrosion, *Corrosion/88*, Paper No. 249, NACE, Houston, TX, 1988.

46. R Prasad, Pros and cons of ATP measurement in oil field waters, *Corrosion/88*, Paper No. 87, NACE, Houston, TX, 1988.

47. HR Rosser, WA Hamilton, *Appl Environ Microbiol* **6**:45 (1983).

48. *Standard Methods for Examination of Water and Wastewater*, 17th ed., American Public Health Association, Washington, DC, 1989.

3

ADSORPTION IN CORROSION INHIBITION

3.1 ADSORPTION OF INHIBITOR AT THE METAL SURFACE

It is generally agreed that corrosion inhibition is due to the adsorption of the inhibitor molecule at the metal–solution interface, which is accompanied by a change in potential difference between the metal electrode and the solution due to the non-uniform distribution of electric charges at the interface. The metal–electrolyte interface is characterized by an electrical double layer, sometimes by a triple layer. A schematic representation of the electrical double layer is given in Fig. 3.1. The first layer is a sheet of charges at the metal surface caused by an excess or deficiency of electrons. The second layer (region A) is formed on the solution side of the interface by specially adsorbed ions. The loci of the centers of these charges form the inner Helmholtz plane of the double layer. These anions lose their coordinated water molecules or water sheaths, displace adsorbed water molecules from the metal surface, and in turn are adsorbed on portions of the bare metal surface. These ions are known as potential-determining ions. The charges are balanced in part by hydrated ions of opposite charge in the outer Helmholtz plane in region B called counterions. Outside this area (i.e., region C in the figure) is known as the Gouy–Chapman diffuse layer, where the concentrations of the counterions decrease toward that of bulk electrolyte and balance the net charge close to the metal surface.

The ions forming the double layer are distributed not only because of their kinetic motion and the surface electric field but also because of specific chemical interactions between the ions and interface, that is, where region A was reached by the

Green Corrosion Inhibitors: Theory and Practice, First Edition. V. S. Sastri.
© 2011 John Wiley & Sons, Inc. Published 2011 by John Wiley & Sons, Inc.

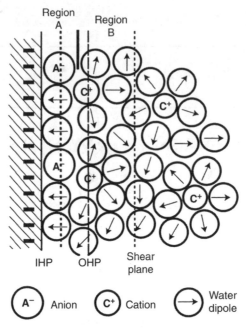

Region A

Region B

IHP   OHP   Shear plane

A⁻ Anion   C⁺ Cation   Water dipole

IHP = inner Helmholtz plane
OHP = outer Helmholtz plane

FIGURE 3.1   Schematic representation of the electric double layer.

ions. These interactions depend upon distance and encompass hydrogen and covalent bonds and π bonds or hydrophobic interactions that do not occur in the outer regions.

The variation of potential that occurs with distance from the interface is depicted in Fig. 3.2. Considering the regions A and B as a parallel plate capacitor, the potential drops linearly from P_a to P_b, and this potential, known as the Stern potential, cannot be measured directly, but is measured with respect to the hydrogen electrode, so that the standard relative electrochemical potentials can be obtained. The addition of a corrosion inhibitor into the electric double layer changes its composition and structure. Hence, measurement of the capacitance of the double layer before and after the addition of the corrosion inhibitor may be used to monitor the adsorption of the inhibitor.

The adsorption of inhibitors can also be monitored by the changes in zeta potential ζ. The zeta potential is the potential required to cause electrokinetic movement within the electrolyte and is considered to be just outside the outer Helmholtz plane. The total potential drop does not change when an indifferent electrolyte is added. However, the thickness of the double layer decreases because the equivalent number of charges of the opposite sign is always required for compensation of the potential-determining ions. Therefore, the potential distribution is changed and the zeta potential is reduced. The interaction of ions or neutral polar molecules with the electric double layer affects

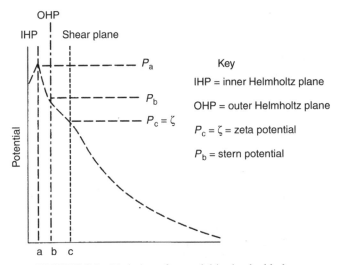

FIGURE 3.2 Variation of potential in the double layer.

its structure as well as its properties. When an inhibitor I approaches and adsorbs at the metal–solution interface, it may be written as:

$$M(nH_2O)_{ads} + I_{(sol)} = MI_{ads} + nH_2O_{(sol)}.$$

In the process of adsorption of the inhibitor, the inhibitor displaces n water molecules initially adsorbed on the metal. The adsorption of the inhibitor on the metal occurs because the interaction energy between the metal and the inhibitor is more favorable than the interaction energy between the metal and the water molecules.

The dielectric properties of solvent water are affected during the adsorption of an inhibitor on the metal. In bulk water the water molecules are disoriented and the dielectric constant is 80. In the electric double layer the dipoles of water molecules are oriented, which causes lowering of the dielectric constant. Dielectric constants of 6 and 40 have been estimated in solvent water in the inner and outer Helmholtz planes, respectively. Ions or molecules with high dipole moment, on adsorption at the metal surface, affect the electric double layer by causing a change in the dielectric properties of water molecules in the Helmholtz double layer. This effect has been observed in the corrosion inhibition of iron in 10% hydrochloric acid by quinolinium compounds (1). The corrosion inhibition of the iron by quinolinium compounds is attributed to the strong ordering effect by the π electrons of the inhibitor molecules on the water molecules at the interface, in addition to electronic and geometric factors.

3.2 CORROSION INHIBITORS

A corrosion inhibitor is a chemical substance that, upon addition to a corrosive environment, results in reduction of corrosion rate to an acceptable level. Figure 3.3

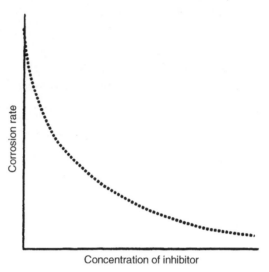

FIGURE 3.3   Corrosion rate as a function of inhibitor concentration.

shows the relationships between the concentration of an inhibitor and corrosion rate
and Fig. 3.4 shows the concentration of inhibitor and degree of inhibition. These
relationships were studied by Sieverts and Leug, and the figures resemble adsorption
isotherms. Corrosion inhibitors are generally used in small concentrations.
A corrosion inhibitor should not only mitigate corrosion but also be compatible with
the environment. Usually the corrosion inhibitor is rated in terms of inhibition
efficiency I and is defined as

$$I = \frac{(CR)_o - (CR)_i}{(CR)_o} \times 100$$

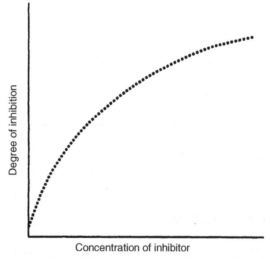

FIGURE 3.4   Curve showing inhibitor efficiency as a function of inhibitor concentration.

where CR is the corrosion rate and the subscripts i and o refer, respectively, to the presence and absence of the inhibitor.

A corrosion inhibitor can mitigate corrosion in two ways. In some cases, the corrosion inhibitor can alter the corrosive environment into a noncorrosive or less corrosive environment through its interaction with the corrosive species. In other cases, the corrosion inhibitor interacts with the metal surface and renders protection of the metal from corrosion. Thus, based on the mode of interaction, there are two broad classes of inhibitors: environment modifiers and adsorption inhibitors.

In the case of environment modifiers, the action and mechanism of inhibition is a simple interaction with the corrosive species in the environment, thereby preventing the attack of the metal by the aggressive species. This is exemplified by oxygen scavengers such as hydrazine or sodium sulfite together with cobaltous nitrate and biocides, used in inhibiting microbiological corrosion. In the case of corrosion in neutral and alkaline solutions, oxygen reduction is the cathodic reaction, which can be mitigated by the oxygen scavengers and thus inhibit the corrosion.

$$\text{Cathodic reaction}: O_2 + 2H_2O + 4e^- \rightarrow 4OH^-.$$

$$\text{Oxygen scavenging}: 5O_2 + 2N_2H_4 \rightarrow 2H_2O + 4H^+ + 4NO_2,$$
$$2SO_3^{2-} + O_2 \rightarrow 2SO_4^{2-}.$$

In the case of inhibitors that adsorb on the metal surface and inhibit the corrosion, there are two steps: (i) transport of inhibitor to the metal surface and (ii) metal–inhibitor interactions. The process is analogous to the transport of a drug molecule in the body to the required site followed by its interaction with the site to provide relief from an ailment. The most important step involves the interaction of the metal with the inhibitor molecule. These are chemical interactions and will be dealt with later.

Depending upon whether the cathodic reaction or the anodic reaction is suppressed by the corrosion inhibitor, the inhibitors have been further classified as follows:

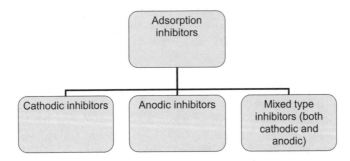

Cathodic inhibitors inhibit the hydrogen evolution in acidic solutions or the reduction of oxygen in neutral or alkaline solutions. It is also observed that the cathodic branch of the polarization curve is affected when a cathodic inhibitor is added to a system, as shown in Fig. 3.5. Distinct differences are seen in the cathodic

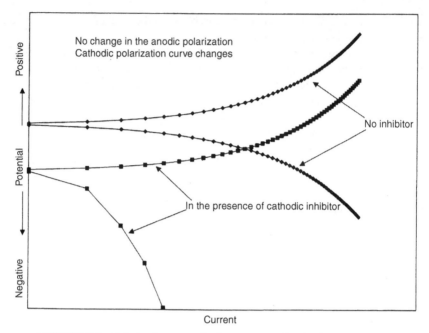

FIGURE 3.5   Polarization curve in the presence of cathodic inhibitor (2).

branch of the uninhibited or inhibited systems (2). Substances with high overpotential for hydrogen in acidic solutions and those that form insoluble products in alkaline solutions are, in general, effective cathodic inhibitors. Some examples of the inhibitors are inorganic phosphates, silicates, or borates in alkaline solutions, which inhibit the oxygen reduction at the cathodic sites. Other substances such as carbonates of calcium and magnesium, due to limited solubility, block the cathodic sites.

Anodic inhibitors are generally effective in the pH range of 6.5–10.5 (near neutral to basic). Oxyanions such as chromates, molybdates, tungstates, and also sodium nitrite are quite effective anodic inhibitors. These oxyanions are thought to play a role in repairing the defects in the passive iron oxide film on the metallic iron surface. In the case of chromate or dichromate, the concentration of the inhibitor used is critical. A potentiodynamic polarization diagram showing the effect of the concentration of the inhibitor is depicted in Fig. 3.6. The figure clearly shows the effect of the concentration of the inhibitor. It is clear from the figure that protection is rendered when a sufficient amount of the inhibitor is present. It is also seen from the figure that corrosion protection is poor when the inhibitor concentration is insufficient. This behavior is displayed when dichromate is used as an inhibitor. Sometimes dichromate is also considered a "dangerous inhibitor" when insufficient concentration of the inhibitor is used.

Mixed-type inhibitors affect both anodic and cathodic branches of a polarization curve. Organic compounds function as mixed-type inhibitors. The organic inhibitors adsorbed on the metal surface provide a barrier to dissolution at the anode and a barrier to oxygen reduction at the cathodic sites, as shown in Fig. 3.7. The protective

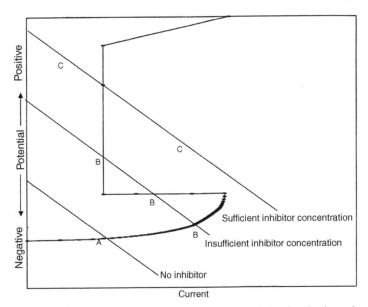

FIGURE 3.6 Polarization diagram of an active–passive metal showing the dependence of the current on the concentration of passivation-type inhibitor (2).

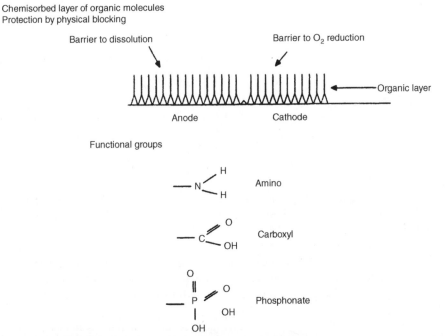

FIGURE 3.7 Protection from corrosion by organic inhibitors.

functional groups in the organic mixed-type inhibitors can be amino, carboxyl, or phosphonate.

3.3 ADSORPTION ISOTHERMS

The adsorption of inhibitors on the metal surface is governed by the residual charge on the metal and the chemical structure of the inhibitor. The two types of adsorption of an organic inhibitor on a metal surface are physical or electrostatic and chemisorption. The addition of a corrosion inhibitor in increasing amounts results in progressive decrease in corrosion rate or progressive increase in corrosion inhibition. The relationships between corrosion rate and concentration of inhibitor and corrosion inhibition and concentration of inhibitor shown in Figs. 3.3 and 3.4 were investigated by Sieverts and Leug (3). The figures resemble adsorption isotherms.

Langmuir showed the classical relationship between the concentration of the adsorbate and the amount of adsorption. The fractional surface S covered by adsorption is related to the concentration C of the adsorbed species in solution by the equation

$$S = \frac{aC}{1 + aC},$$

where a is a characteristic constant for the specific adsorbate. In the context of corrosion, the relationship can be written as

$$S = k\frac{m_0 - m}{m_0}C.$$

The above equation may be written as

$$\frac{m_0}{m_0 - m} = A\left(\frac{1}{C}\right) + B.$$

Figure 3.8 is a plot of $m_0/(m_0 - m)$ versus $1/C$, and the straight lines obtained for three inhibitors show the validity of Langmuir isotherms in the case of the three inhibitors.

Adsorption isotherms are often used to demonstrate the performance of organic adsorbent-type inhibitors, and the surface coverage rates determined by capacitance measurements give a good correlation with the adsorption isotherm plot. The various types of adsorption isotherms along with the verification plots are given in Table 3.1.

The adsorption of an inhibitor affects the Helmholtz double layer structure and hence the potential in the double layer. The adsorption of an inhibitor can be

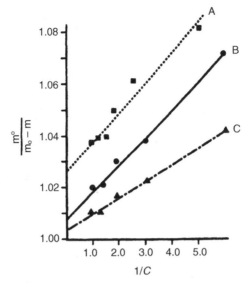

FIGURE 3.8 Verification plot of Langmuir isotherms for inhibitors.

monitored by determining the electrokinetic potential or by determining the capacitance before and after the adsorption of the inhibitor, by considering the system to be a parallel plate capacitor. For effective adsorption of an inhibitor on a metal surface the forces of interaction of metal and inhibitor must be greater than the interaction forces between the metal and water molecules.

Physical adsorption is due to electrostatic attraction between inhibiting ions or dipoles and the electrically charged metal surface. The surface on the metal is defined by the position of the free corrosion potential E_{corr} of the metal with respect to its potential of zero charge, PZC. Some values of potentials of zero charge are given in Table 3.2. At PZC the net charge on the electrode is zero. At potentials more positive than PZC, the electrode is positively charged; at potentials more negative than PZC, the electrode is negatively charged. When $\Phi = E_{corr} - E_{q=0}$ is negative, cations are

TABLE 3.1   Adsorption Isotherms (3–6)

Name	Isotherm[a]	Verification Plot
Langmuir	$\theta/(1-\theta) = \beta C$	$\theta/(1-\theta)$ versus log C
Frumkin	$[\theta/(1-\theta)]e^{f\theta} = \beta C$	θ versus log C
Bockris–Swinkels	$\theta/(1-\theta)^n[\theta + n(1-\theta)^{n-1}/n^n = Ce^{-\beta}/55.4$	$\theta/(1-\theta)$ versus log C
Tempkin	$\theta = (1/f) \ln KC$	θ versus log C
Virial Parson	$\theta e^{2f\theta} = \beta C$	θ versus log(θ/C)

[a]θ, % $P/100$, surface coverage; β, $\Delta G/2.303RT$; ΔG, free energy of adsorption; R, gas constant; T, temperature; C, bulk inhibitor concentration; n, number of water molecules replaced per inhibitor molecule; f, inhibitor interaction parameter (either 0, no interaction; +, attraction; or −, repulsion); K, constant; and % $P = 1$ inhibited corrosion rate/uninhibited corrosion rate.

TABLE 3.2   Values of Zero Charge Potentials (7)

Metal	Potential of Zero Charge, mV (SHE)
Ag	− 440
Al	− 520
Au	+ 180
Bi	− 390
Cd	− 720
Co	− 450
Cr	− 450
Cu	+ 90
Fe	− 350
Ga	− 690
Hg	− 190
In	− 650
Ir	− 40
Nb	− 790
Ni	− 300
Pb	− 620
Pd	0
Pt	+ 20
Rh	− 20
Sb	− 140
Sn	− 430
Ta	− 850
Ti	− 1050
Zn	− 630

adsorbed, and when $\Phi = E_{corr} - E_{q=0}$ is positive, anions are adsorbed. This occurs in the case of inorganic or organic ions as well as dipoles. At equal values of Φ for different metals such as iron and mercury, similar behavior for adsorbing organic species is expected as well as observed (8).

The usefulness of the concept and the value of PZC can be illustrated by considering the observed synergism (9) in the corrosion inhibition of iron in sulfuric acid solutions by quaternary ammonium cations in the presence of chloride. At the free corrosion potential of iron in sulfuric acid solution, the surface charge of the metal is positive, that is, E_{corr} is $- 0.2\,V_H$ compared to PZC of $- 0.37\,V_H$. Under these conditions, organic cations do not adsorb. But due to the adsorption of chloride, the PZC shifts to more positive values, and $\Phi = E_{corr} - E_{q=0}$ becomes negative and adsorption of cations is favored.

The forces involved in electrostatic adsorption are generally weak. The inhibitor adsorbed on the metal due to electrostatic force can be desorbed easily. The main feature of electrostatic adsorption is that the adsorbed ions are not in direct physical contact with the metal. A layer of solvent molecules separates the metal from the ions.

The physical adsorption process has low activation energy and is relatively independent of temperature.

Chemisorption is the most important type of interaction between the metal surface and an inhibitor molecule. Unlike physisorption, in chemisorption the adsorbed inhibitor is in direct contact with the metal surface. It is surmised that a coordinate type of bond involving electron transfer from the inhibitor to the metal takes place (10). An opposing view held by Bockris (11) is that there is no chemical bond between the metal and the adsorbed chemical species.

The chemisorption process is slower than electrostatic sorption and has higher activation energy. The temperature dependence shows higher inhibition efficiency at higher temperatures. Unlike electrostatic adsorption, chemisorption is specific for certain metals and is not completely reversible. The nature of the metal and the organic inhibitor has a decisive effect on the bond between the metal and the inhibitor, and charge transfer from the inhibitor molecule to the metal is facilitated when the inhibitor molecule has a functional group with an atom containing a lone pair of electrons. Availability of π electrons due to the presence of multiple bonds or aromatic rings in the inhibitor molecule is thought to facilitate charge (electron) transfer from the inhibitor to the metal.

The organic inhibitors used have reactive functional groups, which are the sites for the chemisorption process. The strength of the adsorption bond depends upon the electron density on the donor atom present in the functional group and its polarizability.

3.4 ANODIC DISSOLUTION AND ADSORPTION

The anodic dissolution of a metal depends on the electrode potential, pH of the solution, electrolyte concentration, and the metal involved. The three mechanisms of anodic dissolution (12) are (i) acid-catalyzed, (ii) base-catalyzed, and (iii) water-catalyzed mechanisms. The characteristic features of the three mechanisms are shown in Fig. 3.9. The dissolution current increases with increase in hydrogen concentration in an acid-catalyzed mechanism, while the dissolution current at constant potential decreases in the case of a base-catalyzed mechanism; dissolution current is independent of hydrogen ion concentration in the case of water-catalyzed mechanism. The anions of the electrolyte can either promote or inhibit the anodic dissolution. The anion-assisted dissolution observed in acid solutions may be attributed to species like $FeCl_4{}^-$ formed in hydrochloric acid or specific adsorption of anions, which reduces the activation energy for metal dissolution.

In the case of base-catalyzed and water-catalyzed mechanisms, anions will compete with water and hydroxyl ions during adsorption on the metal surface. If the adsorbed anions block the adsorption sites for water and hydroxyl ions or increase the activation energy for metal ion transfer across the interface, the anodic dissolution will be inhibited. The pH dependence of the anodic dissolution current of iron at constant potential depicted in Fig. 3.10 shows (13) three distinct regions corresponding to the three mechanisms cited earlier.

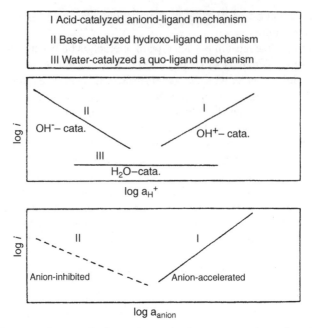

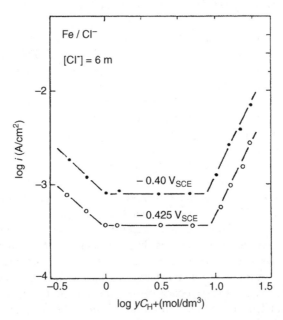

FIGURE 3.9 Dependence on hydrogen ion and anion concentrations of anodic dissolution current at constant potential in the active state of metals for three mechanisms.

FIGURE 3.10 Dependence on hydrogen ion activity of anodic dissolution current of iron in the active state.

The three mechanisms may be given in terms of the following reaction schemes:

(I) Anion-ligand mechanism:

$$Me + A_{aq}^- \rightarrow MeA_{ad} + e^-,$$
$$MeA_{ad} \rightarrow MeA_{aq}^+ + e^-.$$

(II) Hydroxo-ligand mechanism:

$$Me + H_2O \rightarrow MeOH_{ad} + H_{aq}^+ + e^-,$$
$$MeOH_{ad} \rightarrow MeOH_{aq}^+ + e^-.$$

(III) Aquo-ligand mechanism:

$$Me \rightarrow Me_{ad}^+ + e^-,$$
$$Me_{ads}^+ \rightarrow Me_{aq}^{2+} + e^-.$$

Subscripts "ad" and "aq" refer to species adsorbed on the metal surface and those dissolved in bulk solution, respectively. All three mechanisms are operative simultaneously, and under certain conditions one mechanism will predominate. The predominance of any one mechanism depends on the pH, concentration of anion, and the electrode potential of the metal. It is clear that water, hydroxyl ion, and other anions such as chloride play a decisive role in the anodic dissolution by adsorption on the metal surface as well as formation of metal complexes such as $FeCl_4^-$.

Organic inhibitors used to inhibit corrosion of metals in acid media are known to adsorb on the metal surface and inhibit anodic dissolution. The strong adsorption of organic inhibitors can result in sufficient increase in activation energy to inhibit anodic dissolution as shown in Fig. 3.11. The potential of zero charge of the metal has to be favorable for the adsorption of the inhibitor. When the electrode potential of the metal is more positive than the potential of zero charge, anions preferentially adsorb on the metal due to the electrostatic forces. The specific adsorption likely changes the PZC itself. The PZC of iron changes to a more positive value upon adsorption of chloride ions. It is thought that the change in PZC due to specific adsorption depends on whether the adsorbed species is in the inner Helmholtz plane as counterion without any charge transfer or whether it reacts with the metal surface forming a surface chemical compound involving partial or full charge transfer to the metal. The shift in PZC likely causes redistribution of the potential drop at the metal–solution interface, which affects the dissolution rate.

3.4.1 Formation of Passive Films

When a metal electrode subjected to active dissolution in a particular medium is polarized in the positive potential direction by means of a potentiostat, the anodic dissolution current will decrease rapidly at certain critical potential and the metal

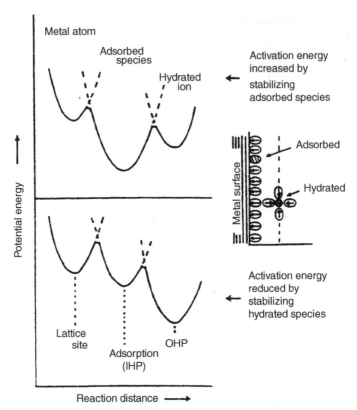

FIGURE 3.11 Schematic potential energy diagrams for the anodic dissolution that proceeds through an adsorption intermediate (IHP, inner Helmholtz plane; OHP, outer Helmholtz plane).

becomes passive. The passive state of a metal reached either by the applied potential or by the addition of a strong oxidizing agent to the solution is due to the formation of a monatomic or a polyatomic oxide film on the surface of the metal.

The three possible mechanisms of formation of passive films are (i) direct film formation, (ii) dissolution–precipitation, and (iii) anodic deposition. In the direct film formation mechanism, the interaction of the metal surface with the solution leads to an oxygen adsorption film (MeO_{ad}) or a compact oxide film (MeO_{ox}).

$$Me + H_2O_{aq} \rightarrow MeO_{ad} + 2H^+_{aq} + 2e^-,$$
$$Me + H_2O_{aq} \rightarrow MeO_{ox} + 2H^+_{aq} + 2e^-.$$

The dissolution–precipitation mechanism involves precipitation to form an oxide film on the metal surface

$$Me \rightarrow Me^{2+}_{aq} + 2e^-,$$
$$Me^{2+}_{aq} + H_2O_{aq} \rightarrow MeO_{ox} + 2H^+_{aq}.$$

In the anodic deposition mechanism, the anodic oxidation of metal in solution leads to formation of an oxide film of higher oxidation state:

$$Me \rightarrow Me_{aq}^{2+} + 2e^-,$$

$$2Me_{aq}^{2+} + 3H_2O_{aq} \rightarrow Me_2O_3 + 6H_{aq}^+ + 2e^-.$$

The formation of a passive film changes the potential distribution across the metal–solution interface. The potential distribution across the metal–solution interface is schematically shown in Fig. 3.12. When the bare metal surface is exposed to the solution, the anodic dissolution current is controlled by the potential difference $\Delta\Phi_{m/s}$, which is located across the Helmholtz double layer on the solution side. When the metal surface is covered with the film, the total potential difference $\Delta\Phi_{m/s}$ is partitioned into three parts, namely, $\Delta\Phi_{m/s}$, $\Delta\Phi_f$, and $\Delta\Phi_{f/s}$. As a consequence, the metal dissolution occurs in consecutive reactions consisting of transfer of metal ions from metal phase to the surface film, migration of ions through the film, and transfer of metal ions across the Helmholtz layer at the film–solution interface. At steady state, the rate of metal dissolution from the film is equal to the rate of ion migration through the film.

The ion conduction current depends on the potential difference in the film $(\Delta\Phi_f)$ and the metal dissolution current is controlled by $\Delta\Phi_{f/s}$, the potential difference at the film–solution interface. At steady state, the ion conduction current is equal to the metal dissolution current and the potential difference $\Delta\Phi_{f/s}$ at the film–solution interface is determined by the electrochemical reaction:

$$H_2O_{aq} = O_f^{2-} + 2H_{aq}^+,$$

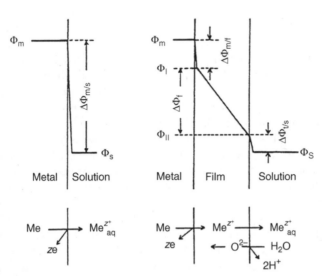

FIGURE 3.12 Schematic potential distribution across metal–solution and metal–film interface.

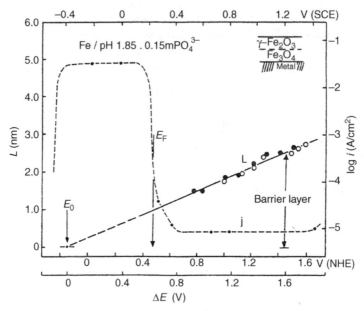

FIGURE 3.13 Anodic polarization and thickness of passive film on iron by potentiostatic 1 h oxidation in phosphoric acid.

and the dissolution current becomes a function of solution pH and activity of oxygen ions in the outermost layer of the film. The passivation of the metal occurs due to the passive film, since the potential difference $\Delta\Phi_f$ across the film accounts for most of the potential difference $\Delta\Phi_{m/s}$. It is to be noted that the passive film acts as a barrier for ion migration. The anodic polarization curve of iron and thickness of the passive film formed on iron in phosphoric acid solution (14) are shown in Fig. 3.13. It is clear from the figure the dissolution current (passivity maintaining current) is nearly constant irrespective of the potential, while the thickness of the passive film increases linearly with increasing potential. This indicates that potential difference $\Delta\Phi_{f/s}$ at the film–solution interface remains constant, and the increase in electrode potential is equivalent to the increase in potential difference across the film, $\Delta\Phi_f$. From the linear relation between the film thickness and the electrode potential, the electric field strength in the film is estimated to be 10^6 V/cm. In contrast to iron, the dissolution current of nickel increases with increasing potential as shown in Fig. 3.14. In this case, the potential difference $\Delta\Phi_{f/s}$ at the film/solution is not constant and the increase in electrode potential contributes to an increase in $\Delta\Phi_f$ and $\Delta\Phi_{f/s}$. The increase in $\Delta\Phi_{f/s}$ is due to an increase in oxygen ion activity in the passive film on nickel (15).

The surface charge of a passive film is directly related to the potential difference $\Delta\Phi_{f/s}$ at the film–solution interface. The pH at which the surface charge of the oxide film equals zero is known as ZPC. When the solution pH is greater than ZPC of the oxide film, the film surface is positively charged due to adsorption of hydrogen ions. A decrease in pH results in increases in $\Delta\Phi_{f/s}$ and hence increased dissolution current. An example of the pH dependence of the dissolution current of iron in phosphate

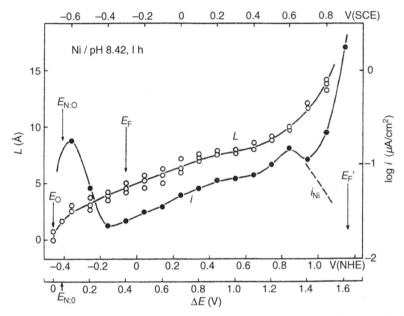

FIGURE 3.14 Anodic polarization curve and thickness of passive film on nickel by potentiostatic 1 h oxidation in pH 8.4 borate buffer solution.

medium is shown in Fig. 3.15. The dissolution current increases linearly in the pH range 4 to 1. Thus, the adsorption of cations or anions on the oxide film results in a change in ZPC to low or high pH value and consequent change in potential difference $\Delta\Phi_{f/s}$ that would increase or inhibit the dissolution (16).

The acid–base nature of the oxide film and the ions in solution play an important role in adsorption on the film. The metal ions act as acids and oxide and hydroxyl ions act as bases. Anions such as chloride can replace oxide and hydroxyl ions. Chloride ions are known to cause breakdown of passive film on iron and cause pitting corrosion.

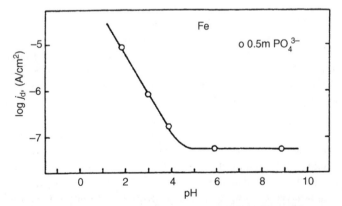

FIGURE 3.15 pH dependence of dissolution current of passive iron in phosphate solution.

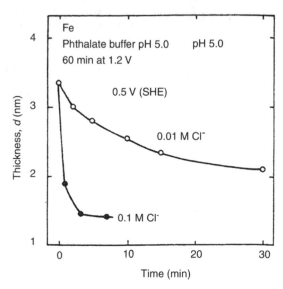

FIGURE 3.16 Variation of thickness of passive film on iron on introduction of chloride ions in pH 5 buffer solution.

The chloride ions adsorb on the passive film to form species such as $FeCl^{2+}$ or $FeCl_2^+$, which are soluble. An increased dissolution results in reduced thickness of the passive film (17) as shown in Fig. 3.16. Surface analysis of the outer layers of the iron samples with passive films formed showed the presence of anions such as borate that is indicative of the anodic dissolution and deposition of the dissolved species on the metal surface. The deposition of insoluble metal salts or their incorporation in the outer layers contributes to thickness and stability of passive films.

3.5 ROLE OF OXYANIONS (PASSIVATION) IN CORROSION INHIBITION

Oxyanions such as nitrite, chromate, molybdate, and tungstate passivate iron and steel in slightly acidic or near neutral solutions. Nitrite and chromate passivate iron both in aerated and deaerated solutions, while molybdate and tungstate passivate only in aerated solutions. Nitrite and chromate inhibit the anodic reaction and accelerate the cathodic reaction. Molybdate and tungstate inhibit only the anodic metal dissolution reaction. Anodic inhibition by these oxyanions is a result of adsorption of the oxyanions on the metal surface followed by passivation.

Surface analysis of films formed on iron (18) exposed to nitrite, chromate, molybdate, and tungstate showed the presence of chromium, tungsten, and molybdenum. No significant amount of nitrogen due to the nitrite inhibitor was detected. Detailed coordinated surface analysis showed (19) several monolayers of chromium, about one monolayer of tungsten, and a fractional monolayer of molybdenum oxides on the surfaces of samples exposed to the inhibitors. Mixed iron and chromium

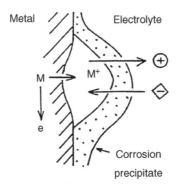

FIGURE 3.17 Corrosion precipitate that has ion selectivity.

oxide were detected by X-ray photoelectron spectroscopy (20). Both nitrite and chromate adsorb on the metal surface and inhibit the anodic dissolution. When passivation occurs due to anodic inhibition, adsorbed chromate ions probably react directly with iron to form the passive film consisting of mixed hydrous oxide. In the case of molybdate, tungstate, and nitrite, the anodic dissolution reaction is inhibited, and the inhibitors probably take part in repairing the defects in the passive iron oxide film.

The corroding metal surface is often covered with a porous corrosion-produced precipitate film of hydrated metal oxides or insoluble salts. The corrosion precipitate film can either promote or inhibit further corrosion of the metal under the precipitate film. When anodic dissolution occurs under a deposit of precipitate film with concurrent cathodic reaction elsewhere, there will be a migration of cations from the inside occluded solution to the outside bulk solution, while anions will migrate from outside into the inside region (21) (Fig. 3.17).

The hydrated iron oxide can be either anion selective in acidic and neutral chloride solutions or cation selective in alkaline solutions. Considering the hydrated iron oxide as a membrane, the ion-selectivity property of the precipitate will depend upon the sign and concentration of the fixed charge of the preferentially adsorbed ions in the micropores of the membrane. When the fixed charge on the corrosion precipitate membrane is positive, it is anion selective and acts as cation selective when there is adsorbed anion such as molybdate (Fig. 3.18). When the corrosion precipitate film is anion selective, chloride anions will migrate from outside the bulk solution into the inside occluded solution of the precipitate.

With time, the occluded solution will become enriched in chloride and hydrogen ions resulting from hydrolysis of iron chloride and leading to favorable conditions for localized corrosion and, more specifically, underdeposit corrosion. This phenomenon is most likely to occur with respect to anion-selective precipitate.

In the case of cation-selective corrosion precipitates, migration of chloride anions into the inside solution is inhibited; relatively mobile ions like hydrogen ions migrate outward, leaving dissolved metal ions in the occluded solution. Due to lack of inward flow of water, the metal ions inside the occluded solution will form metal hydroxide, leading to corrosion inhibition under a cation-selective precipitate film.

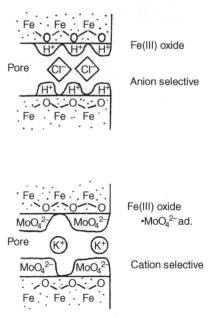

FIGURE 3.18   Ion selectivity in hydrated iron oxide corrosion precipitate.

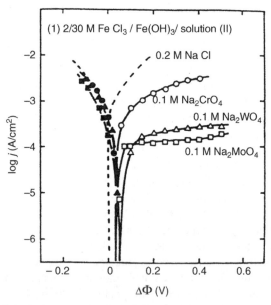

FIGURE 3.19   Polarization curves of ferric hydroxide precipitate membrane in ferric chloride solution on one side and inhibitors on the other side.

The potentiodynamic polarization curves of ferric hydroxide precipitate membranes in ferric chloride solution on one side and sodium chromate, sodium molybdate, and sodium tungstate solutions on the other side (22) are shown in Fig. 3.19. These membranes have bipolarity, and the anodic current is due to positive charge flow from ferric chloride solution to the outer solution. The passivation effect is nearly the same in the case of sodium molybdate and sodium tungstate, while the degree of passivation due to chromate is considerably less. This feature has been attributed to weak adsorption of chromate ions (23).

3.6 INHIBITION OF LOCALIZED CORROSION

Localized corrosion occurs on metals covered with passive films due to defects in the passive film or breakdown of the passive film. Aggressive anions such as chloride cause the breakdown of the passive oxide film at weak points on the surface of the metal. A major area of metal surface is in the passive state, while small localized areas on the metal are in an active state, due to inhomogeneities in the passive film. Localized corrosion begins to occur after a lapse of time during which the metal interacts with the aggressive environment. This time lapse is known as the induction time for pit nucleation. The induction time for pit nucleation depends upon the concentration of the aggressive anion and the electrode potential of the metal. Thus, localized corrosion such as pitting involves two steps; namely, pit nucleation followed by pit growth.

The origin of pits depends on the presence of weak points in the surface oxide layer. The corrosion rate is generally high at the weak points compared to the rest of the passive surface. The high corrosion current at the weak points leads to a local increase in the surface salt concentration due to transport modes such as migration and diffusion and adsorption of the anion on the electrode. The high concentration of anions displaces the water molecules from the Helmholtz double layer and suppresses the passivation reaction

$$M + H_2O \rightarrow MeO + H^+ + e^-$$

and promotes the oxide film dissolution

$$MeO + A^- + H^+ \rightarrow MeA + H_2O.$$

As a result of these processes, the oxide film thickness decreases and an increase occurs in the electric field across the film. Under the effect of the high electric field, anions penetrate the oxide film. The density of lattice defects increases, which together with the high electric field causes an increase in ionic conductivity of the oxide layer. The oxide layer loses its passivating properties and becomes a non-passivating oxide, which can sustain high corrosion current densities. The passive film breakdown is then a simple transformation of the passive oxide, $MeO(p)$, into the nonpassive oxide, $MeO(np)$:

$$MeO(p) \underset{H_2O}{\overset{A^{n-}}{\rightleftharpoons}} MeO(np).$$

By decreasing the anion concentration, the equilibrium shifts in favor of the passivating oxide, and hence there is a decrease in the rate of localized corrosion.

The three steps involved in the propagation of pits are (i) reaction at the metal–oxide interface, (ii) transport of ions through the oxide layer, and (iii) reactions at the oxide–electrolyte interface.

Electrochemical oxidation of the metal occurs at the metal–oxide interface

$$M \rightarrow M^{n+} + ne^-.$$

The metal cation then fills the vacancy, V_M, in the oxide lattice (24),

$$M^{n+} + V_M \rightarrow M_{CL}.$$

The ion then moves from metal–oxide interface to the oxide–electrolyte interface as denoted by the equation

$$(M_{CL})_{n-1} + (V_M)_n \rightarrow (V_M)_{n-1} + (M_{CL})_n,$$

where the subscripts n and $(n-1)$ denote the respective more distant and less distant positions of the vacancy (V) or metal cation (M_{CL}), with respect to the metal–oxide interface.

The kinetics of localized corrosion depends upon the electrolyte concentration and the potential due to the electrochemical reaction at the oxide–electrolyte interface. The reaction at the oxide–electrolyte interface may be considered the rate-determining step of metal dissolution in the pits. The metal cation traversing through the oxide layer reacts at the oxide–electrolyte thus

$$M_{CL} \rightarrow V_M + nh + M_o,$$

where h denotes holes. This is followed by the reaction

$$M_o \rightarrow M^{n+} + ne^-.$$

The electrons generated in the above reaction react with positive holes in the lattice

$$ne^- + nh \rightarrow o.$$

The overall reaction at the oxide–electrolyte interface is

$$M_{CL} \rightarrow V_M + M^{n+}.$$

The reaction that produces holes,

$$M_{CL} \rightarrow V_M + nh + M_o,$$

can also be a source of vacancies in the oxide layer: One such case is that in which the reaction of holes with electrons consumes only a fraction of the electrons and the remaining fraction escapes through the oxide–metal interface into the metal, then a corresponding amount of vacancies and positive holes are left behind in the oxide lattice. The overall process can be represented by competitive reactions:

$$M_o + H_2O \rightarrow MO + 2H + ne^-.$$

The low amount of water and the high concentration of salt in the electrolyte present in the pit favor the reaction

$$M_o A_{ads}^{n-} \rightarrow MA + ne^-,$$

which may be written in terms of hydrated species as

$$M_o[A^-(H_2O)]_{ads} \rightarrow MA + H_2O + ne^-.$$

An effective inhibitor should be capable of inhibiting both pit nucleation and pit growth. Pit nucleation is a complex phenomenon, so it is difficult to predict the effect of inhibitors on the kinetics and mechanism of pit nucleation and repassivation of pits. Although the mechanism of pit growth is relatively better known, the prevailing conditions in the pits, such as low pH, high concentrations of anions such as chloride, and chloride salt layer in pits, are conducive to rapid metal dissolution. The conditions prevailing in pit growth are hard to control, and hence suitable inhibitors are sought for inhibition of the pit nucleation process.

The quality and breakdown of passive film are governed by (i) the ability of the film to adsorb aggressive anions such as chloride from solutions, (ii) the composition and structure of the film, (iii) film thickness, (iv) film history, (v) time, (vi) temperature, (vii) electrode potential and the composition of the electrolyte used for anodic passivation, (viii) semiconductive properties of the film, and (ix) the presence of mechanical and electrostatic stresses.

Inhibition of pit nucleation can be achieved by (i) improving the protective nature of the film by changing the composition and structure of the film; (ii) reducing the imperfections in the film; and (iii) reducing the extent of anion adsorption on the film by competitive adsorption of other chemical species.

Inhibitors used for prevention of general corrosion in neutral media may be applied for combating localized corrosion. The selection of inhibitors for localized corrosion requires knowledge of the behavior of inhibitors in neutral and acid solutions.

3.7 ADSORPTION OF HALIDE IONS

It is generally found that adsorption of chloride ions on a passive metal leads to pitting corrosion. The dissociative adsorption of water on a passivated metal leads to a reaction of the type

$$M_p + H_2O \rightarrow M_{pH^+}^{OH^-}.$$

At pH values of zero charge, that is, pH_0, $[OH^-]_{sur} = [H^+]_{sur}$. At pH values greater than zero charge (i.e., $pH > pH_0$), the surface of the metal is negatively charged

$$M_p + H_2O \rightarrow M_pOH^-$$

and adsorption of the negative chloride ion is prevented. At pH values lower than pH_0, adsorption of anions is favored:

$$M_p(H_2O) + A^{n-} \rightarrow M_p(H_2O)A^{n-}.$$

The adsorbed anion can be either corrosive or inhibitive, as is the case in inhibition of iron by sulfate anion (25).

The pH_0 values of iron oxide and aluminum oxide are 8.8 and 9.1, respectively (26). Chloride was found to adsorb on iron and aluminum oxides below the pH_0 values. Adsorption of chloride on passive iron gave a Tempkin type adsorption isotherm, indicating surface inhomogeneity and variable adsorption energy. Adsorption of aggressive and inhibitive ions gave an isotherm corresponding to Frumkin, Tempkin, and Langmuir type isotherms (27–29). Adsorption was found to depend on the nature of the metal, structure of the inhibitor, solution composition, pH, potential, and temperature. The potential dependence for aggressive ions and inhibitors is different. The concentration ratio of the aggressive ion to that of the inhibitor plays an important role in inhibition of localized corrosion, due to the competitive adsorption of ions.

Studies based on exposure of iron to solutions of sodium chloride and sodium nitrite, and to solutions of sodium sulfate and sodium nitrite, led to the empirical equation (30)

$$\log C_A = a + b \log C_p$$

where C_A and C_p are the concentrations of aggressive ion and passivating ion respectively, and a and b are constants. The ratio $\log C_A/\log C_p$ is found to be the decisive factor in whether or not pitting occurs. Similar dependence was observed in systems consisting of aluminum (31), iron, and carbon steel (32). The pitting behavior in these systems has been ascribed to the competitive adsorption of the aggressive chloride ion and the passivating nitrite ion. Other combinations of aggressive anions and passivating inhibitors in which the empirical relationship was valid are noted in Table 3.3.

An important relationship between the pit nucleation potential E_{np} and the concentration of aggressive ion

$$E_{np} = a - \log C_A$$

has been found to describe the pitting behavior of metals (35–37). An inhibitor that shifts E_{np} to more positive values is considered to be an effective inhibitor for pitting

TABLE 3.3 Anion/Inhibitor Combinations

C_A/C_i	Metal	Reference
Chloride/molybdate		33
Chloride/chromate		
Chloride/acetate	Aluminum	34
Iodide/benzoate		
Chloride/sodium phenylanthranilate		31
Nitrate/nitrite	Nuclear waste-carbon steel	32
Sulfate/nitrite		

corrosion. The effect of an inhibitor on the pit nucleation potential is given by the equation (38)

$$E_{np} = a + b + \log\left(\frac{C_A}{C_{In}}\right),$$

where C_A and C_{In} are the concentrations of anion and the inhibitor, respectively, and a and b are constants. The equation was found to describe the behavior of both inorganic and organic inhibitors.

Attempts were made to describe competitive adsorption of ions in terms of Langmuir (39) and Tempkin (40) isotherms. The Tempkin treatment was found to be more appropriate because of the inhomogeneity of the metal surface and consequent increase in lateral repulsive interaction with coverage. Assuming the adsorption of chloride anions in clusters, the adsorption of inhibitor and chloride to be independent of each other, the surface coverage θ_A by aggressive ions A^- has been described as

$$\theta_A = \frac{2.303RT}{r_A}\log K_A + \frac{2.303RT}{r_A}\log a_A + \frac{\gamma_A Z_A FE}{r_A},$$

where θ_A is the surface coverage of aggressive anions A^-, a_A is the bulk activity of aggressive anion A^-, Y_A is a factor that takes into consideration that specifically adsorbed anions of charge Z_A move through only a part of the potential drop across the interface, r_A is the Tempkin parameter, and

$$K_A = K'_A e^{\Delta H_A^\circ/RT},$$

where ΔH_A° is the initial heat of adsorption.

For the coverage by the inhibitor, we have

$$\frac{\theta_A}{\theta_{inh}} = \theta_{crit}; E = E_{pit}.$$

At constant activity a_i of inhibitor anion *N-n*-dodecanoyl sarcosine (NLS), the following relationship may be written

$$\frac{dE_{\text{pit}}}{d\log(a_A - a_{\text{inh}})} = \frac{2.303RT}{F} \frac{1}{\theta_{\text{crit}}(r_A/r_i)(Y_i Z_i - \gamma_A Z_A)}.$$

This equation is similar to the pit nucleation potential relationship discussed earlier.

3.8   INFLUENCE OF ENVIRONMENTAL FACTORS

The effect of temperature on corrosion is a complex function. A change in temperature can affect (i) kinetics and mechanism of metal dissolution and oxide formation and (ii) adsorption and desorption. Thus, change in temperature can increase or decrease or have no effect on corrosion inhibition. Changes in temperature can cause decomposition of the inhibitor or changes in the metal dissolution process. The data on the effective activation energy for the corrosion of iron in 1 N HCl solution as a function of inhibitor concentration have been attributed to the nature of adsorption, changes in the metal dissolution mechanism, and the thermal stability of the inhibitor (41).

The data obtained on the corrosion of iron in 5 mM sodium chloride and 2 mM sodium phenylanthranilate showed (42) a clear trend of higher nucleation potential corresponding to higher activation energy for the inhibition reaction. This trend is indicative of a stronger metal/inhibitor bond, resulting from increased electron density at the metal/inhibitor center due to the electron-donor substituents in the inhibitors.

The pH of the solution can influence the effectiveness of the inhibitor. In neutral solutions containing dissolved oxygen, there is a critical pH value above which passivation is favored. Some pitting corrosion inhibitors are effective in neutral solutions and ineffective in acid solutions. Other inhibitors effective in acid solutions can be ineffective in neutral solutions. The surface coverage of an inhibitor and, in some instances, even monolayer coverage has proven to give adequate protection.

The extent of inhibition usually increases with an increase in concentration of the inhibitor. In some cases, the corrosion may decrease with increase in inhibitor concentration, which may be attributed to the formation of soluble metal complexes, change in the mode of adsorption or the inhibitor may act as a cathodic depolarizer at

TABLE 3.4   Effective Activation Energy for Corrosion of Iron as a Function of (C_4H_9O) PSe Inhibitor Concentration

Concentration of Inhibitor (mol/L)	Effective Activation Energy (kcal/mol)
1×10^{-4}	20
5×10^{-4}	24
1×10^{-3}	42
1×10^{-2}	15

higher inhibitor concentrations. In the case of some inhibitors used in small concentrations, the inhibitors can act as cathodic depolarizers and increase the corrosion rate. For example, in a solution containing 30 mg/L of NaCl and 70 mg/L of Na_2SO_4, chromate inhibitor at less than 50 mg/L increased the pitting rate of steel.

3.9 ADSORPTION INTERACTIONS

The bond formed between the metal or an oxide surface due to the adsorption of the inhibitor has been rationalized in terms of the hard and soft acid and base (HSAB) principle (43). Hard bases have donor atoms such as N, O, and F of high electronegativity and low polarizability. Soft bases contain P, S, Cl, and I as donor atoms of low electronegativity and high polarizability. Hard acids have high positive charge and are small in size. Soft acids are large in size, with small positive charge and available electrons for sharing. According to the HSAB principle, hard acids prefer to react with hard bases and soft acids with soft bases. In accordance with the principle, hard bases containing O and N donor atoms should be adsorbed on iron and aluminum oxides that contain hard acids; namely, trivalent iron and aluminum. Passivated iron and aluminum are known to adsorb hydroxide, phosphate, borate, and sulfate on their surface. All these anions are good inhibitors, with the exception of sulfate, which is known to accelerate (32) and to inhibit (44) pitting corrosion.

A relationship was found to hold between inhibition efficiency and the electronegativity of the donor atom. The relationship may be written as

$$\log \frac{I}{1-I} = \beta x + \acute{\alpha},$$

where I is the inhibitor efficiency, x is the electronegativity of the donor atom in the inhibitor, and $\acute{\alpha}$ and β are constants.

Some systems in which the inhibition efficiency was in accord with the HSAB principle are noted in Table 3.5.

It is necessary that the inhibitor be more nucleophilic than the aggressive anion and the product formed between the inhibitor and passive oxide not be soluble in the medium.

TABLE 3.5   Inhibition Efficiency and HSAB Principle

Metal	Medium	Inhibitor
Iron Nickel	Deaerated perchloric acid and borate solutions	R_3N, R_3P, R_3Sb, R_2O, R_2Se, and R_2Te R = propyl or butyl group (45)
Iron oxide		Hard bases containing O and N atoms
Iron oxide Aluminum oxide		Bases I^-, Br^-, and Cl^- Adsorption order: $I^- < Br^- < Cl^-$

Corrosion inhibition by substituted organic compounds in acid solutions showed the role of the nature of the bond formed between the metal and the inhibitor. The nature of the bond depended on the functional group in the inhibitor. For example, the adsorption of the inhibitor increased with decreasing electronegativity, $O < N < S < Se < P$; the adsorption also increased with increase in electron density of the functional group; increased dipole moment of the inhibitor molecule; and increased hybrophobicity of the inhibitor molecule.

3.10 PASSIVATION OF METALS

In neutral solutions, inhibitor efficiency is determined by the ability of the inhibitor to passivate metals. Passivation of metals by the inhibitor is determined by the effect of the inhibitor on the critical current for passivation (i_p). In deaerated 0.1 M solutions of anions, the passivation current for iron was in the order:

$$NO_2^- < OH^- < CrO_4^{2-} < BO_3^{3-} < PO_4^{3-} < CO_3^{2-} < HCO_3^- < NO_3^-,$$

and the passivation ability of the ions decreased along the series from nitrite to nitrate (46). The nitrite ion can readily passivate iron at concentrations greater than critical concentration in the presence of dissolved oxygen. When the oxide film is protective, the nucleation of pits becomes difficult because of the higher value for nucleation potential (E_{np}). Passivation of existing pits is not possible due to the acidic conditions and the high salt content of the pits.

Properties such as electric field strength, growth rate, and surface analytical data (47) on films formed on iron are given in Table 3.6.

The films formed in sodium phosphate and sodium nitrite had high field strength, low growth rates, and the best protective properties. The films formed in chromate, molybdate, and tungstate are less protective, as evidenced by lower field strength and higher growth rates. Low field strength and high growth rate were observed with sodium biphosphate (Na_2HPO_4). The inhibition is due to coverage of the defects in the passive film by iron phosphate and the buffering capacity of the inhibitor. The thin film formed in sodium phosphate solution protects the surface by the buffering capacity and high alkalinity of the inhibitor.

TABLE 3.6 Properties of Films Formed on Iron

Solution	pH	Field Strength (mV/Å)	Growth Rate (Å/Decade)	Auger
0.1 N Na_3PO_4	12	35	5	O, Fe, P only on surface
0.01 N $NaNO_2$	7	32	10	O, Fe, small amount N
0.01 N Na_2CrO_4	8–12	3–28	20–33	O, Fe, Cr, near the substrate lack of Cr
0.01 N Na_2MoO_4	9.6	13.5	15	O, Fe, small amount of Mo
0.05 N Na_2WO_4	8.5	13.5	15	O, Fe, substantial amount of W
0.01 N Na_2HPO_4	9.1	0.02	>1000	O, Fe, P

TABLE 3.7 Effect of Film Modification on Corrosion Potential and Pit Nucleation Potential (47)

pH	Molybdate		Tungstate		Chromate	
	E_{corr} (mV)	E_{np} (mV)	E_{corr} (mV)	E_{np} (mV)	E_{corr} (mV)	E_{np} (mV)
4.5	-50	447	-16	507	$+96$	514
9.0	$+22$	393	-1	460	$+38$	446

The effect of film modification on the corrosion potential and pit nucleation potential for iron in 3% sodium chloride with oxyanions inhibitors such as chromate, molybdate, and tungstate yielded data given in Table 3.7.

The pit nucleation potentials are more positive than $+300\,$mV, which is the value for samples with unmodified air films.

3.11 INHIBITION OF LOCALIZED CORROSION

Some anions such as nitrate, sulfate, chromate, dichromate, phosphate, and hydroxyl ion act as inhibitors in the localized corrosion of iron and nickel. The inhibitors decrease the localized corrosion rate and increase the breakdown potential.

Based on breakdown potential, E_b°, for iron, the effectiveness of inhibitors follows the order

$$NO_3^- \approx SO_4^{2-} < H_2PO_4^- < OH^- < CrO_4^{2-},$$

and for Ni, we have

$$NO_3^- < SO_4^{2-} < H_2PO_4^- < CrO_4^{2-} < HPO_4^{2-} < Cr_2O_7^{2-}.$$

The inhibiting properties are in keeping with the order of molar polarization values

$$NO_3^- < SO_4^{2-} < PO_4^{3-} < OH^- < CrO_4^{2-}.$$

The molar polarization of an anion is assumed to be directly proportional to the degree of adsorption and to the degree of inhibition (48). The adsorption of the inhibiting ion expels the aggressive anion from the electrode surface, thus decreasing the surface catalyst concentration, and increases the activation energy for metal dissolution. Both factors decrease the metal dissolution rate and increase the breakdown potential. This mechanism is operative only at low potentials. At higher potentials a stable passive oxide is formed, due to the passivating effect of weakly adsorbed hydrated ions.

The weakly adsorbed inhibitors in the solution and their inhibitive capacity depends upon the electrode potential and gives rise to polarization curves whose shape is sensitive to the concentration ratio of aggressive ion to that of the inhibitor. The potentiodynamic polarization curves of iron in 0.5 M potassium nitrate and

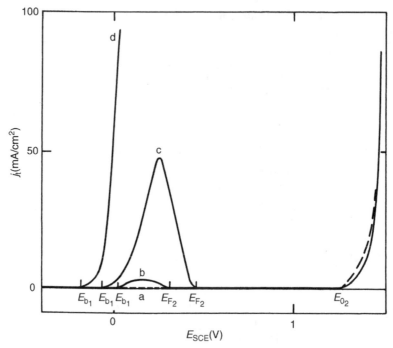

FIGURE 3.20   Potentiodynamic polarization curves of iron in 0.5 M KNO_3 + KCl (a = 0.01 M KCl, b = 0.02 M KCl, c = 0.03 M KCl, d = 0.1 M KCl).

variable concentration of potassium chloride (Fig. 3.20) gave rise to polarization curves indicating (i) absence of localized corrosion at 0.01 M KCl (curve a), (ii) the presence of first breakdown potential and second passivation potential at 0.02 and 0.03 M KCl (curves b, c), and (iii) only the first breakdown potential at 0.1 M KCl (curve d).

The measured potential as a function of chloride/nitrate ratio is plotted in Fig. 3.21. The line a lies to the left side of the diagram and shows the absence of localized corrosion. At the potential (point 1) active metal dissolution takes place. At potential 2, an oxide is formed, and at points 3 and 4 the passivation of the metal occurs (E_{F1}). No differences in polarization curves are observed in neutral solutions between points 1 and 4. At potential 5, the stability of the passive state begins to increase. Hence, at potential G, no localized corrosion occurs.

With an increase in chloride concentration, the connecting line is shifted to the right (connecting line b, polarization curve B). There is no change up to point 4. The first breakdown potential E_{b1} (point 5) lies under curve E and is lower than the second passivation potential, E_{F2}. Hence, localized corrosion occurs at E_{b1}. Then increased polarization leads to point 7 (potential E_{F2}), the nonpassive oxide is transformed into passive oxide, and the electrode passivates again. This passive state is stable only up to point 10, the second breakdown potential, which cannot be achieved since oxygen evolution occurs at point 9.

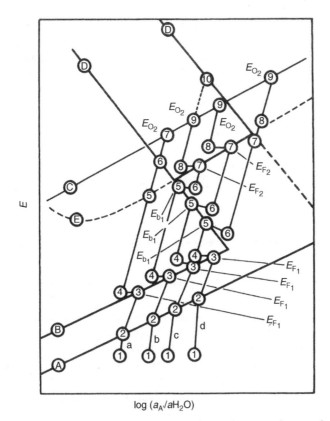

FIGURE 3.21 Plot of potential versus potassium nitrate and potassium chloride concentration.

The connecting line c represents higher chloride/nitrate concentration. From the intersection points of connecting line c into curves D and E, it is clear that with increasing chloride concentration, the first breakdown potential is shifted to lower values and the second passivating potential to higher values. This leads to a wider region of localized corrosion and a narrower passive region. At higher chloride concentration (connecting line d), the first breakdown potential is low and the second passivating potential (point 8) is higher than the second breakdown potential (point 7). This may be a possible reason why the second passivating potential is missing from the polarization curve D.

The different inhibition mechanisms operating at high and low potentials may be the reason for the classification of nitrates as both strong and weak inhibitors. According to Rozenfeld, nitrate is more effective than chromate in the inhibition of localized corrosion of steel. On the other hand, chromate was found to be more effective than nitrate in the corrosion inhibition of pure iron.

In general, the effect of anions on the localized corrosion rate varies with the concentration, potential, and the nature of the metal involved. Anions that inhibit localized corrosion under certain conditions may promote localized corrosion under different conditions. Some examples of this type of anion are nitrate, sulfate, and perchlorate. Based on the effect on localized corrosion kinetics, anions fall into three groups:

(i) Aggressive anions that cause localized corrosion:

$$Cl^-, Br^-, I^-, ClO_4^-, NO_3^-, SO_4^{2-}.$$

(ii) Anions that accelerate localized corrosion, induced by aggressive ions, but do not by themselves cause localized corrosion:

$$NO_3^-, SO_4^{2-}.$$

(iii) Anions that inhibit the localized corrosion rate and increase the breakdown potential:

$$NO_3^-, SO_4^{2-}, CrO_4^{2-}, PO_4^{3-}, ClO_4^-, OH^-.$$

Inorganic inhibitors used in inhibition of localized corrosion can be passivators or species that form insoluble salts on the metal surface. Passivation of the metal can be achieved either by increasing the efficiency of the cathodic reaction or by decreasing the extent of anodic reaction or both. Nitrite and chromate inhibitors passivate the metal by shifting the electrode potential to more positive potentials. In the corrosion, inhibition of iron by chromate trivalent chromium in the outer layer and trivalent iron in the inner layer of the surface film were detected (49). It is thought that the chromate is initially adsorbed on the metal surface followed by reduction of chromate to trivalent chromium. However, in a neutral solution containing dissolved oxygen, reduction of oxygen is faster than chromate, and oxygen is the actual inhibitor while chromate reduces the critical current for passivation due to its adsorption on the surface. Nitrites like chromates are good passivators and can passivate iron even in the absence of oxygen.

During the course of corrosion inhibition by oxyanions (MO_4^{2-}) such as chromates and molybdates, hydroxyl ions are formed (50).

$$CrO_4^{2-} + 4H_2O + 3e^- \rightarrow Cr(OH)_3 + 5OH^-, \quad \gamma = \frac{5OH^- - 3e^-}{3e} = 0.66,$$

$$MoO_4^{2-} + 2H_2O + 2e^- \rightarrow MoO_2 + 4OH^-, \quad \gamma = \frac{4OH^- - 2e^-}{2e} = 1.0.$$

When $\gamma > 0$, the chemical species functions as an inhibitor. The inhibition is due to changes in the Helmholtz double layer due to the adsorption of hydroxyl ions on the surface of the metal. The preferential adsorption of hydroxyl ions precludes the migration and adsorption of chloride ions on the metal surface. It is also thought the hydroxyl ions adsorb on the metal surface in preference to oxidizing anions, in contradiction to the generally held view.

Molybdates and tungstates are weaker oxidizing agents than chromates. The corrosion inhibition of mild steel by molybdate in a solution of sodium hydroxide and sodium sulfate consisted of passivation of the metal in the presence of oxygen. An increase in molybdate concentration resulted in increased adsorption, decrease in critical current density for passivation, and more positive pitting potential (51). This result is probably due to the ability of molybdate to repair defects in the passive film since a large amount of molybdate is adsorbed on the metal surface. Polymolybdates also act as inhibitors for iron in acid media (52).

In deaerated solutions, sodium hydroxide, sodium phosphate, and sodium carbonate form insoluble compounds and give protection of iron (53). Hydroxyl ions in high concentration passivate metals in the absence of oxygen. The nucleation potential of iron (E_{np}) is more positive in solutions of pH 10–14 when the chloride to hydroxyl ion ratio is low due to the competitive adsorption of hydroxyl and chloride anions.

The mechanism of inhibition of localized corrosion by organic inhibitors consists of screening the active sites on the metal by adsorption in preference to aggressive ions. Depending upon the conditions, organic anions such as benzoates, phthalates, and acetates can either inhibit or accelerate localized corrosion. The organic anions accelerate corrosion of iron in deaerated solutions but inhibit corrosion in the presence of oxygen at concentrations higher than critical values. The pH of the iron surface in aerated solutions is presumed to be shifted to higher values by a phenomenon known as surface alkalization. The critical pH_e is given by the relationship (54)

$$pHe = \log C_s + \log C_{O_2} + pK_a - 0.67,$$

where pH_e is the critical pH, C_s is the salt concentration, C_{O_2} is the oxygen concentration, and pK_a is the negative logarithm of the acid dissociation constant. The above equation was found to give results in agreement with experimentally determined pH_e values for the passivation of iron in the pH range 8–9.3.

Alkali metal salts of organic acids such as formic, acetic, propionic, succinic, benzoic, and salicylic acids function as pitting corrosion inhibitors for iron by shifting the nucleation potential of iron to more positive values (55).

The pitting behavior of aluminum in borate buffer solutions (31) containing aggressive ions such as chloride, bromide, and iodide with and without added sodium phenyl anthranilate inhibitor was studied. The pitting nucleation potential was measured. The inhibitor was adsorbed more readily than chloride and ΔE was found to correlate with basicity H.

Pitting corrosion inhibition efficiency of substituted orthoaryl carboxylates in solutions containing halides, thiocyanate, and trichloroacetate anions was determined and correlated with Hammet sigma constants. The results yielded v-shaped curves with electron-donor substituents falling on the left side and the electron-withdrawing substituents occupying the right branch of the curve. The electron-withdrawing substituents increased the pitting potential of aluminum (56).

Metals have a characteristic pitting potential in a given electrolyte. The change in pitting potential involving protection was correlated with effective electronegativity of the metal for sodium anthranilate and sodium phenylanthranilate (57). The effective electronegativity of a metal is related to the potential of zero charge (E_{q_0}) and work function:

$$X_{\mathrm{eff}} = 3/2(E_{q0} - W + 6.1).$$

Studies on the pitting inhibition behavior of substituted benzoates on iron and aluminum in borate buffer solutions of pH range 6–9 containing 0.005 M sodium chloride were shown to obey the equation (58)

$$\Delta E = a + b\pi + c\sigma,$$

where π is the hydrophobicity parameter, σ is the electron-donor parameter, and a, b, and c are constants. In the case of aluminum, hydrophobicity made a significant contribution to inhibition, while both hydrophobicity and electron-donating ability of the substituent played a significant role in the inhibition of pitting corrosion of iron. The protective effect depended on the pH of the medium. The decrease in ΔE is probably related to the pK values of the substituted acids.

Based on anodic polarization data, pit nucleation potentials for AISI 304 stainless steels in the pH range 2–8, it was shown that NLS and N(carboxymethyl)-N-dodecyl glycine (LADA) were effective inhibitors at high pH values, while dodecylbenzene sulfate (DBS) and dodecylsulfonate (LS) were effective at lower pH values (59). Mixtures of the two types of inhibitors showed synergism. The synergism can be attributed to the pK values of these inhibitors. Other studies on these inhibitors showed them to increase the nucleation potential E_{np} but to be ineffective in repassivating existing nucleated pits.

Pitting corrosion of AISI 304 L stainless steel in 0.1 M NaCl and 0.2 M sodium acetate at pH 5.2 was effectively inhibited by N-acylsarcosine. The carboxyl group of the inhibitor attaches to the metal surface. The inhibitor also has buffering effect on pH at pit nucleating sites (60).

In pitting corrosion, the anodic reaction is confined to small active areas and the cathodic reaction occurs on the whole metal surface. The corrosion principle that the total anodic current equals the cathodic current is valid since, when the anodic current density in the pits is high, metal dissolution takes place. In general, it is difficult to choose inhibitors for pits that are already nucleated. Although cathodic inhibition is uncommon, in some cases organic inhibitors have been used to minimize growth of pits filled with iron sulfide. In this case, the inhibitor probably chemisorbs on the iron sulfide surface and forms a barrier to the transport of water, which prevents a hydrogen evolution reaction (61).

REFERENCES

1. G Schmidt, K Bedbur, *Werkst Korros* **36**:273 (1985).
2. N Hackerman, ES Snavely. In: LSV Delinder (ed.), *Corrosion Basics*, NACE, Houston, TX, 1984.
3. ZS Smialowska, G Wieczorek, *Corros Sci* **11**:843 (1971).

4. S Trassate, *J Electroanal Chem* **53**:335 (1974).

5. JOM Bockris, DA Swinkels, *J Electrochem Soc* **111**:736 (1964).

6. I Langmuir, *J Am Chem Soc* **39**:1848 (1947).

7. S Trassati, *J Electroanal Chem Interf Electrochem* **33**:351 (1971).

8. LI Antropov, *Proceedings of the 1st International Congress on Metallic Corrosion 1961*, Butterworths, London, 1961, p. 109.

9. IL Rozenfeld, *Corrosion Inhibitors*, McGraw-Hill, New York, 1981.

10. N Hackerman, RM Hurd, *Proceedings of the International Congress on Metallic Corrosion*, Butterworths, London, 1962, p. 166.

11. JOM Bockris, N Conciocat, F Gutman, *An Introduction to Electrochemical Science*, Wykeham Publ. Ltd., London, 1974, p. 44.

12. N Sato, *Proceedings of the Japan–USSR Corrosion Seminar*, ISCE, Tokyo, Japan, May 1980, p. 35.

13. E McCafferty, N Hackerman, *J Electrochem Soc* **119**(8):999 (1972).

14. N Sato, K Kudo, T Noda, *Z Phys Chem* **98**:217 (1975).

15. H Gobrecht, W Paatsch, R Thull, *Ber Bunsenges Phys Chem* **75**:1353 (1971).

16. SM Ahmed, In: JW Diggle (ed.), *Oxide and Oxide Films*, Vol. 1, Marcel Dekker, New York, NY, 1972, p. 319.

17. W Khalil, S Haupt, HH Strehblow, *Werkst Korros* **36**:16 (1985).

18. J Lumsden, Z Szklarska-Smialowska, *Corrosion* **34**:169 (1978).

19. VS Sastri, RH Packwood, JR Brown, JS Bednar, IE Galbraith, VE Moore, *Br Corros J* **24**:30 (1989).

20. E McCafferty, MK Bernett, JS Murday, *Corros Sci* **28**:559 (1988).

21. N Sato, *Corros Sci* **27**:421 (1987).

22. M Sakashita, N Sato, *Corrosion* **35**:351 (1979).

23. D Kyriacou, *Surf Sci* **8**:370 (1967).

24. FA Kröger, HJ Vink, *Solid State Phys* **3**:307 (1956).

25. G Palumbo, PJ King, KT Aust, *Corrosion* **43**:37 (1987).

26. PM Natishan, E McCafferty, GK Hubler, *J Electrochem Soc* **133**:1061 (1986).

27. Z Sklarska-Smialowska, G Wieczorek, *Corros Sci* **11**:843 (1971).

28. R Bailey, JE Castle, *J Mater Sci* **12**:2049 (1977).

29. VV Ekilik, NV Sergeeva, *Zasch Met* **22**:538 (1986).

30. S Matsuda, HH Uhlig, *J Electrochem Soc* **111**:156 (1964).

31. Yu I Kuznetsov, LJ Popova, *Zasch Met* **19**:727 (1983).

32. JW Congdon, *Mater Perf* **5**:35 (1988).

33. K Sugimoto, Y Sawada, *Corrosion* **32**:347 (1976).

34. H Böhni, HH Uhlig, *J Electrochem Soc* **116**:906 (1969).

35. H Kaesche, *Z Phys Chem* **26**:138 (1960).

36. F Hunkeler, H Böhni, *Werkst Korros* **32**:129 (1987).

37. HP Leckie, *J Electrochem Soc* **117**:1152 (1970).

38. KJ Vetter, HH Strehblow, *Localized Corrosion NACE-3*, NACE, Houston, TX, 1974, p. 240.

39. HH Strehblow, B Titze, *Corros Sci* **17**:461 (1979).

40. E McCafferty, *Proceedings of the 6th European Symposium on Corrosion Inhibitors*, Sez. V, Ann. University Ferrara, 1985, p. 1535.

41. Z Sklarska-Smialowska, B Dus, *Proceedings of the 3rd International Congress on Metal Corrosion*, Vol. 2, Moscow, 1967, p. 111.

42. Yu I Kuznetsov, JA Valuev, *Zasch Met* **23**:822 (1987).

43. RG Pearson, *J Am Chem Soc* **85**:3533 (1963).

44. IL Rozenfeld, JS Danilov, *Corros Sci* **7**:129 (1967).

45. K Aramaki, H Nishihara, *Proceedings of the 6th European Symposium on Corrosion Inhibitors*, Sez. V, Ann. University Ferrara, 1985, p. 67.

46. JGN Thomas, TJ Nurce, *Br Corros J* **2**:13 (1967).

47. JB Lumsden, Z Sklarska-Smialowska, *Corrosion* **34**:169 (1978).

48. JO Bockris, AKN Reddy, *Modern Electrochemistry*, McDonald, London, 1998.

49. E McCafferty, *Proceedings of the 6th European Symposium on Corrosion Inhibitors*, Sez. V, Ann. University Ferrara, 1985, p. 533.

50. Yu N Mikhailovskii, *Zasch Met* **20**:179 (1984).

51. MA Stranick, *Corrosion* **40**:296 (1984).

52. K Ogura, T Ohama, *Corrosion* **40**:47 (1984).

53. JEO Mayne, JH Menter, *J Chem Soc* 103 (1954).

54. W Forker, G Reinhard, D Rahner, Corros Sci, **19**:11 (1979).

55. SM Abd El Haleen, MGA Khedr, AA Abdel Fatah, *Proceedings of the International Congress on Metallic Corrosion*, Vol. 4, Toronto, Canada, 1984, p. 62.

56. Yu I Kuznetsov, *Zasch Met* **30**:359 (1984).

57. Yu I Kuznetsev, JG Kuznetsova, *Zasch Met* **22**:474 (1986).

58. JGN Thomas, *Proceedings of the 5th European Symposium on Corrosion Inhibitors*, Sez. V, no. 2, Ann. University Ferrara, 1980, p. 453.

59. F Zucchi, JH Omar, G Trabanelli, *Proceedings of the 6th European Symposium on Corrosion Inhibitors*, Sez. V, Ann. University Ferrara, 1985, p. 1535.

60. DW De Berry, A Viehbeck, *Corrosion* **44**:299 (1988).

61. RG Asperger, P Hewitt, *Corrosion/86*, Paper no. 135, Houston, TX, National Association of Corrosion Engineers 1986.

4

CORROSION INHIBITION: THEORY AND PRACTICE

Corrosion can be described as the undesirable reaction of a metal or alloy with the aggressive environment. Corrosion inhibition consists of the addition of chemicals known as corrosion inhibitors, which minimize the extent of corrosion. It should be borne in mind that other techniques, such as the removal of dissolved oxygen and adjustment of pH, in some cases can combat corrosion.

The use of inhibitors is concerned with four types of environment:

(i) Natural waters, supply waters, industrial cooling waters in the near-neutral, 5–9 pH range.
(ii) Aqueous acid solutions used in pickling of metals, postservice cleaning of metal surfaces.
(iii) Primary and secondary production of oil and refining.
(iv) Miscellaneous environments.

There are many classifications of inhibitors. Some types, classified as near-neutral pH solutions, are (i) "safe" or "dangerous" inhibitors. The inhibitor must be present above a certain minimum concentration in order for the inhibitor to be effective (safe). An inhibitor present below the critical concentration may result in localized attack, and the inhibitor in this case is called a dangerous inhibitor.

Inhibitors are also known as anodic or cathodic, depending upon whether the inhibitor affects the anodic metal dissolution reaction or the cathodic oxygen

Green Corrosion Inhibitors: Theory and Practice, First Edition. V. S. Sastri.
© 2011 John Wiley & Sons, Inc. Published 2011 by John Wiley & Sons, Inc.

reduction (near-neutral solutions) or hydrogen discharge reaction (acid solutions).

Oxidizing and nonoxidizing inhibitors are characterized by their capacity to passivate the metal. Nonoxidizing inhibitors require dissolved oxygen for maintaining the passive film, while oxidizing inhibitors do not require dissolved oxygen.

Organic or inorganic inhibitors are classified according to whether the chemical is, respectively, organic or inorganic in nature. However, sodium salts of carboxylic acids behave like inorganic inhibitors. Organic inhibitors are further classified as soft, hard, and borderline inhibitors, depending upon the donor atom in the organic inhibitor. Organic inhibitors containing oxygen as the donor atom are termed hard inhibitors; inhibitors containing sulfur as the donor atom are soft; and inhibitors with nitrogen as the donor atom are borderline inhibitors.

The two treatments of water or aqueous solutions in order to combat corrosion are (i) oxygen scavengers to remove dissolved oxygen (deaeration) and (ii) adjustment of pH of the solutions to reduce corrosion (pH control). Deaeration of the solutions can be achieved by the addition of hydrazine or a mixture of sodium sulfite and cobalt chloride. The pH adjustment of iron and steel to a value of 9.0 can be done by adding an alkali such as sodium hydroxide. When metals such as aluminum are involved, pH adjustment to a value of 9 is not suitable, since this is the value at which aluminum corrosion occurs. Thus, pH adjustment is not recommended in a mixed metal system.

A vast literature exists concerning many chemical substances for use as corrosion inhibitors. A few citations are noted below:

(i) 70 compounds as inhibitors, *J Appl Chem* **11**:246 (1961).

(ii) 150 compounds as inhibitors, MG Fontana, RW Staehle (eds.), *Adv Corros Sci Technol*, Vol. 1, Plenum Press, New York, London, 1970.

(iii) Inhibitors for Al and its alloys, *Anti-Corros Meth Mater* **18**(4):8–13 (1971); (5):4–10 (1971).

(iv) Inhibitors for copper, *Anti-Corros Meth Mater* **18**(2):19–23 (1971).

(v) Corrosion inhibitors for industry, *Trans Inst Chem Eng* **47**:T177–T192 (1969).

(vi) Inhibitors for copper, *Anti-Corros Meth Mater* **17**(9):9–15 (1970).

In near-neutral solutions, anodic inhibitors such as chromates, dichromates, nitrites, phosphates, borates, and benzoates assist in maintenance, repair, and reinforcement of the natural oxide film on mild steel. Of these inhibitors, nitrites and chromates are oxidizing type inhibitors. In the presence of aggressive ions, chromates and nitrites present in insufficient quantity for complete protection give rise to localized corrosion and hence become dangerous inhibitors. Toxicity and environmental pollution do not favor the use of chromates. A nonoxidizing benzoate inhibitor can also cause limited localized corrosion (1). Other chemical inhibitors such as polyphosphates, silicates, zinc ions, and tannins belonging to the cathodic type of inhibitors may be used in near-neutral solutions. Some of these may also affect the anodic reaction. Films of zinc hydroxide, phosphates, and silicates are formed on the metal surface. These cathodic inhibitors do not give rise to localized corrosion and can be considered safe inhibitors.

The degree of inhibition of corrosion of metals other than mild steel depends upon both the metal and the inhibitor. The cathodic inhibitor is relatively less susceptible than the anodic inhibitor to the nature of the metal. In general, anodic inhibitors are more efficient than cathodic inhibitors and can give 100% inhibition at suitable concentrations.

In acid solutions, the cathodic reaction is hydrogen discharge, hence, the need for inhibitors that adsorb or bond on the metal surface or raise the overpotential for hydrogen ion discharge. Suitable inhibitors are organic molecules with electron donor atoms such as N, S, and O. Some quaternary ammonium compounds may also be used as inhibitors.

Inhibitor formulations used in near-neutral solutions usually consist of two or three chemical compounds. The multicomponent inhibitor is better than the single inhibitor, since the former is effective for protection of multimetal systems; a mixture of cathodic and anodic inhibitors is better than either a cathodic or an anodic inhibitor alone; and the mixture of inhibitors allows the use of lower concentrations. Any mixtures of inhibitors is known as a synergized formulation.

4.1 FACTORS PERTAINING TO METAL SAMPLES

A general guide indicating the relative performance of inhibitors for metals in near-neutral solutions is given in Table 4.1. Corrosion inhibitors are, in general, specific in action toward metals. It is useful to note that chromates are environmentally unacceptable. Dodecamolybdophosphate has been reported to function as well as chromate inhibitor in corrosion inhibition of many metals (2).

Anions such as nitrates can be aggressive and cause corrosion under some conditions. Nitrates can prevent corrosion inhibition of mild steel by benzoate, chromate, or nitrite (3–5). Nitrates used in antifreeze solution reduce the corrosion of aluminum alloys. Inhibitors are required to prevent the corrosion of mild steel in ammonium nitrate solutions (6). The sulfate ions are aggressive toward mild steel in waters but can inhibit chloride-induced pitting of stainless steels (7) and caustic embrittlement in boilers. These are examples of the dual nature of the anions as well as the dependence of inhibition on the nature of the metal.

When dissimilar metals are involved, it is advisable to incorporate two inhibitors at adequate concentration and in correct proportion. Failure to inhibit the corrosion of one metal can intensify the corrosion of the other metal. For example, failure to inhibit corrosion of copper with benzotriazole (BTA) can lead to increased corrosion of aluminum as a result of deposition of copper from cupric ions in solution on the surface of aluminum.

In some instances, inhibitors can lead to polarity-reversal effects. In corrosive environments the zinc coating on galvanized steel acts sacrificially and prevents the corrosion of steel. But in the presence of sodium benzoate (8, 9) or sodium nitrite (10), the steel exposed at breaks in the zinc coating may corrode readily.

TABLE 4.1 Performance of Corrosion Inhibitors in pH Range of 5–9

Metal (Alloy)	Inhibitor						
	Chromates	Nitrites	Benzoates	Borates	Phosphates	Silicates	Tannins
Mild steel	Good	Good	Good	Good	Good	Reasonably good	Reasonably good
Cast iron	Good	Good	Poor	Good	Good	Reasonably good	Reasonably good
Zinc and zinc alloys	Good	Poor	Poor	Good	–	Reasonably good	Reasonably good
Copper and copper alloys	Good	Partially good	Partially good	Good	Good	Reasonably good	Reasonably good
Aluminum and aluminum alloys	Good	Partially good	Partially good	Variable	Variable	Reasonably good	Reasonably good
Lead–tin soldered joints	–	Aggressive	Good	–	–	Reasonably good	Reasonably good

4.1.1 Sample Preparation

Clean, smooth metal surfaces require a lower amount of corrosion inhibitor for corrosion inhibition than do rough or dirty surfaces. Minimum amount of benzoate, chromate, and nitrite required to inhibit the corrosion of mild steel with a variety of surface finishes has been studied (3–5). The effectiveness of inhibitors such as benzoate was particularly sensitive to surface preparation. Thus, it is difficult to relate laboratory data to a practical system. Other factors, such as oil, grease, or corrosion products, will affect the amounts of inhibitor required, along with possible depletion of inhibitor during service due to its reaction with the contaminants. Rust on the sample can be removed mechanically before addition of the inhibitor. A particular method of removal of rust is phosphate-delayed chromate treatment (11, 12). In this method, the sample is treated with an acid phosphate solution to remove rust, followed by exposure to the chromate inhibitor.

4.1.2 Environmental Factors

Aqueous solutions are the most common environment. Inhibitors are sometimes required for nonaqueous systems, such as aluminum in chlorinated hydrocarbons; oils and greases; and liquid metals, for example, mild steel corrosion in liquid bismuth prevented by addition of Ti, Zr (13). An unusual case is inhibition of corrosion of titanium in methanol by water. In all the cases the chosen inhibitor must be compatible with the system. For example, chromate being a strong oxidizing agent will oxidize glycol and cannot be used as an inhibitor in antifreeze compositions.

The ionic composition of the aqueous near-neutral solutions plays an important role in determination of corrosivity. Chlorides and sulfates are aggressive ions. The concentration of the inhibitor required for protection will depend upon the concentration of the aggressive ions.

The quantitative relationship between concentration of aggressive ions (C_a) present and the required inhibitor concentration (C_i) is

$$\log C_i = K \log C_a + \text{constant},$$

where K is related to the valencies of the ions involved.

The behavior of halides varies with the acidity of the medium. In general, halides are aggressive in near-neutral solutions but improve inhibition in acid solutions. The halides differ in inhibitive action as evidenced by the acceleration of stress-corrosion cracking of titanium in methanol in the presence of chloride ions and inhibition by iodide ions (14).

Both solid and gaseous dissolved impurities can affect the pH of the system and result in lower corrosion inhibition efficiency. In cooling systems of industrial plants, sulfur dioxide, hydrogen sulfide, or ammonia can change the water pH and cause corrosion. Another example, exhaust gases in an automobile can enter into the engine coolant and cause corrosion.

The control of pH is very important for the action of inhibitors. The effective pH range of some common inhibitors is as follows:

Inhibitor	pH Range
Nitrites	5.5–6.0
Polyphosphates	6.5–7.5
Chromates	8.5
Silicates	Wide pH range

In general, inhibition at higher temperatures requires the use of higher concentrations of inhibitors. At higher temperatures, inhibitors may decompose and lose their effectiveness as in the case of decomposition of polyphosphate to give orthophosphate. The presence of calcium will result in precipitation of calcium phosphate.

4.1.3 Concentration of Inhibitor

For effective corrosion inhibition, the inhibitor must be present above a certain minimum concentration. Insufficient concentration of inhibitor can lead to severe corrosion, as mentioned earlier, through safe and dangerous inhibitors. Loss of inhibitor due to film formation or reaction with contaminants in the system must be taken into account when one determines amount of inhibitor required. Loss of inhibitor can also occur due to mechanical effects such as windage losses in cooling towers, blowdown in boilers, and general leaks.

Maintenance of correct concentration level is particularly important in cases where low-level treatments such as 100 ppm are involved. In closed systems such as automobile engines about 0.1% is commonly used; in this application, a good amount of inhibitor is included in the recommended concentration. In any case gross depletion of inhibitors can lead to enhanced corrosion, since the inhibitors involved are the dangerous type.

Inhibitor monitoring can be done by chemical analysis and inspection of test samples or instrumental methods. The instrumental methods are based on the linear polarization method and electric resistance probes. These methods have the advantage in that the data from various areas of the plant can be brought together at a central control point.

4.1.4 Process Conditions

Corrosion can be intensified by mechanical factors. Some factors are inherent or applied stress, fatigue, fretting, or cavitation. Inhibitors cannot be successfully used in these conditions.

Aeration and agitation of medium have a significant effect in corrosion inhibition in near-neutral solutions by causing oxidization of inhibitors. Agitation of the solution helps to transport oxygen and inhibitor to the metal surface. Stagnant solutions require higher amounts of inhibitor than do circulating solutions, as can be observed in the case of polyphosphates (15). In some cases, excessive aeration may be deleterious, since the inhibitor may be depleted by its interaction with oxygen. Tannins serve as an example of this behavior.

For effective protection by inhibitors, it is necessary that crevices at joints, dead ends in pipes, gas pockets, and corrosion product deposits not be present in the system.

Microorganisms can cause (i) general and pitting corrosion, resulting in deposits of corrosion product; (ii) accumulation of fungal growth that impedes water flow; and (iii) degradation and depletion of inhibitors. Bactericides incorporated in the inhibitor formulation can counter the effects of microorganisms.

Polyphosphates and silicates provide inhibition by controlled scale deposition. Uncontrolled scale deposition prevents the inhibitor from reaching the bare metal surface, loss of inhibitor by incorporation into the scale, and reduced heat transfer in the cooling system. The inhibitors may also react with constituents in the water and produce scales. Some examples are the reaction of silicates and phosphates with calcium. The scale formation can be negated by pH control and scale-controlling agents.

Environmental regulations require that treated waters be free from toxic chemicals such as chromates and phosphates. Because of the environmental considerations, low-chromate-phosphate types of formulation have been advanced. For some applications, low-chromate-phosphate formulation is not acceptable and biodegradable chemicals are being introduced.

4.2 INHIBITORS IN USE

There is a vast array of inhibitors available under trade names, and their precise compositions are not known. Thus, a discussion of the use of inhibitors is restricted to inhibitor formulations in chemical terms. Potability and toxicity are the limiting factors in the choice of inhibitors in potable waters. The choice of inhibitors is limited to calcium carbonate scale deposition, silicates, polyphosphates, and zinc salts (16). The silicates do not inhibit corrosion completely; their use is limited to soft waters. The molar ratio of 2 Na_2O:2.1 SiO_2 is preferable for inhibition. Polyphosphates in conjunction with Ca^{2+} or Zn^{2+} provide a good degree of inhibition. The Ca^{2+}: phosphate ratio of 1:5 at a level of 10 ppm has been found to give inhibition in the pH range 5–7. Addition of zinc salts improved the performance of polyphosphates as inhibitors. The concentration of 5 ppm for small towns and 1 ppm for cities has been found to suffice. Higher levels of 10 and 50–100 ppm are suitable for clearing deposits and cleaning old mains, respectively.

4.3 COOLING SYSTEMS

Evaporation is the main source of cooling in a recirculating system. As cooling proceeds, evaporation results in an increase in the dissolved solids in the water until solubility considerations necessitate its limitation by blowdown. The intimate contact of the water in the towers ensures saturation of the system with oxygen. Since different metals are used throughout the network of the system, metal pickup in one site can result in deposition and the setting-up of galvanic couples in another. Inhibition in cooling circuits requires more than the simple addition of inhibitors; the conditions must be favorable for the inhibitor to function effectively.

The factors that affect corrosion inhibition in cooling systems are (i) oxygen saturation at the cooling towers, (ii) ingress of water-soluble gases such as sulfur dioxide due to the scrubbing action of the towers, (iii) pH of the medium, (iv) total dissolved solids of the makeup water, (v) algae spores, which may proliferate readily (biocides, chlorine, chloramines, chlorophenates, and quaternary ammonium salts are effective for combating algae spores), (vi) contamination due to leakage, (vii) sulfate-reducing bacteria (SRB), which produce hydrogen sulfide under slime deposits and cause localized corrosion, and (viii) silt and solids.

Silt and solids coagulate at high temperatures, and their settlement results in attack as a result of differential aeration. Dispersants are often added to keep the solids in solution, but pretreatment of the water is the preferred procedure.

For an inhibitor treatment to be effective it should meet certain criteria, such as (i) it must protect the entire metal from corrosion; (ii) low concentrations of the inhibitor must be effective; (iii) the inhibitor must be effective under a wide range of conditions such as pH, temperature, heat flux, and water quality; (iv) it should not produce deposits on the metal surface that impede heat transfer; (v) the treatment must be environmentally acceptable toxicity for discharge; (vi) it must prevent formation of carbonate and sulfate scales; and (vii) it must combat biological activity due to microorganisms.

Once-through cooling systems made of mild steel can be treated with lime and soda when the chloride content is low (17). Polyphosphates at 2–10 ppm together with a small amount of zinc ions will reduce tuberculation but not necessarily the corrosion rate (18). Other inhibitors used are 9 ppm of an organo-activated zinc–phosphate–chromate (19) and less toxic and environment friendly polyphosphate at 10 ppm. Effluent and economic problems limit the choice of inhibitors and the solution may be in material selection such as cupronickel or aluminum brass. The addition of 1 ppm of ferrous sulfate thrice daily for 1 h to power-house intakes extended the life of condenser tubes by 25–30% (20, 21). High molecular weight, water-soluble nonionic polyacrylamide inhibits corrosion of cupronickel condenser tubes (22).

Open recirculating systems are more suited to corrosion inhibition. The controlled deposition of adherent calcium carbonate film is the most economical method of corrosion inhibition. Local variations in pH and temperature affect the nature and extent of film deposition. In early times, chromates and nitrites were used. An initial dose of 1000 ppm of chromate, lowered to 300–500 ppm in later stages, was used (23). This treatment suffers from drawbacks such as the possibility of localized corrosion in the presence of chloride or sulfate and the environmental problem concerned with chromate disposal. Sodium nitrite at 500 ppm and pH range of 7–9 has been advanced, which is also susceptible to chloride- and sulfate-induced (localized) attack in addition to bacterial decomposition, giving rise to ammonia that causes stress-corrosion cracking of copper alloys. Bacterial decomposition can be controlled by shock treatment with 100 ppm of 2,2′-methylene-bis-(4-chloro phenol) and weekly treatment with sodium pentachlorophenate to control algae (24).

Polyphosphates along with adequate amounts of calcium, magnesium, or zinc ions were found to give adequate inhibition. Polyphosphate:calcium (P_2O_5:Ca) ratio of 3.35:1 is recommended at a level of 15–37 ppm of P_2O_5. The initial dose can be as high as 100 ppm. For a corroded steel system an initial dose of 20 ppm lowered to 10 ppm has been suggested (25). Polyphosphates can revert to orthophosphates and give rise to calcium phosphate scale (26).

Silicates have been used as anodic inhibitors for over 60 years and exist in solution with variable compositions xNa_2O:$ySiO_2$. The inhibition by silicates is affected by pH, temperature, and solution composition. Silicates are ineffective at high ionic strength and are effective in solutions containing salts at 500 ppm or less. The usual effective concentration range is 25–40 ppm of SiO_2. Pitting attack is less severe than in the case of chromates and nitrites. The ratio of SiO_2/Na_2O in the range of 2.5–3.0 is effective for low-carbon steel. The protective film produced by silicates is considered to be a hydrated gel of silica and metal oxide.

For the sake of economy and avoidance of environmental pollution, mixtures of inhibitors at low concentrations have been used. The four components in use are chromates, polyphosphates, zinc salts, and organic compounds (18, 27). A cathodic inhibitor (zinc ions) or polyphosphate is combined with chromate (anodic) ions. These combinations may be used in the pH range of 6–7. Zinc chromate is a heavy metal system and is most effective. This inhibitor impedes the cathodic reduction of oxygen. This mixture inhibits corrosion of copper-base metals, aluminum alloys, and galvanized steel. The concentrations usually used in recirculating cooling water systems are 10 mg/L of zinc and chromate in the pH range of 5.5–7.0 with 7.0 as the optimum value. Formation of calcium sulfate and calcium carbonate scales, insoluble zinc salts above pH 7.5, and the requirement of a metal surface free of deposits are some of the disadvantages that can be overcome by the addition of aminomethylene phosphonic acid (AMP). Another important feature is that zinc chromate is not a nutrient for biological growth, which simplifies biocidal measures.

The zinc dichromate system can be improved by the addition of phosphates and some organics such as lignosulfonates and synthetic polymers. Commercial inhibitors containing chromate:zinc ratio from 0.92 to 3.0 and zinc:phosphate ratio from 0.1 to 3.24 mixed with organic compounds have been documented. In zinc-phosphate formulation, organic compounds such as mercaptobenzothiazole have been incorporated (18). The compound mercaptobenzothiazole inhibits the corrosion of copper. Polyphosphate at 5–10 ppm was found to improve the inhibitive action of 20–40 ppm of silicate inhibitor (18). A mixture of 100 ppm orthophosphate, 40 ppm chromate, and 10 ppm of polyphosphate has been used (28).

In view of environmental concerns, low-chromate formulations are also not acceptable. New formulations involving organic compounds together with zinc ions have been advanced (29). Zinc ions have been combined with AMP and used in the corrosion inhibition of steel with a minimum corrosion rate at 60% of AMP (18). Other phosphonic acids used include 1-hydroxyethylidene-1,1'-diphosphonic acid (HEDP), nitrolo-triphosphonic acid (NTP), phosphono-butane-tetracarboxylic acid (PBTC). A chart for selection of inhibitors showing the potential problems and the inhibitors used, as well as the Ryznar Index, is shown in Fig. 4.2.

Closed recirculating systems are encountered in the cooling of internal combustion engines. Inhibitors are required for engine coolant to prevent corrosion and blockage of coolant flow by corrosion products and to maintain heat transfer efficiency. Antifreeze solutions (ethylene glycol–water) should contain inhibitors such as nitrite, benzoate, borax, phosphate, and copper-specific inhibitors such as sodium mercaptobenzothiazole (NaMBT) and benzotriazole. Triethanol-ammonium orthophosphate (50% solution) and 0.2–0.3% NaMBT mixture has been used (30) in the protection of ferrous metals and copper, respectively. Another inhibitor mixture consists of 5–7.5% sodium benzoate and 0.45–0.55% sodium nitrite. The nitrite protects cast iron while the benzoate protects other metals and soldered joints against the deleterious effect of nitrite. British Standard 3152 incorporated 2.4–3.0% borax ($Na_2B_4O_7 \cdot 10H_2O$) into the nitrite, benzoate mixture. Other formulations (31) consist of 3% borax, 0.1% mercaptobenzothiazole, 0.1% sodium metasilicate, and 0.03% lime (CaO). Standards for inhibited engine coolants are given in British Standard 6580 along with test methods in British Standard 5117.

Alkali metal phosphonates have been used in antifreeze mixtures and these mixtures have a deleterious effect on aluminum alloys. Soluble oils have also been used that can cause ill effects on rubber connections.

Larger volumes of coolant are involved in locomotive engines and the cost of inhibition is high. Another factor is the cavitation attack of cylinder liners. A 15:1 borate–metasilicate at a concentration of 1% has been used with satisfactory performance (32). Nitrate may be added to borate-silicate formulation to combat aluminum alloy corrosion. Tannins and soluble oils may be used. British Standard 3151 benzoate–nitrite mixture may be used (33).

A number of formulations based on nitrites, borates, and phosphates are used in marine diesels. Typical formulations are 1:1 nitrite:borax at 1250–2000 ppm and 1250–2000 ppm of nitrite with sodium phosphate. Inhibitors used in these systems should be nontoxic, due to the possible leakage and contamination of drinking water.

Hot water tanks consist of cast iron or steel boiler, copper piping, steel or cast iron radiators, and copper tanks. Usually the corrosion problems are minimal. Perforation of the radiators and the release of hydrogen gas from the radiators produce magnetite leading to blockage of the pump.

$$3Fe(OH)_2 \rightarrow Fe_3O_4 + H_2 + 2H_2O.$$

The excess magnetite problem can be alleviated by the addition of 0.01% benzotriazole, which complexes copper ions and prevents the catalytic effect of cupric ions on the formation of magnetite. The prevention of copper dissolution reduces magnetite formation. For general corrosion inhibition, 1% sodium benzoate with 0.1% of sodium nitrite has been used with success (34–36). Sodium metasilicate has been successfully used in soft waters. Other inhibitors used are a mixture of silicate and tannic acid, as well as the four-component mixture of sodium benzoate, sodium nitrite, sodium dodecamolybdophosphate, and benzotriazole (37).

The main corrosive agents in steam-condensate lines made of steel and/or copper-base alloys are carbon dioxide and oxygen (38). Neutralization to maintain pH of

TABLE 4.2 Corrosion Inhibition in Refrigerating Brines

System	Inhibitor
Refrigerating brines	Chromates at pH 8–8.5; 56.7 and 90.7 kg dichromate per 28.3 m^3 of calcium or sodium chloride brine
50% seawater (40)	10% sodium nitrite
Desalination system (41)	5 ppm chromate + 30–45 ppm of disodium hydrogen phosphate
Mild steel in oxygen-saturated seawater (42)	Dichromate and phosphate at 50 ppm; chromate, phosphate, zinc, and iodide at 100 ppm
Corrosion of motor vehicles (43)	Corrosion inhibitors in deicing salts
Inhibitors in deicing salts (44)	Polyphosphates
Refrigerating brines (45)	2-Ethyl ethanolamine with benzotriazole at pH 9
Refrigerating brines (46)	Hydrazine hydrochloride with benzotriazole
Refrigerating brines (47)	Gelatin, triethanolamine, and potassium dihydrogen phosphate
Refrigerating brines (48)	Lithium hydroxide with benzotriazole and sodium molybdate
Refrigerating brines	Resorcinol plus sodium nitrate; glycerine plus sodium nitrate; lithium hydroxide plus sodium tungstate
Cooling system of propylene (49) glycol with potassium bicarbonate	Mixture of silicate, polyphosphate, and sucrose (low-toxic formulation)

8.5–8.8 has been advocated (39). Some of the neutralizing agents are ammonia, cyclohexylamine, morpholine, and benzylamine. Ammonia is not suitable since incomplete neutralization will give rise to pockets of unneutralized condensate, and ammonia can attack copper alloys. The amines function favorably by condensing at the rate of steam, but this is expensive since 3 ppm of amine is required to 1 ppm of carbon dioxide (39).

Long-chain amines such as octadecylamine and its acetate salt is a well-suited inhibitor. This inhibitor at a level of 1–3 ppm is used to offset corrosion due to carbon dioxide and oxygen as well as exfoliation of 70 Cu–30 Ni tubes (38).

Total corrosion inhibition in saline medium is difficult to achieve, but a marked decrease in corrosion has been achieved in some systems containing chloride as shown in Table 4.2.

4.4 PROCESSING WITH ACID SOLUTIONS

Acids are widely used in acid pickling, industrial acid cleaning, acid descaling, oil well acidizing, and manufacturing processes in different industries. Inhibitors are used in acid solutions in order to minimize corrosive attack of the metal or alloy. The selection of a suitable inhibitor depends on the type of acid, concentration of the acid, temperature, flow velocity, the presence of dissolved inorganic or organic substances, and the type of metallic material exposed to the acid solution. The most commonly used acids are hydrochloric, sulfuric, nitric, hydrofluoric, citric, formic, and acetic acid.

Hydrochloric acid is most commonly used in a pickling bath. Large-scale continuous treatment, such as metal stripping and wire pickling and regeneration of depleted pickling solutions, is one of the advantages in using hydrochloric acid

TABLE 4.3   Corrosion Inhibition in Acid Solutions

Medium	Inhibitor	Reference
Pickling steel in sulfuric acid	Sulfur-containing organic compounds	50
Pickling steel in hydrochloric acid	Nitrogen-containing organic compounds	50
Pickling of steel	72 compounds and 32 mixtures of nitrogen- and sulfur-containing compounds better than either type alone	51
Pickling in sulfuric acid	112 compounds; sulfur-bearing compounds performed well: 12 out of 14 effective compounds are sulfur-bearing compounds	52
Pickling in sulfuric acid	0.003–0.01% inhibitor—phenyl thiourea, di-*ortho*-tolyl thiourea, mercaptans, and sulfides gave 90% inhibition	
Pickling steel	Pyridine, quinoline, and other amines	
Stainless steel in hydrochloric acid	Decylamine, quinoline, phenyl thiourea, and dibenzyl sulfoxide	53
Sulfuric and hydrochloric acid media	Coal tar base fractions were used; 0.25 vol% distilled quinoline base with 0.05 M sodium chloride in 4 N sulfuric acid at 93°C; sodium chloride acted synergistically	54, 55
Sulfuric acid	Phenylthiourea and potassium iodide	56
Acid solutions	Acetylenic compounds; 2-butyne-1,4-diol-1-hexyne-3-o1; 4-ethyl-1-octyne-3-ol	57–59
Acid solutions	Tetraalkyl ammonium bromides with alkyl having C > 10 atoms	60, 61
Iron in H_2S-saturated sulfuric acid	Tetrabutyl ammonium sulfate	62
PB class inhibitor	Reacting butyraldehyde with ammonia and polymerizing the resultant complex	
PB-5	0.01–0.15% arsenic salt in 20–25% HCl	
ChM inhibitor	Mixture of hexamine with potassium iodide, a regulator and a foaming agent	
BA-6	Condensation product of hexamine with aniline	
Katapin series	*p*-Alkyl benzyl pyridine chlorides	

instead of sulfuric acid. Other acids, such as nitric, phosphoric, sulfamic, oxalic, tartaric, citric, acetic, and formic acids, are used for special applications. In the case of pickling with HCl, up to 200 g/L are commonly used at 60°C with a pickling time of 30 min. Sulfuric acid at 200–300 g/L level at a temperature up to 90°C may also be used. Such severe conditions require effective inhibitors.

The inhibitors used in pickling should (i) be effective at inhibiting metal dissolution, (ii) unlikely to overpickle, (iii) be effective at low inhibitor concentration, (iv) be effective at high temperatures, (v) possess thermal and chemical stability, (vi) be effective at inhibiting hydrogen entry into the metal, (vii) possess good surfactant properties, and (viii) have good foaming properties.

Usually, inhibitors have poor surfactant and foaming properties, and as a result, wetting agents, detergents, and foaming agents are generally added to the commercial inhibitor formulation. Wetting agents help the pickling acid to penetrate into cracks and fissure in the scale and also remove the scale. Wetting agents are known as pickling accelerators. Wetting agents that have degreasing properties are known as pickling degreasing agents. Wetting agents and detergents help produce a clean metal surface after the pickling acid drains from the sample. When the foaming ability of wetting agents and detergents is insufficient, additional foaming agent is added to the pickling acid formulation. A typical pickling inhibitor formulation consists of a mixture of active inhibitor, wetting agent, detergent, foaming agent, solvent, and sometimes a cosolvent.

The inhibitors used in acid pickling are generally mixtures of nitrogen-bearing organic compounds, acetylenic alcohols, and sulfur-containing organic compounds. The surface-active compounds may be anionics, such as alkyl or alkyl phenyl sulfonates or alkyl sulfates with C_{10}–C_{18} alkyl chains, or nonionics, such as ethoxylated derivatives of higher alcohols, phenols, or alkyl phenols. Nonionics are used when precipitation of iron is to be prevented.

Commercial inhibitors are generally available in the form of liquids. The solvent may be water or the acid in which the inhibitor is to be used. Depending upon the amount of solvent or cosolvent, about 5–50 g of the inhibitor is used for 1 L of pickling acid. Some of the inhibitors used in pickling operations are given in Table 4.3.

4.5 CORROSION PROBLEMS IN THE OIL INDUSTRY

The four main processes involved are (i) primary production, (ii) secondary production, (iii) refining, and (iv) storage. Corrosion problems arise in the primary production due to the presence of water that accompanies the oil. The water can contain various corrosive agents such as carbon dioxide, hydrogen sulfide, organic acids, chlorides, and sulfates. The wells containing hydrogen sulfide are known as sour wells and the wells devoid of hydrogen sulfide are known as sweet wells. Sour wells are very corrosive. When the oil:water ratio is suitable, the crude oil by itself can be protective against corrosion.

Most inhibitors consist of organic compounds containing nitrogen such as (i) aliphatic fatty acid derivatives, (ii) imidazolines, (iii) quaternary ammonium compounds, (iv) complex amine mixtures based on abietic acid, (v) petroleum sulfuric acid salts of long-chain diamines, (vi) other salts of diamines, and (vii) fatty amides of aliphatic diamines. The compounds used consist of oleic and naphthemic acid salts of n-tallow propylenediamine; diamines, $RNH(CH_2)_nNH_2$, where R = C_8–C_{22} and $n = 2$–10.

The method of application of inhibitor varies since many factors are involved. Some factors are oil to water ratio, types of oil and water composition, fluid velocity, temperature, type of geologic formation, emulsion formation, economics, solubility, and specific gravity of the inhibitor. There are many methods of injecting the inhibitor into the well. Some of these methods are slug treatment, batch treatment, weighted treatment (63), microencapsulation method (64), and squeeze method (65).

Water is forced down into the deposits to obtain oil in secondary recovery. Certain corrosion problems are due to oxygen and bacteria. Oxygen may be removed by sulfite addition, but there is the possibility of precipitation of calcium sulfite. Organic compounds containing nitrogen may be used (66). Some of the inhibitors used are given below:

- Organic nitrogenous compounds (67).
- 40% methanol solution oleyldiamine adduct of sulfur dioxide (68).
- Zinc-glassy phosphate type inhibitor at 12–15 ppm and pH 7.0–7.2 used (69).
- Silicates at 100 ppm.
- Sodium arsenite with a surfactant used in acidizing oil wells (50).
- Acetylenic type compounds hexynol and ethyl octynol combined with ethylene diamine, dimethyl formamide, urea, or ammonia (70).
- Acetylenic compounds and organic nitrogenous compound performed well in deeper wells in which 28–30% acid is used in acidizing (71).
- Hydrochloride salt of aliphatic amine with an amine number of 15.75 obtained by nitration of paraffins at 0.1–0.15% was used in 20% HCl, which lowered the corrosion rate of steel at 43°C by 20 times (72).

Inhibitors are required in the oil refining stage and the corrosivity is increased by the presence of H_2S, CO_2, O_2, HCl, and naphthenic acids. The acids in the medium can be neutralized by the addition of alkali to attain a final pH of 7.0–7.5. The neutralization process is not only expensive but also can cause scale formation.

Imidazolines are used in conjunction with ammonia. Imidazoline at 6 ppm in conjunction with ammonia to pH 7.5 provided good protection of crude topping unit. Another inhibitor consists of 4 ppm of amino alkyl aryl phosphate in light hydrocarbon medium. The choice of an inhibitor in refining should also be made to include not only the extent of corrosion, but other factors such as the adverse effect of inhibitors on catalysts in refineries.

The main source of corrosion in oil storage tanks is the water that is present at the bottom part of the tanks. Inhibitors such as nitrites, silicates, and polyphosphates have been used. Other inhibitors such as imidazolines, itaconic salts, oleic acid salts of some amines, and polyalkene glycol esters of oleic acid have been used.

TABLE 4.4   Corrosion Inhibition in Carbon Dioxide and Hydrogen Sulphide Media

Material	Medium	Inhibitor	Reference
Steel	CO_2	Monoethanolamine saturated with CO_2 and 20 ppm of bismuth(III) citrate	75
Steel	CO_2	Alkanolamines with thionitrogen compound 18–25% monoethanol amine with 200 ppm of ammonium thiocyanate	76
Steel	CO_2	Potassium carbonate solution containing 0.25% copper carbonate and antimonyl potassium tartrate	77
Steel	CO_2	Potassium carbonate solution containing 0.5% ferric sulfate, 0.6% EDTA, and 0.25% antimonyl potassium tartrate	78
Steel	CO_2	0.2% sodium metavanadate	79
Steel	H_2S, CO_2, 3% NaCl, 50 wt% gasoline	100 ppm N-tridecylmaleamic acid and tridecylamine salt	80
Steel	H_2S, CO_2, 5% NaCl, pH 3.6	Diethylaminoacetonitrile (0.05 g/L)	81
Steel	5% NaCl, 0.5% acetic acid saturated with H_2S	Dodecyl trimethyl ammonium chloride	82
Steel	5% NaCl, 0.5% acetic acid saturated with H_2S	Reaction product oleyl pyridine and butyl chloride	83
Steel	5% NaCl, 0.5% acetic acid saturated with H_2S	20 ppm of polyamide by the reaction of lauric acid with triethylene tetramine	84
Steel	Gas condensate and water saturated with H_2S	Mixture of 2,3- and 2,4-dimethyl and 2,4,6-trimethyl pyridines	85

Contamination of gasoline with moisture can lead to corrosion problems in fuel distribution systems. Inhibitors are used to counter these corrosion problems. The effective inhibitors are esters of carboxylic or phosphoric acids. Corrosion inhibition in water-contaminated gasoline is described in NACE TM-01-72 and ASTM D6651 (IP135). Contamination with moisture of the order of 0.3% in gasoline alcohol mixtures can lead to a significant amount of corrosion as described in the literature (73). Triazoles such as 3-amino-1H-1,2,4-triazole with polyisobutylene and maleic acid anhydride have been used to counter corrosion (74).

Some of the inhibitors used in carbon dioxide and hydrogen sulfide environments are given in Table 4.4.

4.6   CORROSION INHIBITION OF REINFORCING STEEL IN CONCRETE

In general, added corrosion inhibitor in small amounts should reduce corrosion rate to an acceptable level and not adversely affect the properties of concrete. Since the corrosion inhibitor is added only once to concrete and no further replenishing of the inhibitor is involved, the criteria for an effective inhibitor for concrete are (i) the inhibitor should be soluble to allow homogeneous distribution and not readily leachable from concrete; (ii) the inhibitor must be compatible with the aqueous cement phase, to render its full protection potential; (iii) the inhibitor should not affect the properties of concrete; (iv) the amount of consumption of the inhibitor must be low; and (v) there should be no detrimental effects of the inhibitors such as setting time, strength, and durability of the concrete.

Calcium nitrite is an extensively used inhibitor in concrete. Calcium nitrite is also used on a large scale. Calcium nitrite provides protection in the presence of chlorides, does not affect the properties of concrete, and is readily available for commercial use in concrete. The principal anodic reaction occurring in concrete may be written as

$$Fe \rightarrow Fe^{2+} 2e^{-}.$$

Other reactions lead to the formation of compounds such as $Fe(OH)_2$, $Fe_3O_4 \cdot nH_2O$, or $\gamma FeO \cdot OH$ at the surface of reinforcing steel. Calcium nitrite, an anodic inhibitor is thought to be responsible for the constant repair of the weak spots in the oxide film as well as the following reactions with ferrous ions (86, 87):

$$2Fe^{2+} + 2OH^{-} + 2NO_2^{-} \rightarrow 2NO + Fe_2O_3 + H_2O,$$

$$Fe^{2+} + OH^{-} + NO_2^{-} \rightarrow NO + \gamma FeO \cdot OH.$$

The nitrite inhibitor competes with chloride and hydroxyl ions in the reaction with ferrous ions. Since nitrite helps to repair the flaws in the oxide film, the probability of chloride ion attack at the flaws of the oxide film and subsequent dissolution of the oxide film in the form of soluble chloro complexes of iron is reduced. Since nitrite is primarily involved in repairing the oxide film, and the oxide films are of the order of monolayers, the nitrite inhibitor is not consumed to a great extent.

The best method is entails coating the reinforcing steel bars with a strong inhibitor slurry instead of incorporating the inhibitor into the concrete mixture. The inhibitors that have been found effective are as follows:

- Sodium benzoate (2–10% in slurry coating) (88)
- Sodium nitrite (89, 90)
- Sodium benzoate and sodium nitrite (91)
- Mixture of grease, Portland cement, sodium nitrite, casein, and water
- Sodium mercaptobenzothiazole

- Stannous chloride
- Calcium nitrite and microsilica (Pozzolan)
- Calcium nitrite and stannous chloride (92)
- Sodium nitrite, potassium dichromate, and formaldehyde (93)
- Sodium nitrite, sodium borate, and sodium molybdate (94)
- Calcium nitrite and sodium molybdate (95)
- Calcium nitrate (96)
- Organic amines and esters (97)
- Alkanolamine (98)

4.7   CORROSION INHIBITION IN COAL–WATER SLURRY PIPELINES

A comprehensive account of the causes, control, and economics of internal corrosion of slurry pipelines has been given in the literature (99). The performance of inhibitors such as chromate, polyphosphate, organic phosphonates, and catalyzed sodium sulfite has been evaluated, and some data on their performance are given in Table 4.5. Hexavalent chromium has been found to be effective in mitigating both primary and residual types of corrosion.

Potassium chromate in conjunction with coal adsorption additives (CAA) such as gelatin and Triton X-100 have been used in coal–water slurries for corrosion inhibition of steel. The resulting data (Table 4.6) show that the coal adsorption additive adsorbs on the coal and allows the chromate inhibitor to protect the pipeline steel from corrosion.

Since chromate is environmentally unacceptable, other inhibitors such as molybdate were studied (100) as a function of variables such as pH, inhibitor concentration, and time. About 100 ppm of sodium molybdate at pH of 8.5 proved to be effective. More extensive studies on the corrosion inhibition of mild steel by single component systems such as potassium chromate, sodium molybdate, sodium tungstate, sodium dihydrogen phosphate, HEDP, sodium nitrite, and sodium phosphate and binary combinations of potassium chromate/sodium molybdate/sodium tungstate with HEDP/sodium phosphate/sodium nitrite with and without coal, showed the binary systems to be effective, giving more than 90% inhibition (101) (Table 4.7).

4.8   CORROSION INHIBITION IN THE MINING INDUSTRY

Corrosion in mining operations can be characterized as an electrochemical attack enhanced by abrasion. The conditions favoring corrosion are highly corrosive mine water, grinding media, dissimilar metals, dissolved oxygen, wide pH range, and corrosive species in solution.

TABLE 4.5 Characteristics of Corrosion Inhibitors

Inhibitor	Corrosion Inhibition Effectiveness	Cost/ Performance	Environmental Acceptability	Ease of Handling and Feeding	Typical Feed Concentrations (mg/L)	Typical Corrosion Rates to Mild Steel (mpy)
High chromate	E	P	P	E	500–1000	0–1
Low chromate	F	G	P	E	20–80	3–5
Chromate–zinc	E	E	F	E	2–30	1–2
Polyphosphate	F	F	G	F	10–75	5–15
Polyphosphate–zinc	G	G	G	F	5–50	3–10
Organic phosphonate	P	UA	G	E	5–20	5–20
Organic phosphonate–zinc	F	F	G	E	2–15	5–15
Catalyzed sodium sulfite	F	F	G	P	10–80 variable with oxygen content	Minimal data

E, excellent; G, good; F, fair; P, poor; UA, unacceptable.

TABLE 4.6 Corrosion Rates of Various Coal–Water Slurries (50% V Coal)

Inhibitor (K_2CrO_4) (ppm)	CAA	CAA Concentration (ppm)	Corrosion Rate (mil/year)
Coal and water	–	–	7.5
Coal and water	–	–	8.3
0.5	–	–	7.8
0.5	Triton X-100	5	6.8
0.5	Gelatin	10	4.7
12.5	–	–	3.4
12.5	Triton X-100	5	<1.0
12.5	Triton X-100	5	<1.0
12.5	Gelatin	10	1.5
12.5	Gelatin	100	<1.0

Corrosion of mining equipment is due to exposure to (i) mine air, (ii) mine water, and (iii) mine dust. The composition of mine waters varies from mine to mine. The pH values vary from 2.8 to 12.3, chloride from 5 to 10,500 ppm, and sulfate from 57 to 5100 ppm. Low pH and high chloride and sulfate content of mine waters are conducive to severe corrosion.

The presence of significant amounts of chloride and sulfate in mine waters leads to the corrosivity of the mine waters. In addition to this, the pH of mine waters contributes to the corrosivity due to the wide pH range of 2.8–12.3 in the mine waters. The high pH of the mine water may be due to the use of cement in the backfill. The low pH of mine water is probably due to the oxidation of pyrite.

$$2FeS_2 + 7O_2 + 2H_2O \rightarrow 2FeSO_4 + 2H_2SO_4.$$

The water in abandoned underground mines may be acidic due to the reaction

$$FeS_2 + 8H_2O \rightarrow Fe^{2+} + 2SO_4^{2-} + 2H^+ + 7H_2.$$

Acidic mine water can also be produced by bacteria such as thiobacillus thiooxydans and thiobacillus ferrooxydans. The rate of production of sulfuric acid is about four times greater in the presence of bacteria than the rate in the absence of bacteria.

Mine air causes atmospheric corrosion of mining equipment to an extent determined by the composition of mine air, humidity, and temperature. Mine air contains significant amounts of oxides of sulfur and nitrogen. Oxides of sulfur result from burning of coal contaminated with pyrite and the nitrogen oxides originate from the use of underground explosives. The oxides of sulfur and nitrogen coupled with high humidity and temperature cause corrosion of mining equipment.

Mine dust usually consists of coal dust, pyrite present in coal seams, and gypsum, which is used to dust mine to reduce the amount of combustible particles. The dust in

TABLE 4.7 Corrosion Inhibition by Oxyanions, Medium 2.45 g/L NaCl; 0.45 g/L Na$_2$SO$_4$; pH 8.58 (8)

| | | Corrosion Rate (mpy) | | |
| | | | With Coal | |
Reagent	Concentration	Without Coal	15%	30%
No inhibitor		51.9		
Kr$_2$CrO$_4$	0.01	1.1		
	0.001	5.0		
Na$_2$MoO$_4$	0.01	1.1		
	0.001	4.5		
Na$_2$WO$_4$	0.01	0.1		
	0.001	4.8		
NaH$_2$PO$_4$	0.01	2.5		
	0.001	16.4		
Na$_2$C$_2$O$_4$	0.01	45.4		
	0.001	56.1		
NaNO$_2$	0.01	2.1		
	0.01	2.1		
	0.001	6.8		
HEDP	0.01	28.9		
	0.001	8.2		
	0.0001	9.6		
	0.00001	31.8		
No inhibitor			56	64
K$_2$CrO$_4$ + HEDP	0.005 + 0.005	3.0	0.9	5.2
K$_2$CrO$_4$ + NaNO$_2$	0.005 + 0.005	1.9	2.4	1.3
K$_2$CrO$_4$ + Na$_3$PO$_4$	0.005 + 0.005	3.6	1.5	3.1
Na$_2$MoO$_4$ + HEDP	0.005 + 0.005	11.4	0.6	3.8
Na$_2$MoO$_4$ + NaNO$_2$	0.005 + 0.005	0.2	0.6	1.9
Na$_2$MoO$_4$ + Na$_3$PO$_4$	0.005 + 0.005	0.1	0.5	4.9
Na$_2$WO$_4$ + HEDP	0.005 + 0.005	0.8	1.7	4.8
Na$_2$WO$_4$ + NaNO$_2$	0.005 + 0.005	0.1	0.4	3.1
Na$_2$WO$_4$ + Na$_3$PO$_4$	0.005 + 0.005	0.4	0.5	4.4

coal mines with high sulfur content is very corrosive to mining equipment. The dust settling on metals promotes atmospheric corrosion.

Miscellaneous agents such as flotation reagents and other reagents such as xanthates, sulfuric acid, and sulfur dioxide are corrosive. Cooling waters, process water, and wastewater in mills can be corrosive and cause severe internal corrosion of pipes, valves, process equipment, and storage tanks.

The inhibitors used in mining operations are (i) passivation inhibitors, (ii) precipitation inhibitors, and (iii) adsorption inhibitors. Typical examples of passivation type inhibitors are chromates and phosphates. These inhibitors prevent the anodic reaction. They are incorporated in the oxide film on the metal, thereby stabilizing it and preventing further dissolution. The concentration of the inhibitor used is critical

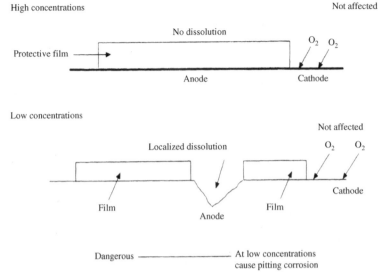

FIGURE 4.1 Passivation (anodic) inhibitors.

because at low concentrations underprotection results in severe pitting. Pitting corrosion is due to the large cathode-to-anode surface areas when some active anodic sites are not protected. As the name implies, anodic inhibitors inhibit the anodic process:

$$Fe \rightarrow Fe^{2+} + 2e^-$$

The action of passivation inhibitors is illustrated in Fig. 4.1.

Precipitation inhibitors block the cathodic reaction by precipitation at the cathode that occurs due to high local pH values. Zinc salts are an example of precipitation inhibitors. The cathodic inhibitors suppress the cathodic reaction

$$O_2 + 2H_2O + 4e^- \rightarrow 4OH^-$$

by precipitation of insoluble species at the cathodic sites

$$Zn^{2+} + 2OH^- \rightarrow Zn(OH)_2\downarrow,$$

which forms a barrier for oxygen reduction as illustrated in Fig. 4.2.

In general, organic inhibitors containing functional groups such as amine, carboxyl, and phosphonates adsorb on the metal surface and form a barrier to metal dissolution as well as oxygen reduction, as illustrated in Fig. 4.3.

The performance of both inorganic (passivating) and organic (adsorption) inhibitors with three types of mine waters was studied (102) and it was found that a 1:1 mixture of phosphonates and phosphate gave 90% inhibition.

Inhibitors for the prevention of corrosion of mild steel in acid mine waters of pH 2.5 were evaluated (103), and the effectiveness of the inhibitors was in the order: *n*-butylamine, urea, guanyl urea sulfate < benzotriazole, hexadecyl pyridinium chloride < thiourea < potassium oxalate. Potassium oxalate gave 93% inhibition.

High concentrations

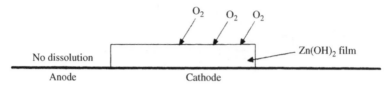

Low concentrations

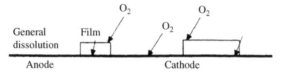

FIGURE 4.2   Precipitation (cathodic) inhibitors.

Potassium oxalate forms tris(oxalate)iron(III) complex, which decomposes on exposure to sunlight and precipitates ferrous oxalate on the steel sample:

$$Fe^{3+} + 3C_2O_4{}^{2-} \rightarrow Fe(C_2O_4)_3{}^{3-} \rightarrow FeC_2O_4 + CO_2.$$

Corrosion inhibitors have been found to reduce ball wear in grinding sulfide ore (104) and hematite ore (105). The performance of both single and binary components of inhibitors in grinding ball corrosion has been evaluated (106).

Chemisorbed layer of organic molecules
Protection by physical blocking

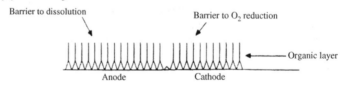

Functional groups

$$-N\begin{smallmatrix}H\\H\end{smallmatrix} \quad \text{Amino}$$

$$-C\begin{smallmatrix}O\\OH\end{smallmatrix} \quad \text{Carboxyl}$$

$$-\overset{O}{\underset{OH}{P}} \begin{smallmatrix}O\\OH\end{smallmatrix} \quad \text{Phosphonate}$$

FIGURE 4.3   Adsorption inhibitors.

4.9   ATMOSPHERIC CORROSION INHIBITION

Atmospheric corrosion accounts for more failures in terms of cost and tonnage than any other form of corrosion and any single environment. The cost in the United States alone is several billion dollars. Atmospheric corrosion varies considerably within geographical areas and with local conditions. Climate conditions can be temperate, tropical, or arctic in nature. There are marine, industrial, urban, and rural atmospheres. The corrosion rate in an industrial area can be 100 times the rate in a desert or arctic region.

Atmospheric conditions depend upon the geographic orientation. In general, the eastern and southern sides of a structure corrode to a lesser extent than the western and northern sides, since the former sides, exposed to rain and dew, dry up faster than the western and northern sides. General uniform corrosion is the common form of atmospheric corrosion, with localized corrosion being rare.

The conditions required for atmospheric corrosion are (i) the relative humidity greater than 70% and (ii) the presence of an electrolyte. Atmospheric gaseous pollutants such as sulfur dioxide dissolve in water and upon oxidation produce sulfuric acid that corrodes carbon steel. This is also known as the acid rain effect.

Other atmospheric pollutants that cause atmospheric corrosion are NO_2, chloride, and fluoride. The values of critical humidity for various pollutants are different, and the value is 60% for sulfur dioxide. In general, a combination of high humidity, high temperature, and industrial pollutants increase the atmospheric corrosion rates. Severe atmospheric corrosion occurs in marine, tropical, or semitropical atmospheres.

Among the contaminants present as salts on the metal surface, chloride has the greatest effect on atmospheric corrosion (107). Chloride has more profound effects when associated with porous corrosion products on steels (108). Once adsorbed, the tenacity of adherence of chloride is so great that it cannot be desorbed by the surface cleaning techniques used in practice.

The effect of microbiological organisms on atmospheric corrosion is not well understood (109). The microbiological effect occurs when carbonate and sulfate deposits are present on the metal surface, leading to the formation of hydrogen, which can affect the mechanical properties of the metal.

Coating or painting the structure is the oldest method of protection from atmospheric corrosion. Coatings for various applications have been documented in the literature (110).

Another common method of combating atmospheric corrosion is by the use of vapor phase corrosion inhibitors (VPCI). Atmospheric corrosion of small and intricate objects can be prevented by vapor phase inhibitors. Vapor phase inhibitors are particularly suitable when the sample has hidden crevices and remote areas.

Corrosion inhibitors may be applied at the beginning when a new structure is installed or at any other time the structure is exposed to the atmosphere. Application of atmospheric corrosion inhibitors may be grouped as (i) chemical conversion coatings, (ii) rust modifiers or stabilizers, (iii) application of oil base inhibitors, and (iv) chemical modification of adsorbent.

Some examples of inhibitors used in chemical conversion coatings are given in Table 4.8.

TABLE 4.8 Applications of Corrosion Inhibitors

Inhibitor	Function	Reference
Nitrites; chromates; phosphates	Passivating type; nitrite is used in concrete structures	111
Phosphates	Applied on steel before painting	112
Polyphosphate with Ca^{2+} ions	At pH 6–7 transparent film formed	113
5% organophosphate with 8–16% zinc phosphate	Uniform protective film	114
Red lead	Deactivates the rust	115
Tannins with chromate	Passivation of iron	116
Benzoic acid with tannic acid and/or phosphate	Enhanced protection	117
Rust stabilizers or modifiers		
Tannins along with phosphates	Protection of steel; ferric tannate complex formed on the surface	118
Phosphoric acid	Protection of steel	
Oil-based inhibitors		
25 parts triethanolamine; 30 parts laurylsarcosine; 10 parts benzotriazole	Rust inhibition	119
Salts of C_8–C_{14} fatty acids with C_2–C_4 dialkylamine	Rustproofing agent	114
Dibutylamine caprylate in mineral oil	Inhibition	120
Tannic acid, glyoxal, and chloroacetic acid in a lubricant	Protection from corrosion and friction	121
Amines, benzotriazole in combination with nitrites, phosphates	Inhibition	114
Sulfonates in oil base	Inhibition	122
Emulsified iron tannin complex in oil	Inhibition	123
N-Oleoylsarcosine with primary and tertiary amine	Inhibitor	

TABLE 4.9 Mechanism of the Corrosion Inhibition

Inhibitor	Mechanism
Nitrite; phosphate; chromate	Improved passivation and oxide growth; Cr, P incorporated in the film
Benzoate	Adsorbs and provides a barrier
Tannins	Protection due to iron tannate complex, which forms a barrier
Amines, imines	Formation of protective metal surface amine complexes, which are protective
Benzotriazole	Formation of complexes such as Cu(I)BTA, which are protective

TABLE 4.10 Corrosion Inhibition in Miscellaneous Environments

Medium	Inhibitor	Reference
Fertilizer (NH_4NO_3–NH_3–H_2O)	500 ppm mercaptobenzothiazole, thiourea, and ammonium thiocyanate	124
NH_4NO_3–urea–H_2O	Ammonium hydrogen phosphate and polyphosphate	125
Solids through steel pipelines	Sodium-zinc polyphosphate glass	126

The mechanism of protection of metals in atmospheric corrosion of metals may be summarized as shown in Table 4.9.

Other miscellaneous environments and the inhibitors used are given in Table 4.10.

REFERENCES

1. DM Brasher, AD Mercer, *Br Corros J* **3**:120–129 (1968).

2. (a) DM Brasher, JE Rhoades-Brown, *Br Corros J* **4**:74 (1969); (b) DM Brasher, JE Rhoades-Brown, *Br Corros J* **8**:50 (1973)

3. DM Brasher, AD Mercer, *Br Corros J* **3**:120 (1968).

4. AD Mercer, IR Jenkins, *Br Corros J* **3**:130 (1968).

5. AD Mercer, IR Jenkins, JE Rhoades-Brown, *Br Corros J* **3**:136 (1968).

6. D Gherhardi, L Rivola, M Troyll, G Bombara, *Corrosion* **20**:73 (1964).

7. IL Rozenfeld, VP Maksimchuk, *Proc Acad Sci (USSR), Phys Chem Soc* **119**:986 (1956).

8. F Wormwell, AD Mercer, *J Appl Chem* **2**:150 (1952).

9. PT Gilbert, SE Hadden, *J Appl Chem* **3**:545 (1953).

10. JGN Thomas, AD Mercer, *4th International Congress on Metallic Corrosion, Amsterdam*, 1969, NACE, Houston, Texas, 1972, p. 585.

11. RN Ride, *J Appl Chem* **8**:175 (1958).

12. WT Edwards, JE Le Surf, PA Hayes, *2nd European Symposium on Corrosion Inhibitors, Ferrara*, 1965, University of Ferrara, 1966, pp. 679–700.

13. P Spinedi, G Signorelli, *1st European Symposium on Corrosion Inhibitors, Ferrara*, 1960, University of Ferrara, 1961, pp. 643–652.

14. F Mazza, S Trasatti, *3rd European Symposium on Corrosion Inhibitors, Ferrara*, University of Ferrara, 1971, pp. 277–291.

15. G Butler, D Owen, *Corros Sci* **9**:603 1969.

16. GB Hatch, *Mater Prot* **8**:31 (1969).

17. W Stericker, *Ind Eng Chem* **37**:716 (1945).

18. CS Cone, *Mater Prot Perf* **9**:32 (1970).

19. AV Fisher, *Mater Prot* **3**:8 (1964).

20. TW Bostwick, *Corrosion* **17**:12 (1961).

21. WB Brooks, *Mater Prot* **7**:24 (1968).

22. BC Edwards, *Corros Sci* **9**:395 (1969).

23. M Darrin, *Ind Eng Chem* **38**:368 (1946).

24. JF Conoby, TM Swain, *Mater Prot* **6**:55 (1967).

25. AV Fisher, *Mater Prot* **3**:8 (1964).

26. J Becher, EA Savinelli, *Mater Prot* **3**:15 (1964).

27. RV Comeaux, *Hydrocarbon Process* **46**:129 (1967).

28. KM Verma, MP Gupta, AK Roy, *Technol Quart (Bull Fertil Corp, India)* **5**:98 (1968).

29. CM Hwa, *Mater Prot Perf* **9**:29 (1970).

30. APTB Squires, The protection of motor vehicles from corrosion, *Society of Chemical Industry*, Monograph no. 4, 1958.

31. J Dulat, *Br Corros J* **3**:190 (1968).

32. British Standards Institution, BS6580, Milton Keynes, 1985.

33. *Cavitation Corrosion, Its Prevention in Diesel Engines, Symposium*, Nov 10, 1965, British Railways Board, 1966.

34. AM Spivey, *Chem Ind* 22 April:657 (1967).

35. J Venczel, G Wranglen, *Corros Sci* **7**:461 (1967).

36. CW Drane, *Br Corros J* **6**:39 (1971).

37. JA Von Fraunhofer, *Br Corros J* **6**:28 (1971).

38. MF Obrecht, *2nd International Congress on Metallic Corrosion*, New York, 1963, NACE, Houston, Texas, 1966, pp. 624–645.

39. RB Maase, *Mater Prot* **5**:37 (1966).

40. TP Hoar, *J Soc Chem Ind* **69**:356 (1950).

41. BD Oakes, JS Wilson, WJ Bettin, *Proceedings of the 26th NACE Conference*, 1969, NACE, Houston, Texas, 1970, p. 549.

42. RA Legault, WJ Bettin, *Mater Prot Perf* **9**:35 (1970).

43. Motor Vehicle Corrosion, Influence of De-Icing Chemicals, OECD Report, Oct 1969.

44. P Asanti, *Proc Inst Mech Eng* **182**(Part 3J):73–79 (1967/1968).

45. Japanese Patent 5891176, *Chem Abstr* **99**:179935 (1983).

46. Japanese Patent 57209981, *Chem Abstr* **98**:145984 (1983).

47. S Gisela, East German Patent 211357, *Chem Abstr* **102**: 97635 (1985).

48. M Itoh, *Corros Eng* **36**:139 (1987).

49. M Hanazaki, N Harada, K Shimada, US Patent 4,655,951, *Chem. Abstr* **107**:10537 (1987).

50. W Machu, *3rd European Symposium on Corrosion Inhibitors*, 1970 Ferrara, University of Ferrara, 1971, pp. 107–119.

51. RL Every, OL Riggs, *Mater Prot* **3**(9):46 (1964).

52. HH Uhlig, *The Corrosion Handbook*, John Wiley & Sons, Inc., *New York*, 1948, p. 910.

53. V Carassiti, G Trabanelli, F Zucchi, *2nd European Symposium on Corrosion Inhibitors*, Ferrara, 1965, University of Ferrara, 1966, pp. 417–448.

54. RM Hudson, QL Looney, CJ Warning, *Br Corros J* **2**:81 (1967).

55. RM Hudson, CJ Warning, *Mater Prot* **6**:52 (1967).

56. M Alfandary, *2nd European Symposium on Corrosion Inhibitors*, Ferrara, 1965, University of Ferrara, 1965, pp. 363–375.

57. M Froment, A Desestret, *2nd European Symposium on Corrosion Inhibitors*, Ferrara, 1965, University of Ferrara, 1966, 223–236.

58. JG Funkhouser, *Corrosion* **17**:283t (1961).

59. IN Putilova, EN Chislova, *Zasch Met* **2**:290–294 (1966).

60. RJ Meakins, *J Appl Chem* **13**:339 (1963).

61. RJ Meakins, *Br Corros J* **6**:109 (1971).

62. IL Rozenfeld, VP Persiantseva, TA Damaskina, *Zasch Met* **9**:690 (1973).

63. CC Patton, DA Deemer, HM Hillard, *Mater Prot* **9**:37 (1970).

64. JE Haughin, B Mosier, *Mater Prot* **3**:42 (1964).

65. JK Kerver, FA Morgan, *Mater Prot* **2**:10 (1963).

66. JI Bregman, *3rd European Symposium on Corrosion Inhibitors*, Ferrara, 1970, University of Ferrara, 1971, pp. 339–382.

67. LW Jones, JP Barrett, *Corrosion* **11**:217t (1955).

68. AK Dunlop, RL Howard, PJ Raifsnider, *Mater Prot* **8**:27 (1969).

69. HB Hateh, PH Ralston, *Mater Prot* **3**:35 (1964).

70. RJ Tedeschi, PW Natali, HC McMahon, *NACE 25th Conference*, 1969, NACE, Houston, Texas, 1970, pp. 173–179.

71. AW Coulter, CM Smithey, *Mater Prot* **8**:37 (1969).

72. IN Rybachok, MA Mikhailov, NA Tarasova, *Korr Zasch V NettegProm* (7):7–10 (1971).

73. *7th Symposium Int. Carbur. Alcool.*, Technip, Paris, France,1986.

74. RL Sung (Texaco), US Patent 4,282,007 (1981).

75. BD Oakes, MS Dupart, DC Cringle, US Patent 41420 (1983, 1984).

76. LS Krawczyk, CW Martin, RL Pearce, US Patent, 4,431,563 (Feb 14, 1984).

77. M Okubo, K Nagai, Japanese Patent 7853540 (May 16, 1978).

78. M Okubo, K Nagi, Japanese Patent 7853539 (May 16, 1978).

79. BF Mago, Ger. Offen. 2518827 (Oct 30, 1976).

80. K Oppenlaender, J Stark, K Barthold, Ger. Offen. DE 3237108 (Apr 12, 1984).

81. IL Rozenfeld et al., Canadian Patent 1,114,594 (Dec 22, 1981).

82. Nippon Steel Corp., Japanese Patent 8190986 (July 23, 1981).

83. Nippon Steel Corp., Japanese Patent 8190985 (July 23, 1981).

84. Nippon Steel Corp., Japanese Patent 8190984 (July 23, 1981).

85. KM Akhmrov, A Kuchkarov, A Ikramov, VY Mudrakova, *Chem Abstr* **91**:76458e (1979).

86. NS Berke, TG Weil, *Adv Concrete Technol*, 899–924 (1992).

87. B El-Jazairi, NS Berke, *Corrosion of Reinforcement in Concrete*, Elsevier Science Publishers Ltd., Wishaw, Warwickshire, UK, 1990, pp. 571–585.

88. North Thames Gas Board, British Patent 706319 (Mar 31, 1954).

89. VM Moskin, SN Alexseev, *Beton Zhelez* (1):28 (1957).

90. SN Alexseev, LM Rozenfeld, *Beton Zhelez* (2):388 (1958).

   (a) KW Treadway, AD Russell, *Highways Public Works* **36** (1704): 19–21 (1968); (b) KW Treadway, AD Russell, *Highways Public Works* **36** (1705): 40–41 (1968)

92. BB Hope, AKC Ip, *Corrosion Inhibitors for Use in New Concrete Construction*, Research & Development Branch, Ontario Ministry of Transportation, 1987.

93. CA Loto, *Corrosion* **48**:759–763 (1992).

94. KK Sagoe-Krentsil, VT Yilmaz, FP Glasser, *Adv Cem Res* 4–15, 91–96 (1991, 1992).

95. SV Thompson, Thesis, Queen's University, Kingston, Ontario, Canada, 1991.

96. H Justnes, EC Nygaard, *Corrosion and Corrosion Protection of Steel in Concrete*, Sheffield, Academic Press, Sheffield, England, 1994, pp. 491–502.

97. CK Nmai, SA Farrington, GS Bobrowski, *Concrete International*, Apr 1992, pp. 45–51.

98. B Miksic, L Gelner, D Bjegovic, L Sipos, *Proceedings of the 8th European Symposium on Corrosion Inhibitors*, Vol. 1, University of Ferrara, 1995, p. 569.

99. RB Jacques, WR Neil, *Proceedings of the 2nd International Technical Conference on Slurry Transportation*, Las Vegas, Nevada, Mar 2-4 1977, p. 124.

100. VS Sastri, R Beauprie, M Desgagne, *Mater Perf* **25**:45 (1986).

101. VS Sastri, J Bednar, *Mater Perf* **29**:42 (1990).

102. RT White, MINTEK Report no. M180, Council for Mineral Technology, Randburg, South Africa, Apr 13 1985.

103. DV Subramanyam, GR Hoey, Research Report R274, CANMET Report, Ottawa, Ontario, Canada.

104. GR Hoey, W Dingley, C Freeman, *CIM Bull* **68**:120 (1975).

105. AW Lui, VS Sastri, J McGoey, *Br Corros J* **29**:140 (1994).

106. AW Lui, VS Sastri, M Elboujdaini, J McGoey, *Br Corros J* **31**:158 (1996).

107. S Yasukawa, K Katoh, M Yasuda, H Imaizumi, *Boshoku Gijutsu*, **29**:12 (1980).

108. AJ Rostron, *Corros Sci* **19**:321 (1979).

109. M Walker, Microbiological Ecology of Metal Surfaces, PhD dissertation, Harvard University, 1986.

110. CG Munger, *Corrosion Prevention by Protective Coatings*, NACE, Houston, Texas, 1984, p. 369.

111. PHG Draper, *Corros Sci* **7**:**91** (1967).

112. M Harsy, L Ludanyi, J Emri, Hungarian Patent 40470 (1986).

113. OL Sarc, L Kastelan, *Corros Sci* **16**:25 (1976).

114. A Raman, *Reviews on Corrosion Inhibitor Science, Technology*, Paper 1. 14, NACE, Houston, Texas, 1993.

115. G Lineke, WD Mahu, *Deut Farbanzeit* **28**:423 (1974).

116. E Ivanov, Yu I Kuznetsov, *Zasch Met* **24**:36 (1988).

117. L Sharma, M.Sc. thesis, Louisiana State University, Baton Rouge, 1986.

118. RE Cromarty, *Corros Coat,* South Africa, **12**:3, 6, 8, 11, 14 (1985).

119. KK Daiwa Kasei Kenkyusho, Japanese Patent JP58130284 (1983).

120. U Ploog, KH Koch, German Patent GO2355007 (1975).

121. N Nichimura, S Masaki, T Nakanishi, Japanese Patent JP77141440 (1977).

122. W Akada, T Kurashima, M Nakamura, *Bosoi Kanri* **23**:10 (1979).

123. MDS Bretes, European Patent EP140179 (1985).

124. H Koesche, H Laengle, J Rueckert, *Werk Korr* **22**:673 (1971).

125. WP Banks, *Mater Prot* **7**:35 (1968).

126. Anonymous, *Mater Prot* **6**:61 (1967).

5

CORROSION INHIBITION MECHANISMS

Corrosion inhibition mechanisms operating in an acid medium differs widely from one operating in a near-neutral medium. Corrosion inhibition in acid solutions can be achieved by halides, carbon monoxide, and organic compounds containing functional group heteroatoms such as nitrogen, phosphorus, arsenic, oxygen, sulfur, and selenium, organic compounds with multiple bonds, proteins, polysaccharides, glue, bitumen, and natural plant products such as chlorophyll and anthocyanins (1). The initial step in the corrosion inhibition of metals in acid solutions consists of adsorption of the inhibitor on the oxide-free metal surface followed by retardation of the cathodic and/or the anodic electrochemical corrosion reactions.

5.1 INTERFACE CORROSION INHIBITION

Adsorption of inhibitors on corroding metals has been studied by direct methods involving radioactive isotopes (2–5) and solution depletion monitoring; these methods provide definitive data, unlike indirect methods, such as double layer capacitance (6), coulometry (6), ellipsometry (7), and reflectivity (7). Qualitative correlation has been observed between the reduction in corrosion rate and the amount of inhibitor adsorbed in the case of iron, nickel, and stainless steels with inhibitors such as iodide, carbon monoxide, organic amines, thiourea, sulfoxides,

Green Corrosion Inhibitors: Theory and Practice, First Edition. V. S. Sastri.
© 2011 John Wiley & Sons, Inc. Published 2011 by John Wiley & Sons, Inc.

sulfides, and mercaptans. For complete coverage, $\theta = 1$ and incomplete coverage, $\theta < 1$. In the case of polyvinyl pyridine adsorption on iron in hydrochloric acid at $\theta < 0.1$ monolayer, the corrosion rate was reduced by 80%.

Adsorption of the inhibitor on a metal surface means that the inhibitor-adsorbed area is resistant to corrosion. By the same token, the inhibitor-free area of the metal surface is prone to corrosion. The inhibitor efficiency should then be proportional to the fraction of the surface covered with the adsorbed inhibitor. This assumption has given consistent results leading to clarification of factors influencing inhibition and adsorption. The relationship between the fractional surface covered with inhibitor and the degree of inhibition is not always reliable. At low surface coverage ($\theta < 0.1$), inhibitor efficiency may be greater than at high surface coverage (8). In some cases, acceleration of corrosion by inhibitors at low concentration has been observed (9, 10). The differences in mechanisms must also be considered along with the effectiveness of inhibitors (11). The data on the inhibitor effectiveness as obtained from corrosion rate data along with the information on adsorption from solution show the dependence of adsorption on the following factors (12): (i) surface charge on the metal, (ii) the nature of the functional group and the structure of the inhibitor, (iii) interaction of inhibitor with water molecules, (iv) lateral interactions of adsorbed inhibitor, (v) reaction of adsorbed inhibitor, and (vi) surface charge on the metal is the electric charge on the metal at the metal–solution interface. In solution, the charge on the immersed metal is denoted by the potential with respect to the zero-charge potential. This potential, referred to as Φ potential, is considered to be more important than the potential on the hydrogen scale. When Φ is positive, anions are adsorbed and when Φ is negative, cations are adsorbed. The adsorption of dipoles in neutral molecules on metals also depends upon Φ potential. The adsorption of ions or dipoles on metals with the same value of Φ potential will be independent of the metal. The difference in behavior of an inhibitor toward different metals can be rationalized on the basis of differences in Φ potentials.

5.2 STRUCTURE OF THE INHIBITOR

The structure of the inhibitor, especially in the case of organic inhibitors, plays a very important role in corrosion inhibition of metals. Organic inhibitors such as amines (aliphatic amines, pyridines, imidazoles) and thiazoles form a coordinate bond with metals such as iron, copper, and zinc. Corrosion inhibition of metals by organic inhibitors containing electron donor atoms such as oxygen, nitrogen, sulfur, and selenium is in keeping with the decreasing electronegativity order $O < N < S < Se$ (13, 14). The metal–inhibitor bond is also formed with inhibitors containing π electrons associated with triple bonds or aromatic rings. The stability of the metal–inhibitor complex is reflected in the degree of corrosion inhibition. Thus, it is useful to consider the determination of the stability constants of a metal–inhibitor complex.

5.2.1 Stability Constants of Zinc–Triazole Complexes (15)

The addition of triazole to aqueous zinc salt solution results in the displacement of the coordinated water molecules bound to zinc ions by the triazole inhibitor followed by the formation of a zinc–triazole complex. The stability of the zinc–triazole complex plays a vital role in corrosion inhibition.

When a solution of zinc ions is added to an inhibitor solution, the pH decreases, which may be explained by the overall complex formation reaction

$$Zn(H_2O)^{2+} + 2H(Inh) \rightleftharpoons Zn(Inh)_2 + 2H_3O^+.$$

The formation of $Zn(Inh)_2$ complex is facilitated by the removal of hydronium ions through the reaction with hydroxyl ions. Thus, the effect of the complex formation on inhibition efficiency of triazole compounds should increase with pH.

The stability constants K_1, K_2, and $\beta = K_1 K_2$ of zinc–triazole complexes, benzotriazole (BT), 3-amino-5-heptyl-1,2,4-triazole (AHT), and bisaminotriazole (BAT4) were determined by potentiometric pH titration of $5 \times 10^{-5}\,M\,Zn^{2+}$ and $10^{-4}\,M$ triazole with $3 \times 10^{-3}\,M\,KOH$. From the pH profile, the stability constants were calculated using the equation

$$\frac{\bar{n}}{(\bar{n}-1)L} = \left\{\frac{2-\bar{n}[L]}{\bar{n}-1}\right\}\beta - K_1$$

where $\bar{n}$ is the average number of triazole molecules, $[L]$ is the activity of triazole anion, and β is the overall stability constant $= K_1 K_2$.

The stability constants obtained are given in Table 5.1.

The stability constants are in the order BAT4 > AHT > BT, which correlates well with the corrosion inhibition efficiency observed at pH 9 and inhibitor concentration of 1 ppm. The influence of stability of the complex formed is of great importance at low concentrations of the inhibitor. At higher concentrations of the inhibitor, solubility and adsorbability appear to play an important role.

The structure of the organic inhibitor and the substituents in the inhibitor has profound effect on corrosion inhibition of metals. The electron density on the donor atom in the functional group of the organic inhibitor depends upon the substituents present in the organic inhibitor. Some of the inhibitors with substituents that serve as examples in corrosion inhibition are (i) substituted pyridines (16–19), (ii) anilines (17, 20, 21), (iii) aliphatic amines (8), (iv) amino acids (22), (v) benzoic acids (23), and

TABLE 5.1 Stability Constants of Zinc–Triazole Complexes

Complex	log K_1	log K_2	log β
Zn–BT	5.0	4.8	9.8
Zn–AHT	8.3	8.1	16.4
Zn–BAT4	13.1	12.8	25.9

aliphatic sulfides (24). Thus, it is logical to examine the structure–activity relationships applicable to corrosion inhibitors.

5.3 STRUCTURE–ACTIVITY RELATIONSHIPS

The Hammett equation may be written as

$$\log \frac{k_R}{k_H} = \rho\sigma,$$

where k is an equilibrium or rate constant, ρ is the reaction parameter, assumed to be 1.0 for *meta*- and *para*-substituted benzoic acids, and σ is the substituent constant reflecting its total electronic effect on the reaction center. Since ρ increases with ionic character of the reaction (i.e., the polarity of the transition state), it is reasonable to assume that electrochemical processes involving charge transfer would be sensitive to electronic effects of the substituent.

Application of the Hammett equation in corrosion inhibition by a blocking mechanism was proposed, and the following equation was advanced (25):

$$\log \frac{1 - \gamma_R}{1 - \gamma_H} = \rho\sigma$$

where γ_R and γ_H are the inhibition efficiencies calculated as the ratios of the corrosion rates in blanks (K_0) and inhibited solutions (K_{Inh}), and ρ and σ are the constants defined earlier. It was later shown that linear free energy relationships (LFER) could be applied to acid corrosion inhibition by organic substances and rate-determining electrochemical reactions without need for the inhibitor-blocking mechanism.

Hammett substituent constants for various *meta* and *para* substituents are given in Table 5.2.

Logarithm of corrosion rates of Armco iron obtained with *para*-substituted benzonitriles as inhibitors showed good correlation with σ values (26). Corrosion rates obtained with substituted pyridine correlated well with Hammett σ values. The corrosion rate decreased with increasing negative σ value, indicating an increase in electron density at the pyridine nitrogen atom and increased interaction between the inhibitor and the metal (27). Figure 5.1 shows the correlation of percent inhibition of iron and Hammett σ values with substituted pyridine as the inhibitor.

The corrosion rate data obtained with *para*-substituted benzonitrile compounds as inhibitors given in Table 5.3 show the dependence of corrosion rates on the σ values.

The *para*-methyl benzonitrile showed reduced corrosion rate while the other substituted benzonitriles gave higher corrosion rates in keeping with the σ values. Positive σ values for substituents indicate electron withdrawal, and negative σ values indicate electron-donating tendencies of the substituents. Thus, the bonding of the inhibitor with the metal is stronger or weaker depending upon whether the inhibitor is electron rich or electron poor.

TABLE 5.2 Hammett Substituent Constants

Substituent	σ meta	σ para	Substituent	σ meta	σ para
CH_3	− 0.069	− 0.170	O	− 0.708	− 1.00
CH_2CH_3	− 0.07	− 0.151	OH	+ 0.121	− 0.37
$CH(CH_3)_2$	− 0.068	− 0.151	OCH_3	+ 0.115	− 0.268
$C(CH_3)_3$	− 0.10	− 0.197	OC_2H_5	+ 0.1	− 0.24
C_6H_6	+ 0.06	− 0.01	OC_6H_5	+ 0.252	− 0.320
$C_6H_4NO_2$-p	—	+ 0.26	$OCOCH_3$	+ 0.39	+ 0.31
$C_6H_4OCH_3$-p	—	− 0.10	F	+ 0.337	+ 0.062
$CH_2Si(CH_3)_3$	− 0.16	− 0.21	$Si(CH_3)_3$	− 0.04	− 0.07
$COCH_3$	+ 0.376	+ 0.502	PO_3H	+ 0.2	+ 0.26
COC_6H_5	—	+ 0.459	SH	+ 0.25	+ 0.15
CN	+ 0.56	+ 0.660	SCH_3	+ 0.15	0.00
CO_2	− 0.1	0.0	$SCOCH_3$	+ 0.39	+ 0.44
CO_2H	+ 0.35	+ 0.406	$SOCH_3$	+ 0.52	+ 0.49
CO_2CH_3	+ 0.321	+ 0.385	SO_2CH_3	+ 0.60	+ 0.72
$CO_2C_2H_5$	+ 0.37	+ 0.45	SO_2NH_2	+ 0.46	+ 0.57
CF_3	+ 0.43	+ 0.54	SO_3	+ 0.05	+ 0.09
NH_2	− 0.16	− 0.66	$S(CH_3)_2$	+ 1.00	+ 0.90
$N(CH_3)_2$	− 0.211	− 0.83	Cl	+ 0.373	+ 0.227
NHCOCH	+ 0.21	0.00	Br	+ 0.391	+ 0.232
$N(CH_3)_3$	+ 0.88	+ 0.82	I	+ 0.352	+ 0.276
N_2	+ 1.76	1.91	IO_2	+ 0.70	+ 0.76
NO_2	+ 0.710	+ 0.778			

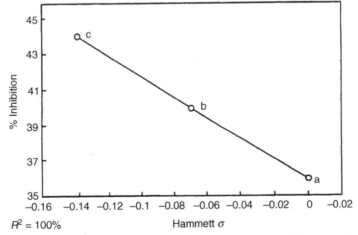

FIGURE 5.1 Percent inhibition for *meta*-substituted methyl pyridines as a function of Hammett σ values: (a) pyridine; (b) 3-methyl pyridine; (c) 3,5-dimethyl pyridine.

TABLE 5.3 Corrosion Rate Data with Benzonitriles

Inhibitor	σ Value	Corrosion Rate log K (mg/(cm^2 h))
Benzonitrile	0.00	-2.65
p-Chlorobenzonitrile	0.227	-1.00
p-Nitrobenzonitrile	0.778	-1.6532
p-Methyl benzonitrile	-0.17	-2.398

Theoretical calculations involving quantum chemical methods such as the extended Hückel molecular orbital (EHMO), modified neglect of differential overlap (MNDO), zero neglect differential overlap approximation (ZNDO), and density functional Theory (DFT) methods were applied to metal–inhibitor systems. These calculations give values of the energies of the highest occupied molecular orbital (E_{HOMO}) and lowest unoccupied molecular orbital (E_{LUMO}) of the inhibitor molecule of interest.

By definition, ionization energy, $I_E = -E_{HOMO}$, electron affinity, $E_A = -E_{LUMO}$, electronic chemical potential, $-\mu = (I_E + E_A)/2 = \chi$, and absolute hardness, $\eta = (I_E - E_A)/2$. The fraction of electronic charge ΔN transferred from the inhibitor (B) to the metal (A) is

$$\Delta N = \frac{\chi_a - \chi_b}{2(\eta_a - \eta_b)}.$$

With these relationships the fractional charge transferred from the inhibitor to the metal was calculated.

The corrosion inhibition data obtained for the various systems were analyzed in terms of the Hammett relationship

$$\log \frac{I}{I_0} = \rho\sigma,$$

where I is the percent inhibition with the substituted inhibitor, I_0 is the percent inhibition with the parent unsubstituted inhibitor, σ is the Hammett parameter, and ρ a constant characteristic of the system. The data on percent inhibition and $\Delta\Delta N$, the fraction of charge due to the substituent (i.e., the fraction of charge of the substituted inhibitor minus the fraction of charge of the parent inhibitor), have been found to obey a Sastri equation of the type

$$\log \frac{I}{I_0} = \kappa\Delta\Delta N,$$

where κ is a constant. It is to be expected that σ and $\Delta\Delta N$ are related to each other in that both signify the electronic effects due to the substituent in the inhibitor molecule.

The analysis of the data on corrosion inhibition of iron by pyridines in HCl, H_2SO_4, and H_2S, of aluminum by pyridines in HCl, and of iron by anilines and benzimidazoles in HCl yielded the data given in Table 5.4.

TABLE 5.4 Data on Iron and Aluminum Inhibition

System	Hammett Constant, ρ	Proposed Constant, κ	Value of Ratio, $\Delta\Delta N/\sigma$
Iron–methyl pyridines			
HCl medium			
meta-Derivatives	0.622	10.01	0.0620
ortho-Derivatives	0.357	8.05	0.0397
H$_2$SO$_4$ medium			
meta Derivatives			
1% v/v inhibitor	1.34	18.75	
2% v/v inhibitor	1.29	18.07	0.0714
3% v/v inhibitor	0.408	5.71	
pH 4.5, 3000 mg/L H$_2$S			
meta-Derivatives	0.423	6.27	0.066
ortho-Derivatives	0.179	4.74	0.038
Aluminum–methyl pyridines			
HCl medium			
meta-Derivatives	6.9	26.8	0.260
ortho-Derivatives	2.3	22.6	0.105
Iron–aniline			
HCl medium			
para-Derivatives	0.681	24.8 (13.6)	0.0299 (0.0385)
Iron–benzimidazole			
HCl medium			
ortho-Derivatives			
10^{-4} M inhibitor		24.5 (34.3, 14.6)	
10^{-3} M inhibitor		10.06 (13.9, 6.4)	
10^{-2} M inhibitor		3.25 (406, 2.43)	

The data obtained by analysis through the equations show the following trends. Inhibition of iron with methyl pyridines:

   (i) $\rho_{H_2SO_4} > \rho_{HCL} > \rho_{H_2S}$ for both *meta*- and *ortho*-substituted inhibitors
   (ii) $\rho_m > \rho_o$ in HCl, H$_2$SO$_4$, and H$_2$S
   (iii) ρ (1%, 2% inhibitor) $> \rho$ (3% inhibitor) in H$_2$SO$_4$
   (iv) $\rho_{Al} > \rho_{Fe}$ for both *meta*- and *ortho*-substituted inhibitors
   (v) $K_{H_2SO_4} > K_{HCL} > K_{H_2S}$ for both *meta*- and *ortho*-substituted inhibitors
   (vi) $K_m > K_o$ in HCl, H$_2$SO$_4$, and H$_2$S solutions
   (vii) $K_{Al} > K_{Fe}$ in HCl for both *meta*- and *ortho*-substituted inhibitors
   (viii) K (1%, 2% inhibitor) $> K$ (3% inhibitor).

The trends noted above for the values of ρ and K are similar since Hammett's parameter σ and $\Delta\Delta N$ are related and signify the electronic density transmitted by the substituent in the inhibitor molecule to the metal. The observed trends in ρ and values in HCl, H_2SO_4, and H_2S may be explained as follows: the chloride ion is readily adsorbed on metallic iron compared to the bisulfate or sulfate anions. This observation is based on the following evidence:

(i) Bockris et al. (28) found that $2.71 \times 10^{-9} \, mol/cm^2$ of chloride and $1.15 \times 10^{-9} \, mol/cm^2$ of bisulfate are adsorbed on a Pt electrode.

(ii) The potentials of adsorption at zero point charge for chloride and sulfate are -461 and $-438 \, mV$, respectively (29).

(iii) Using radioactive isotopes of chloride and bisulfate, the greater degree of adsorption of chloride than bisulfate on iron has been demonstrated (30).

Thus, in HCl solution, displacement of chloride adsorbed on iron by the inhibitor is more difficult to accomplish than of bisulfate, and hence inhibition in H_2SO_4 solution is greater than in HCl solution. The inhibition in H_2S solution can be explained by the operation of a dual mechanism; namely, by pyridines assisted by bisulfide in a synergistic pathway.

The ρ and K values for *meta*-substituted inhibitors are greater than for *ortho*-derivatives due to the electronic charge transfer, the ρ and K values at lower concentrations of the inhibitor being greater than at higher concentrations of the inhibitor. This indicates that the electron charge transfer effects are more significant at lower concentrations of inhibitor, which correspond to lower coverage of the metal surface by inhibitor.

The ρ and K values for the inhibition of corrosion of aluminum in HCl solution by methyl pyridines are greater than those for iron in HCl solution. This can be explained on the basis that the adsorption of chloride on iron is preferred over aluminum. In the case of aluminum (31), the adsorption of sulfate on the metal is a maximum at pH 4.0 and is negligible below pH 2.0. Adsorption of chloride on aluminum is expected to be similar to that of sulfate with a maximum at pH 4.0 and negligible below pH 2.0. Since the data on inhibition of corrosion of aluminum by methyl pyridines refer to a solution of pH < 2.0, the competing adsorption of chloride ion on the metal surface is expected to be negligible, thus facilitating the interaction of the aluminum metal with the methyl pyridines.

The *para*-substituted anilines give a ρ value of 0.68 and K value of 24.8 for the methyl and bromo derivatives. The lower value of 13.6 for K is for the case of thiocyanato-substituted aniline (1) and indicates that the thiocyanato group may be interacting with the iron metal in addition to the iron–aniline interaction. In this case, there are two adsorption modes, namely Fe–NH_2 and Fe–NCS, resulting in the lower K value of 13.6.

The K values of iron-substituted benzimidazoles decrease progressively when the concentration of the inhibitor is increased from 10^{-4} to 10^{-2} M, indicating that the electronic effects due to the substituents are significant at lower concentrations of the inhibitor. The values of K given in parentheses in Table 5.3 refer to 2-hydroxy and 2-amino benzimidazoles derivatives. In the case of 2-amino benzimidazoles, a second interaction, namely Fe–NH$_2$, in addition to the primary Fe–benzimidazole nitrogen might be responsible for the observed degree of inhibition.

Considering the iron inhibitor system, the data show the following trends:

$$\rho_{aniline} \geq \rho_{pyridine},$$

$$K_{aniline} \approx K_{benzimidzoles} > K_{pyridines}.$$

High values for K indicate that the corrosion is sensitive to the substituents in the inhibitor molecule. Lower K values indicate:

(i) lower sensitivity to the substituents;

(ii) operation of an inhibition mechanism other than direct interaction of the inhibitor and the metal, such as the synergistic mechanism in the inhibition of iron by pyridines in H$_2$S medium through Fe–SH, or Fe–NS interaction in the case of the iron–aniline system.

The variation of K values with the inhibitor concentration indicates the variation of the extent of coverage of the metal by the inhibitor and that the substituent effects are prominent at lower concentrations of the inhibitor.

The main advantage of the proposed Sastri equation is its predictive nature for the selection of inhibitors, for example, in predicting that the introduction of a mercapto group (–SH) in aniline or pyridine or benzimidazole inhibitors could produce considerable improvement in the inhibition of corrosion of iron (32).

The effect of substituents in parent organic inhibitor molecules has been correlated with changes in electron densities in the functional groups as noted below:

- Substituted pyridines (16, 17, 19, 25)
- Substituted anilines (17, 33)
- Substituted aliphatic amines (34)
- Amino acids (22)
- Benzoic acids (23)
- Aliphatic sulfides (24).

The electron density at the functional groups in the inhibitor molecules was ascertained from Hammett (25) and Taft constants (24), nuclear magnetic resonance data (17), and quantum chemical calculations (35).

5.4   QUANTUM CHEMICAL CONSIDERATIONS

Quantum chemical calculations give data on (i) energies of highest occupied molecular orbital (E_{HOMO}) and lowest unoccupied molecular orbital (E_{LUMO}), (ii) the fraction of electronic charge on the donor atom of the inhibitor (ΔN),

TABLE 5.5   Diamine $H_2N(CH_2)_nNH_2$

n	E (eV)	Geometry	Percent Inhibition
2	-9.094	Planar	84
3	-8.815	Planar	56.5
4	-9.197	Nonplanar	-42

(iii) heats of reaction between the metal and the inhibitor (ΔH), (iv) degree of softness of the inhibitor (i.e., $E_{HOMO} - E_{LUMO}$).

Hückel molecular orbital theoretical calculations on corrosion inhibitors such as biguanide, dicyanodiamide, guanyl urea, and biuret used in the corrosion inhibition of mild steel in 6% HCl showed a linear relationship between E_{HOMO} and E_{LUMO} with the logarithm of corrosion inhibition efficiency. The inhibitors act as electron donors to iron atoms and prevent corrosion reaction by the formation of a bond with the metal (36).

The linear combination of atomic orbitals (LCAO) self-consistent field (SCF) molecular orbital (MO) study of linear chain diols, diamines, and aliphatic aminoalcohols showed the correlation of inhibition efficiency with E_{HOMO}. Corrosion inhibition is high when the inhibitor is planar and has a high value for E_{HOMO}, as shown in Table 5.5 (37).

The LCAO-SCF-MO method, including on titanium metal atoms, was used to study substituted nitrophenols and nitroanilines, and the degree of corrosion inhibition was correlated with the stabilization energy (38). The higher the stabilization energy the lower was the corrosion rate (Table 5.6).

Quantum chemical calculations using the MNDO method on substituted pyridines and ethane compounds in the corrosion inhibition of iron in HCl solutions (39) gave correlation of percent inhibition with the E_{HOMO}, $E_{LUMO} - E_{HOMO}$ gap (degree of softness) of inhibitor, and the fraction of electrons (ΔN) transferred from the inhibitor to the metal. These correlations are significant since they lead to a method of increasing the degree of corrosion inhibition by incorporating suitable substituents in the parent inhibitor molecule. This leads to selection of corrosion inhibitors based on structural and theoretical considerations.

Theoretical calculations using ZINDO/1 method including both the metal and the inhibitor resulted in data on the heats of formation of the metal–inhibitor complex (30). Some typical data are given in Table 5.7

TABLE 5.6   Corrosion Rates and Stabilization Energies

Inhibitor	Corrosion Rate (mpy)	Stabilization Energy (eV)
Phenol	88.6	1.77
o-Nitrophenol	1.8	4.40
m-Nitrophenol	0.2	4.44
p-Nitrophenol	0.3	4.50
Aniline	86.8	1.11
m-Nitroaniline	0.4	4.53
p-Nitroaniline	0.5	4.57

TABLE 5.7   Heats of Reaction and Percent Inhibition

System	Heat of Reaction (kCal/mol)	Percent Inhibition
Cu–benzimidazole	− 3467.7	92.11
Cu–2-methylbenzimidazole	− 4018.7	93.42
Cu–5,6-dimethylbenzimidazole	− 4547.8	96.71
Fe–benzimidazole	− 3471.1	29.0
Fe–2-methylbenzimidazole	− 4030.7	35.0
Fe–2-phenylbenzimidazole	− 6001.4	95.0

The data show an increase in percent inhibition with increasing negative value for heat of reaction. A linear relationship is observed between heat of reaction and percent corrosion inhibition.

5.4.1   Application of Hard and Soft Acid and Base Principle in Corrosion Inhibition

Pearson enunciated the hard and soft acid and base (HSAB) principle (41). Corrosion inhibitors can be viewed from the HSAB principle and described as hard, soft, or borderline inhibitors (Table 5.8).

In accordance with the HSAB principle, hard acids form complexes with hard bases and soft acids form complexes with soft bases. Borderline acids form complexes with either soft or hard bases.

Corrosion inhibition of iron and aluminum by phosphate is in accord with the HSAB principle, since Fe^{3+} and Al^{3+} are hard acids, which react with phosphate, a hard inhibitor. Corrosion inhibition of Cu^{2+} and Zn^{2+} (borderline acids) by amines (borderline inhibitors) is in keeping with the HSAB principle. Corrosion inhibition of copper, a soft acid by mercaptobenzothiazole (soft inhibitor) obeys the HSAB principle.

The probable mechanism of corrosion of iron in hydrogen sulfide medium involves the formation of a surface complex $Fe(-H-S-H)_{ads}$ and bonding to the nitrogen atom of the inhibitor such as pyridine.

Molecular orbital theoretical calculations of corrosion inhibitors such as methyl pyridines and substituted ethane compounds enabled correlation of corrosion rates with the fraction of electronic charge transferred from the inhibitor to iron metal. The corrosion rate of iron as a function of ΔN (fraction of electronic charge)

TABLE 5.8   Hard, Soft and Borderline Inhibitors

Hard Inhibitor	Soft Inhibitor	Borderline Inhibitor
F^-, PO_4^{3-}, CO_3^{2-}	RSH, RS, R_3P	$C_6H_5NH_2$, C_5H_5N, Br^-
Hard Acids	Soft Acids	Borderline Acids
Fe^{3+}, Cr^{3+}, Al^{3+}	Cu^+, M^0 (metal atoms), bulk metals	Fe^{2+}, Ni^{2+}, Cu^{2+}, Zn^{2+}

transferred from the inhibitor to the metal is shown in Figs. 5.2 and 5.3 for substituted pyridines and substituted ethane compounds, respectively. The corrosion rates decrease with increase in electronic charge transferred from the inhibitor to the metal. These results confirm earlier results obtained on the methyl pyridines (16). The corrosion inhibition of iron by methyl pyridines was in the order: 2,4,6-trimethyl pyridine > 2,4,-lutidine, 2,6-lutidine > 3,5-lutidine, 4-picoline, 2-picoline > 3-picoline > pyridine.

The inhibition of corrosion by aliphatic amines is considered to involve the following equilibria (13):

$$RNH_3{}^+ (sol) \rightleftharpoons RNH_2 + H^+ \rightleftharpoons RNH_2\text{--}Fe^0$$

where $RNH_2\text{--}Fe^0$ denotes the chemisorbed amine. The degree of chemisorption depends on the strength of the metal–amine bond. The relative order of inhibitor effectiveness of aliphatic amines is

$$R_2NH > R_3N > RNH_2 > NH_3,$$

where R is the methyl group, and

$$R_3N > R_2NH > RNH_2 > NH_3,$$

where R is ethyl, propyl, butyl, or amyl group.

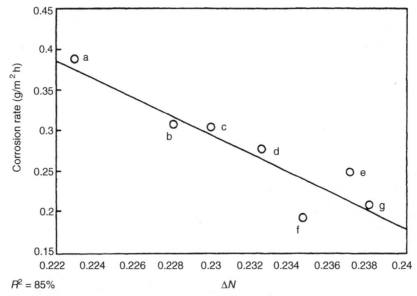

FIGURE 5.2 Correlation of ΔN with corrosion rates of pyridine derivatives: (a) pyridine; (b) 3-methyl pyridine; (c) 2-methyl pyridine; (d) 3,5-dimethyl pyridine (e) 2,6-dimethyl pyridine; (f) 2,5-dimethyl pyridine; (g) 2,4,6-trimethyl pyridine.

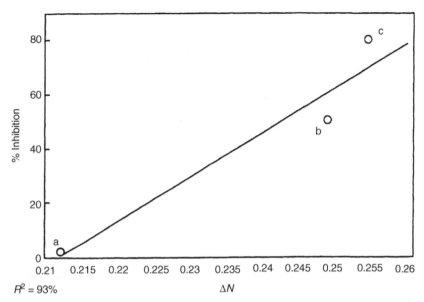

FIGURE 5.3 Correlation of ΔN with percent inhibition of substituted ethane derivatives: (a) $HOCH_2–CH_2–OH$; (b) $HO–CH_2–CH_2–NH_2$; (c) $H_2N–CH_2–CH_2–NH_2$.

Aliphatic amines are better inhibitors than aromatic amines. Heterocyclic amines, such as pyridine, are better inhibitors than aromatic amines but less effective than the corresponding saturated amines. The following is the order of effectiveness of inhibitors:

cyclohexylamine > pyridine > aniline
piperidine > pyridine
dicyclohexylamine > diphenylamine

The introduction of a methyl group in aniline results in increase in the degree of corrosion inhibition due to the hyperconjugation of the methyl group. The degree of inhibition of toluidines is in the order:

o-toluidine > p-toluidine > m-toluidine > aniline.

Sulfur-containing inhibitors are better than nitrogen-containing inhibitors because of the greater polarizability of the sulfur atom over the nitrogen atom. Sulfur is less electronegative then nitrogen (2.5 compared to 3.0) and has two lone pairs of electrons for bonding, compared to the one pair in nitrogen.

The order of effectiveness of mercaptans as inhibitors is

amyl > butyl > propyl > ethyl > methyl

and of sulfides is

$$butyl > propyl > ethyl > methyl.$$

Aromatic sulfides are less effective than aliphatic sulfides. For example, thiophenol is less effective than ethyl mercaptan. Thiocresols are better inhibitors than thiophenol.

Alcohols and phenols are poor inhibitors since they contain more electronegative oxygen and do not form a coordinate bond readily. Based on the same reasoning, selenium compounds are expected to be good inhibitors. Ethyl selenide is found to be a good inhibitor.

The inhibition of corrosion by aldehydes is in the order:

$$butyraldehyde < acetaldehyde < propionaldehyde < formaldehyde.$$

The introduction of a double bond in a compound appears to increase the extent of corrosion inhibition. An example consists of the inhibition effect of allyl alcohol compared to ethyl alcohol. Allyl thiourea is a good inhibitor in contrast to thiourea. Crotonaldehyde is a better inhibitor than butyraldehyde. Thioureas are better inhibitors than urea, and thiocyanates are better than cyanates as inhibitors.

5.5   INHIBITOR FIELD THEORY OF CORROSION INHIBITION

The corrosion inhibitor field theory propounded by Sastri et al. (40, 42) may be described as follows: Corrosion inhibition of metals by organic inhibitors is known to involve the adsorption of the inhibitor on the surface of metals and hence protect the metals from corrosive attack. It is reasonable to assume that the first step of corrosion inhibition is the transport of the inhibitor to the metal surface followed by a second step of adsorption on the metal surface. The adsorbed inhibitor on the metal surface may be viewed as the formation of a metal complex, also known as a coordination complex. The metal–inhibitor complex may be hexacoordinated with one metal–inhibitor bond and five metal–water bonds of quasi-distorted octahedral geometry and the whole complex occupying a lattice site of the metal on the metal surface as shown in Fig. 5.4. Alternatively, the metal ions resulting from the initial stages of corrosion may interact with the inhibitor and form a monoinhibitor pentaaquo metal complex layer in the Helmholtz double layer near the metal, leading to corrosion inhibition as shown in Fig. 5.5. In the metal–inhibitor complex, the metal can be iron, copper, and titanium that are also known as the transition group of metals in which the d-orbitals are populated with electrons. The ions Ti^{3+}, Fe^{2+}, and Cu^{2+} have d^1, d^6, and d^9 electronic configurations, respectively. In other words Ti^{3+}, Fe^{2+}, and Cu^{2+} have 1, 6, and 9 electrons in the d-orbitals. In the terminology of inorganic chemistry, the moiety bonded to the metal is known as a ligand, and the inhibitor may be considered as a ligand.

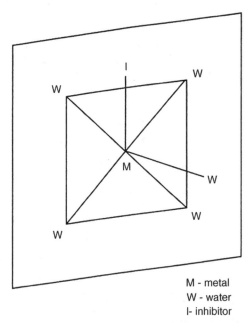

M - metal
W - water
I- inhibitor

FIGURE 5.4   Disposition of monoinhibitor pentaaquo-distorted metal complex adsorbed on the metal surface.

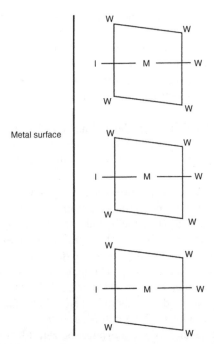

FIGURE 5.5   Disposition of monoinhibitor pentaaquo metal complex in Helmholtz double layer near the metal surface.

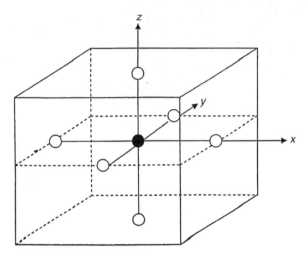

FIGURE 5.6 Octahedral arrangement of a metal complex.

According to the inhibitor field theory propounded by Sastri et al. (42), the metal–inhibitor complex can be considered an aggregate of metal and inhibitor molecules held together by electrostatic forces. In the case of the Ti^{3+} ion, it is surrounded by inhibitor ions or molecules containing a lone pair of electrons available for donation to the metal.

The crystal field model then assumes that the effects of metal–inhibitor complex formation on the transition metal ion is to place the metal ion in the electrostatic field of inhibitor molecules situated at the corners of a regular octahedron as shown in Fig. 5.6. In the crystal field model, the argon core of electrons is neglected and the exterior 3d electrons of the metal are considered. Thus, the electronic properties are ascribed solely to the 3d electrons moving in the field of the coordinated ligands (inhibitor).

In the case of the isolated Ti^{3+} ion, the five distinct 3d states, designated as d_{xy}, d_{yz}, d_{zx}, $d_{x^2-y^2}$, and $d_{3z^2-r^2}(d_{z^2})$, which are degenerate, can be visualized (Fig. 5.7). A look at Figure 5.7 shows that the inhibitor with negative charge will repel an electron in $d_{x^2-y^2}$- and d_{z^2}-orbitals more than an electron in the d_{xy}-, d_{yz}-, and d_{xz}-orbitals. The fivefold degenerate d-orbitals will be split by a regular octahedral environment of ligands (inhibitor) into two groups, namely, d_{xy}, d_{yz}, and d_{xz}, with electron density pointing away from the ligands, and $d_{x^2-y^2}$, d_{z^2} with electron density pointing toward the ligands (inhibitor). The splitting of the d-orbitals in octahedral geometry is shown in Fig. 5.8. The d_{xy}-, d_{yz}-, and d_{xz}-orbitals belong to the t_{2g} level, and d_{z^2}-, $d_{x^2-y^2}$-orbitals to the e_g group. The total energy separation of the t_{2g}- and e_g-orbitals may be represented by Δ or 10 Dq or E_1–E_2 and the energy chosen as the weighted mean energy of the d-orbitals to be zero. Under these circumstances, the energies of t_{2g}- and e_g-orbitals $-2/5\Delta$ and $3/5\Delta$, respectively, satisfy the condition that the weighted mean energy of the d-orbitals be zero [$3(-2/5\Delta + 2(3/5)\Delta = 0$]. The 10 Dq values vary with the nature of the ligand since ligands can have different electron donor atoms. The 10 Dq represents the energy gap between the lower set of

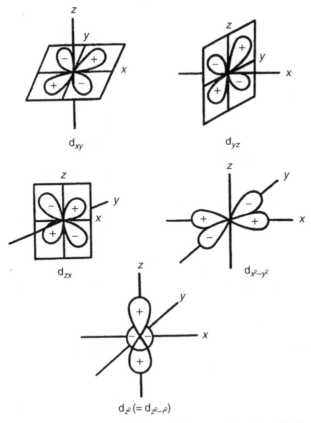

FIGURE 5.7   The angular dependence of the wave function of d electrons.

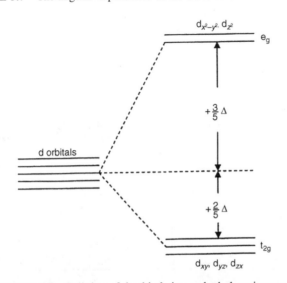

FIGURE 5.8   Splitting of d-orbitals in octahedral environment.

d-orbitals and higher set of d-orbitals in the octahedral configuration of inhibitor. Thus, it also indicates the σ or π bonding of the metal–inhibitor complex.

5.6 APPLICATION TO TYPICAL METAL–INHIBITOR SYSTEMS

The data on the percent inhibition of corrosion of iron and copper by the aniline and benzimidazole inhibitors containing the amine group are given in Table 5.9.

The ratios of percent inhibition of Cu/Fe work out to 1.57, 1.90, with an average value of 1.74 in the case of the aniline inhibitor. The corresponding values in the case of the benzimidazole inhibitor work out to 1.54, 1.65, with the average value of 1.60.

The ligand field parameter, 10 Dq values for ethylenediamine and ammonia ligands given in Table 5.10 give 1.25 and 1.20 for the ratio of Cu/Fe. These values are in the same direction as the percent inhibition of Cu/Fe but lower. The crystal field stabilization energy ratio of Cu/Fe is 6 Dq/4 Dq, and the values 1.88 and 1.80 are obtained with an average of 1.84 comparable with percent inhibition ratios of 1.74 and 1.60 with aniline and benzimidazoles as the inhibitors, respectively.

TABLE 5.9 Data on Corrosion Inhibition of Iron and Copper

System	Percent Inhibition	Ratio of Percent Inhibition (Cu/Fe)
Iron–aniline	52.0	1.57
	43.0	1.90
		Average value 1.74
Copper–aniline	81.82	
Iron–benzimidazole	60.00	
Copper–benzimidazole	92.11	1.54
	99.0	1.65
		Average value 1.60

TABLE 5.10 Ligand Field Parameters for Iron and Copper

System	10 Dq (cm^{-1})	Ratio of 10 Dq	Crystal Field Stabilization Energy Ratio (Cu^{2+}/Fe^{2+}) 6 Dq/4 Dq
Fe^{2+}–ethylenediamine	12,800	1.25	1.88
Cu^{2+}–ethylenediamine	16,000		Average value 1.84
Fe^{2+}–ammonia	12,500	1.20	1.80
Cu^{2+}–ammonia	15,000		

TABLE 5.11   Iron–Inhibitor System (43)

	$HO(CH_2)_nOH$	$H_2N(CH_2)_nOH$	$H_2N(CH_2)_nNH_2$
$n = 2$	1.0	49.0	$84/49 = 1.71$
$n = 3$		40.0	$56.5/40 = 1.41$
$10\,Dq\ (cm^{-1})$	9,400	10,850	$12,500/10,850 = 1.15$

$HO(CH_2)_2OH < HO(CH_2)_2NH_2 < NH_2(CH_2)_nNH_2$
Inhibitor field strength order: O, O < N < N, N

The data on corrosion inhibition of iron with substituted ethylene and propylene inhibitors are given in Table 5.11. The inhibitors studied are dihydroxy, hydroxyamino, and diamino derivatives. It is clear from the data that dihydroxy derivatives are poor inhibitors. As the hydroxyl group is replaced by an amino group, the percent inhibition increases. The inhibition increases in the order: $HO(CH_2)_nOH < HO(CH_2)_nNH_2 < H_2N(CH_2)_nNH_2$.

The order of inhibition is in keeping with increasing $10\,Dq$ values. Thus, we can define "inhibitor field strength" with the following sequence:

$$O, O < O, N < N, N.$$

The ratio of inhibition data of the diamino inhibitor to the monoamino monohydroxy derivative gives the values of 1.71 and 1.41, comparable with the $10\,Dq$ ratio of 1.30.

The data on corrosion inhibition of iron by benzimidazoles, mercaptobenzimidazole, and amino benzothiazole along with $10\,Dq$ values are given in Table 5.12.

The substitution of the mercapto group in benzimidazole increases the effectiveness of the inhibitor. The ratios of the percent inhibition of iron–mercaptobenzimi-

TABLE 5.12   Data on Iron–Inhibitor System

System	Percent Inhibition	
Iron–benzimidazole (Fe–BI)	60.0	$\dfrac{Fe - BISH}{Fe - BI} = 1.47$
Iron–mercaptobenzimidazole (Fe–BISH)	88.0	
Iron–aminobenzothiazole (Fe–ABT)	85.0	$\dfrac{Fe - ABT}{Fe - ABI} = 1.42$

$\dfrac{10\,Dq\ (Fe - BISH)}{10\,Dq\ (Fe - BI)} = \dfrac{13,500}{10,000} = 1.35$
Inhibitor field strength order: BISH > BI; (N, SH) > N
Inhibitor field strength order: O, O < O, N < N, N < N, SH
Order of $10\,Dq$ for iron: $9,400 < 10,850 < 12,500 < 13,500$

dazole/iron–benzimidazole and iron–aminobenzothiazole/iron–benzimidazole work out to 1.47 and 1.42, respectively. These ratios are comparable to the ratio of 1.35, the ratio of 10 Dq values. We can define the inhibitor field strength for iron as BISH, $ABT > BI(N, SH) > N$.

The overall inhibitor field strength sequence for inhibition of iron with different functional groups along with 10 Dq values may be written as O, $O < O$, $N < N, N \leq N, SH$.

The observation that the inhibitor effectiveness order parallels 10 Dq order is useful in the prediction of the effectiveness of the inhibitor with the available values of 10 Dq.

The data on corrosion inhibition of titanium by nitrophenols and nitroanilines are given in Table 5.13. The important point to note is that unlike the case of iron and copper, inhibitors containing oxygen donors are the most effective inhibitors in the inhibition of the corrosion of titanium.

The data show corrosion inhibition to follow the order

$$OH, NO_2 > NH_2, NO_2 > NO_2.$$

The inhibition field strength deduced from the data is

$$O, O > O, N.$$

The 10 Dq values for titanium with the inhibitors containing the functional groups NO_2 and OH are in the order

$$NO_2 > NO_2, NH_2 > NH_2.$$

TABLE 5.13   Data on Titanium Inhibitor Systems

Inhibitor	OH (OH)	Corrosion Rate
Phenol	OH (OH)	1.00
o-Nitrophenol	OH, NO_2 (HO,O)	0.0203
m-Nitrophenol	OH, NO_2 (HO,O)	0.0023
p-Nitrophenol	OH, NO_2 (HO,O)	0.0034
Nitrobenzene	NO_2 (O)	0.015
Aniline	NH_2 (N)	1.000
m-Nitroaniline	NH_2, NO_2 (N,O)	0.0046
p-Nitroaniline	NH_2, NO_2 (N,O)	0.0058
Picric acid	OH, NO_2 (O,O)	0.0035

Order of corrosion inhibition: OH, $NO_2 > NO_2$, $NH_2 > NO_2$
Inhibitor field strength: O, O > O, N
10 Dq order (cm^{-1}): NO_2 (30,450) > NO_2, NH_2 (27,913) > NH_2 (25,375)

TABLE 5.14 Data on Stability Constants and Inhibitor Field Energies

Inhibitor	log K		Δ (cm^{-1})		Inhibition (%)
Fe-trimethylaniline (N) (45)	1.335		2083		51.0
Fe-Schiff bases (N, O) (46)	3.3222 3.7235 3.9191	} 3.65	3750 4167 4167	} 4028	79.0 85.0 90.0
Fe-triazoles (N) (47)	5.193 5.707 6.644	} 5.85	4167 4167 4167	} 4167	90.3 94.8 99.4
Fe-ethylenediamine (2N)	7.65		4170		52, 60
Cu-en (2N)	10.63		5120		81.8, 92.1, 99.0
Cu-MBT (S)	6.30		3440		73

The data on the stability constants, percent inhibition, and inhibitor field energies are given in Table 5.14. The stability constants are plotted against the percent inhibition in Fig. 5.9. The percent inhibition is seen to increase linearly with the stability constants of the formation of metal–inhibitor complex. The corrosion inhibitor field energies also bear a linear relationship with the logarithm of stability constants of the metal–inhibitor complexes as shown in Fig. 5.10.

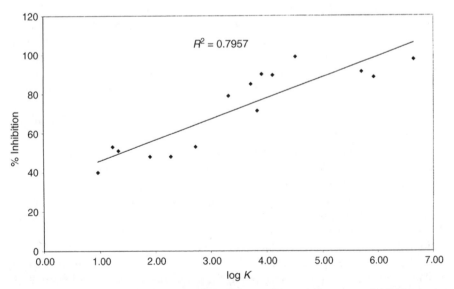

FIGURE 5.9 Plot of logarithm of stability constants against percent inhibition.

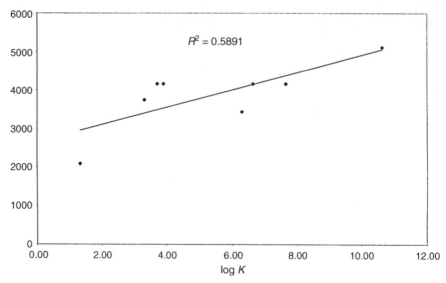

FIGURE 5.10 Plot of logarithm of stability constants versus inhibitor field energy.

These correlations confirm the operation of inhibitor field theory and mechanism in corrosion inhibition.

5.7 PHOTOCHEMICAL CORROSION INHIBITION

Mild steel is an important structural material in mining equipment and is subject to corrosion-related failure while in contact with acid mine waters. The mine waters in sulfide ore mines had a pH range of 2.5–8.0 with significant amounts of ferric and ferrous ions, and ions of Cu^{2+}, Ni^{2+}, Ca^{2+}, Mg^{2+}, Cl^-, SO_4^{2-}, and HCO_3^-. The corrosion rates of mild steel in synthetic mine waters of pH 2.5 containing 0.012 M of ferric and 0.001 M cupric ions with and without the inhibitors were determined (48).

The data on the corrosion rates both in the presence and in the absence of inhibitors are given in Table 5.15.

The deviation in corrosion rates was within 5 mpy. The data show that the corrosion rates are related to the ease of reduction of ferric and cupric ions. In the case of solutions containing n-butylamine or urea or guanylurea sulfate, the corrosion rates are high, paralleling the near complete reduction of cupric and ferric ions in solution. This observation is further supported by the negative shift in corrosion potential. Benzotriazole and hexadecyl pyridinium chloride are nearly equal in corrosion inhibition of steel. Thiourea yielded 77% inhibition, but the ferrous sulfide film formed is conducive to the hydrogen embrittlement damage of steel. Potassium oxalate gave the best protection among the inhibitors tested, resulting in 93% inhibition. The corresponding corrosion potential shifted from -515 to -110 mV.

TABLE 5.15   Data on Corrosion Rates

Inhibitor	Concentration (M)	Corrosion Rate[a] (mpy)	Inhibition (%)
None	–	1437	–
n-Butylamine	0.01	2175	0
Urea	0.125	2575	0
Guanyl urea sulfate	0.0224	1682	0
Benzotriazole	0.01	860	40
Hexadecyl pyridinium	0.01	914	36
Chloride	0.01	326	77
Thiourea	0.048	106	93[b]
Potassium oxalate	0.048	220	85[c]

[a]Accuracy of corrosion rates is within 5.0 mpy.
[b]Exposed to daylight.
[c]Dark.

Corrosion inhibition by potassium oxalate was found to be photosensitive, as evidenced by the lower corrosion rate on exposure to daylight (106 mpy) than darkness (220 mpy). Photochemical decomposition of a trisoxalato iron complex is a well-established reaction, and this system is in use as a chemical actinometer with quantum yields of 1.18 at 3650–3663 Å and yields of 0.93–1.28 at 4700–2537 Å, respectively.

The deposit on the steel sample exposed to the solution containing a potassium oxalate inhibitor was analyzed using X-ray and electron microprobe techniques and identified as hydrated β-ferrous oxalate (β-FeC$_2$O$_4$·2H$_2$O). The complex formed in solution containing potassium oxalate and ferric ion may be written as

$$Fe^{3+} + 3C_2O_4^{2-} \rightarrow Fe(C_2O_4)_3^{3-}.$$

The trisoxalato iron(III) complex formed undergoes decomposition to yield β-ferrous oxalate, which is deposited on the steel sample. The decomposition of the oxalate complex is photosensitive,

$$[Fe(C_2O_4)_3]^{3-} \xrightarrow{h\nu} \beta\text{-}FeC_2O_4 + 4CO_2.$$

It yields a deposit of β-ferrous oxalate on the surface of the steel sample and gives satisfactory corrosion inhibition. Thus, nontoxic potassium oxalate may be added to acid mine water exposed to daylight and allowing it to achieve satisfactory photochemical corrosion inhibition. This is the first example of photochemical corrosion inhibition.

It is generally agreed that when a metal is immersed in an aqueous solution, water molecules are adsorbed:

$$M + H_2O \rightarrow M(H_2O)n.$$

Upon addition of a corrosion inhibitor, the inhibitor is adsorbed:

$$M(H_2O) + I \rightarrow MI + H_2O,$$

and the adsorbed water molecule is removed from the surface. Thus, the metal–inhibitor interaction energy must be greater than metal–water interaction energy. When the inhibitor is an organic compound such as an amine, the adsorption of the amine on the metal becomes more favorable as the length of the hydrocarbon part of the amine increases. Some examples of the order of effectiveness of substituted amines in corrosion inhibition are as follows (49):

- triethylamine > diethylamine > ethylamine
- tri-*n*-propylamine > di-*n*-propylamine > mono-*n*-propylamine
- tri-*n*-butylamine > di-*n*-butylamine > mono-*n*-butylamine
- tri-*n*-amylamine > di-*n*-amylamine > mono-*n*-amylamine
- monoamylamine > monobutylamine > monopropylamine > monoethylamine > methylamine
- triamylamine > tributylamine > tripropylamine > triethylamine > trimethylamine.

Lateral interactions between the adsorbed inhibitor molecules can be significant at high surface coverage since the inhibitor molecules are close to one another. Attractive lateral interactions occur between molecules containing hydrocarbon components such as *n*-alkyl substituents, due to an increase in van der Waals' attractive forces between adjacent molecules. These attractive forces between adjacent molecules lead to stronger adsorption. On the other hand, repulsive interactions between ions or dipoles lead to weaker adsorption.

In the case of ions, repulsive interactions can be altered to attractive interactions through a combination with an ion of opposite charge. Thus, in a solution of inhibitive anions and cations, the adsorption of both ions may be increased and the degree of inhibition greater than in a single component, such as a cation or anion alone. The effects of inorganic anions and organic amines on the corrosion of mild steel in various mineral acids were studied, and the following order of adsorption was observed (50):

$$I^- > Br^- > Cl^- > SO_4^{2-} > ClO_4^-.$$

Synergistic inhibition has been observed in mixtures of anionic and cationic inhibitors. The largest effect on corrosion inhibition was observed with iodide alone and iodide in combination with amines. The results are suggestive of a strong interaction with a steel surface through chemisorption. The inhibition efficiency enhancement by iodide was found to depend upon the molecular structure of the

TABLE 5.16   Conversion of Primary Inhibitor into Secondary Products

Primary Inhibitor	Secondary Product
Sulfoxides (51)	Sulfides
Quaternary phosphonium	Phosphine
Arsonium compounds (52)	Arsine
Acetylenic compounds (53)	Reduction and polymerization to give multimolecular polymeric film
Thiourea (10)	Gives bisulfide, which promotes corrosion

inhibitor. The degree of synergism between iodide ion and organic amines was in the order:

$$octamethyleneimine > di\text{-}n\text{-}butylamine > di\text{-}n\text{-}octylamine.$$

The observed synergistic effects may be attributed in part to the stabilization of the adsorbed anion layer by quaternary organic cations, possibly by some sort of covalent bonding.

The adsorbed corrosion inhibitor may undergo electrochemical reduction to yield a product species that may also act as an inhibitor. Corrosion inhibition due to an initially added inhibitor is known as primary inhibition, and the inhibition by the reaction product produced during primary inhibition is termed secondary inhibition. Thus, the corrosion inhibition effect may increase or decrease with time, depending upon whether the secondary inhibition is more or less effective than primary inhibition. Some of the inhibitors giving rise to products that act as secondary inhibitors are given in Table 5.16.

5.8   INFLUENCE OF INHIBITORS ON CORROSION REACTIONS IN ACID MEDIA

In acid solutions the anodic and cathodic reactions are the oxidation of the metal and reduction of hydrogen ions, respectively.

$$\text{Anodic reaction}: \quad M^0 \rightarrow M^{n+} + ne^-.$$
$$\text{Cathodic reaction}: \quad H^+ + e^- \rightarrow H,$$
$$H + H \rightarrow H_2.$$

In solutions saturated with air, cathodic reduction of oxygen occurs above a pH of 3.0

$$O_2 + 2H_2O + 4e^- \quad \rightarrow \quad 4OH^-,$$
$$O_2 + 2H^+ + 4e^- \quad \rightarrow \quad 2OH^-.$$

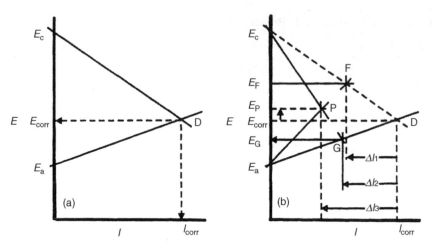

FIGURE 5.11 Anodic and cathodic polarization curves. (a) Illustrates significant terms for freely corroding metal. (b) A schematic diagram showing metallic corrosion, protection, and inhibition.

The added inhibitor may stifle the rate of anodic reaction, the cathodic reaction, or both anodic and cathodic reactions. The change in corrosion potential upon the addition of an inhibitor can often indicate the process inhibited. An inhibitor can inhibit either the cathodic or anodic process or both the anodic and cathodic processes. The change in potential in the positive direction upon addition of the inhibitor indicates inhibition of the anodic reaction, and a change in potential in the negative direction indicates inhibition of the cathodic process. No observable change in the corrosion potential upon the addition of an inhibitor can be interpreted as inhibition of both anodic and cathodic processes. All three scenarios are illustrated in Fig. 5.11. It is clear from the figure that corrosion inhibition is more effective when both the anodic and cathodic processes are inhibited than when either the cathodic or anodic process alone is inhibited.

The adsorption of inhibitors on the metal can be studied electrochemically by determination of the anodic and cathodic polarization curves of the corroding metal (54). A displacement of the polarization curve without a change in Tafel slope upon addition of an inhibitor indicates that the adsorbed inhibitor is blocking active sites and inhibiting the reaction without affecting the mechanism of the corrosion process. An increase in the Tafel slope of the polarization curve upon addition of the inhibitor indicates the effect of the inhibitor on the corrosion reaction mechanism. The Tafel slope is usually determined by polarization of the metal under conditions of potential and current density far removed from normal corrosion. This may result in differences in adsorption and mechanisms of inhibition of polarized metals compared to naturally corroding metals. Thus, the potentiodynamic polarization diagrams obtained may not be conclusive about the mechanism of corrosion. This difficulty can be overcome by the rapid polarization technique as illustrated in Fig. 5.12.

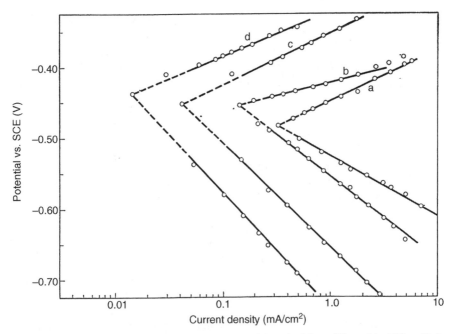

FIGURE 5.12   Polarization curves obtained by rapid method for mild steel in 10% sulfuric acid with dibutyl thioether at 30°C. (a) 0 M; (b) 0.001 M; (c) 0.002 M; (d) 0.005 M.

A more reliable procedure involves the determination of true polarization curves near the corrosion potential by the simultaneous measurement of applied current and potential. This method, although laborious, can be done, but it is little used. Rapid polarization methods may be used (Fig. 5.12). The inhibitors in acid solutions may affect corrosion reactions as follows:

(i) The adsorbed inhibitor may form a diffusion barrier through the development of a film on the metal surface, which acts as a barrier to diffusion of ions or molecules to and from the metal surface and, hence, retards corrosion reactions. Some examples of this type of inhibitor are agar agar, gelatin, and polysaccharides. The films formed by these inhibitors give rise to resistance polarization and concentration polarization affecting both cathodic and anodic reactions. A similar phenomenon occurs when the inhibitor forms a polymeric film from acetylenic (53) and sulfoxide (55) compounds.

(ii) The adsorbed inhibitor prevents metal atoms from participation in both the anodic and cathodic reactions without affecting the basic mechanism of the corrosion reactions. In this case, the Tafel slopes remain unaffected. This type of corrosion inhibition has been observed in the case of 2,6-dimethyl quinoline (10), β-naphthoquinoline (56), and aliphatic sulfides (24).

   The anodic and cathodic processes might be inhibited to different extents (10). The anodic dissolution of metal can occur at dislocations in the metal surface where metal atoms are held together less firmly with neighbors. These weak points are in small proportion in comparison to the total metal surface. The cathodic hydrogen evolution process occurs on most of the metal surface area. The adsorption of the inhibitors at low surface coverage occurs at anodic sites, while at high surface coverage the adsorption of the inhibitor occurs at both anodic and cathodic sites, thus inhibiting both anodic and cathodic corrosion reactions.

(iii) The corrosion reactions involve the adsorbed intermediate species with surface metal atoms, such as adsorbed hydrogen atoms in the hydrogen evolution process and adsorbed FeOH species in the anodic dissolution of iron metal (57). The added corrosion inhibitors interfere with the formation of aforementioned intermediates and form intermediates containing the corrosion inhibitor, which act as catalyst in the sense that they remain unchanged. This type of participation of the inhibitor results in the change of Tafel slope. The formation and the participation of the adsorbed inhibitor species such as (Fe-Inh) or (FeOH·I) have been implicated in the case of inhibitors such as halides, aniline, benzoate, and furoate ion in the anodic dissolution of iron with an increase in Tafel slope for the reactions.

The presence of hydrogen sulfide in acid solutions increases the rate of dissolution of iron and decreases the Tafel slope (9). The anodic dissolution through the $Fe(SH^-)_{ads}$ path is easier than through the $Fe(OH)_{ads}$ path. The corrosion of iron due to hydrogen sulfide is complex, since increased corrosion has been observed in the presence of thioureas and corrosion inhibition in the presence of amines, quaternary ammonium cations due to synergistic enhancement of the interaction of the bisulfide anion and the inhibitor.

Corrosion inhibitors may also inhibit the rate of hydrogen evolution on metal by their effect on the mechanism of the corrosion reaction, as indicated by an increase in the Tafel slopes of the cathodic branch of the polarization curves. This effect is clearly seen in the case of phenyl thiourea (PHTU) (58) as shown in Fig. 5.12. Similar effects and behavior have been observed with acetylenic compounds (59), anilines (60), benzaldehyde derivatives (61), and pryilium compounds (61). The rate-determining step for the hydrogen evolution reaction on iron in acid solutions with pH less than 2.0 is thought to be a recombination of hydrogen atoms to form molecular hydrogen. It has been shown that the addition of anilines, benzaldehydes, and pyrilium compounds to acid solutions (HCl) inhibit the discharge of hydrogen ions to form adsorbed hydrogen atoms on iron and hence arrests the hydrogen evolution process.

In the case of inhibitors such as amines and sulfoxides, protonation of the inhibitors will occur and may accelerate the cathodic hydrogen evolution reaction by the participation of the protonated species. The discharge of protonated species to produce an adsorbed hydrogen atom at the metal surface occurs more readily than the discharge of a hydrogen ion in the case of zinc, since zinc has a higher overvoltage than iron.

The adsorption of inhibitor ions on metal surfaces changes the character of the Helmholtz double layer at the metal–solution interface, resulting in changes in the rates of electrochemical reactions (62). The adsorption of cations such as quaternary ammonium ions or protonated amines confers more positive potential at the surface of the metal, where the ions in the solution tend to approach. Thus, the acquired positive potential negates the discharge of positively charged hydrogen ions. The theoretically calculated corrosion inhibition values for inhibition of hydrogen ion discharge of iron in acidic solutions due to the added pyridine inhibitor were in fair agreement with the experimental values.

In the case of adsorption of anions on the metal surface, the potential will be more negative on the metal side of the Helmholtz double layer, which in turn will accelerate the rate of discharge of hydrogen ions as observed with benzoate (63) and sulfosalicylate anions (57).

5.9   CORROSION INHIBITION IN NEUTRAL SOLUTIONS

The corrosion of metals in near-neutral solutions saturated with air (oxygen) involves the metal covered with films of oxides, hydroxides, and the reduction of dissolved oxygen, as this cathodic reaction shows:

$$\text{Cathodic reaction}: \quad O_2 + 2H_2O + 4e^- \rightarrow 4OH^-.$$

Unlike in acid solutions, the metal is covered with an oxide film. The metal may also be covered with films of hydroxides or other insoluble salts. Thus, the inhibitors used in neutral solutions are different from those in acid solutions, with the exception of compounds such as gelatin, agar agar, and dextrin, which are effective both in acid solutions as well as near-neutral solutions. The inhibition in neutral solutions can be achieved by stabilization of the protective films. The inhibitor may form a surface film of insoluble salt by precipitation or reaction to form insoluble hydroxides of zinc, magnesium, manganese, and nickel at cathodic sites by combining with the hydroxyl ion produced by the reduction of oxygen; soluble calcium may precipitate as calcium carbonate by combining with dissolved carbon dioxide and formation of thin phosphate films of zinc or calcium. These insoluble salt films prevent the diffusion of dissolved oxygen to the metal surface and act as cathodic inhibitors, and oxygen reduction cannot occur on the film surface.

Some inhibitors such as chromate, nitrite, benzoate, silicate, phosphate, and borate, anions of weak acids, render stability to the oxide film on the metal by forming thin protective passive films. The passivating oxide films resist the diffusion of metal ions and hence inhibit the anodic dissolution of the metal. Thus, these inhibitive ions are known as anodic inhibitors. These anodic inhibitors are used to arrest the corrosion of iron, zinc, aluminum, copper, and their alloys in near-neutral media.

The three important properties of the oxide film that determine the protective nature of the oxide film are (i) the Flade potential, (ii) the breakdown potential, and (iii) the corrosion current. The negative potential limit of the stability of the oxide film defines the Flade potential; the oxide film becomes unstable and nonprotective at potentials more negative than Flade potentials. The critical breakdown potential is the positive potential limit of stability of the oxide film. At the breakdown potential and more positive potentials, the oxide film becomes unstable and results in localized corrosion such as pitting corrosion in the presence of chloride ions. The corrosion current due to the diffusion of metal ions through the passive film and the dissolution of metal ions at the oxide–solution interface should be small to allow for protection of the metal from corrosion.

Flade potential, breakdown potential, and corrosion current are affected by pH, the anion, and the oxidizing agent present in the solution. The inhibition of corrosion is possible when the appropriate combination of anions, pH, and oxidizing agent is present, so that the potential lies between the Flade potential and breakdown potential, and the corrosion current is small.

5.10   CORROSION INHIBITION OF IRON: INTERPHASE AND INTRAPHASE INHIBITION

The corrosion of iron can be inhibited by the anions weak acids (64, 65). But anions of strong acids oppose the effect of inhibitive anions of weak acids and instigate the breakdown of the passive protective oxide film. Some such aggressive anions are halides, sulfate, and nitrate. It has been observed that most of the anions exhibit dual characteristics; namely, inhibitive and aggressive behavior toward iron and the balance between inhibitive and aggressive behavior depends upon (i) anion concentration, (ii) pH of the medium, (iii) dissolved oxygen concentration, (iv) nature of the metal surface, and (v) temperature.

The corrosion inhibition of iron can be achieved when the anion concentration exceeds a critical value (64–66). The inhibitive anions become aggressive and cause breakdown of the oxide film when their concentrations are below the critical value. It has been found that the effective inhibitive anions have low critical concentrations for inhibition. The following order of decreasing inhibition toward steel has been observed (64): azide, ferricyanide, nitrite, chromate, benzoate, ferrocyanide, phosphate, tellurate, hydroxide, carbonate, chlorate, o-chlorobenzoate, bicarbonate, fluoride, nitrate, formate.

The inhibitive anions can inhibit corrosion of iron at pH values higher than the critical values. In Table 5.17 are the approximate critical pH values for some anions in inhibition of iron in 0.1 M solution of the anion.

The critical pH value depends on the concentration of the inhibitive anion. In the case of nitrite and azelate anions, the critical pH for inhibition decreases with an increase in anion concentration. In the case of benzoate inhibitor, increase in the concentration of the anion resulted in increase in the critical pH value.

Corrosion inhibition by inhibitive anions requires oxidizing agents, such as dissolved oxygen, usually supplied by aeration. The critical oxygen concentration

TABLE 5.17   Critical pH Values for Corrosion Inhibition

Anion	Critical pH	Reference
Chromate	1.0	62
Azelate	4.5	63
Nitrite	5.0–5.5	64
Benzoate	6.0	65, 66
Phosphate	7.2	67
Hydroxide	~12.0	60

for the inhibition of iron in 0.1 M solution of sodium benzoate at pH 7 has been found to be ~0.3 ppm (73). The passive film becomes unstable when the oxygen concentration is less than 0.3 ppm. Critical oxygen concentration increases for inhibition (74) of iron in 0.1 M sodium benzoate solution of pH < 7 in contrast to the system at pH of 7.0. The nature of the anion has a profound influence on the critical oxygen concentration required for corrosion inhibition. Powerful oxidizing anions such as chromate, nitrite, and pertechnate may not require the presence of dissolved oxygen to function as inhibitors. The critical oxygen concentrations required for some effective inhibitive nonoxidizing anions are low. By increasing the dissolved oxygen concentration above the value in air-saturated solutions, the inhibition of iron corrosion is achieved even in the presence of chloride ions (74). Aeration and rapid stirring help the transport of oxygen to the metal surface and inhibition at lower critical anion concentrations (70). Addition of an oxidizing agent such as hydrogen peroxide lowers the pH and the critical concentrations of sodium benzoate and sodium azelate needed for corrosion inhibition of steel (75).

The critical concentration of inhibitive anions necessary for corrosion inhibition increases with the concentration of aggressive anions present in the solution (76–80). Brasher and Mercer (76–80) have advanced a relationship between the maximum concentration of aggressive anion C_{agg}, which gives full protection by a certain concentration of inhibitive anion C_{inh} as

$$\log C_{inh} = n \log C_{agg} + K,$$

where K is a constant dependent on the nature of inhibitive and aggressive anions and n is approximately the ratio of the valency of the inhibitive anion to the valency of the aggressive ion. This relationship indicates competition between the action of the inhibitive anion and aggressive anion. It is obvious that more aggressive anions translate into tolerance of smaller concentrations of aggressive anion by the inhibitive anion. The order of tolerance of aggressive ions by the inhibitive anion very nearly parallels the order of aggressiveness of the anions in terms of breakdown of the passive film on iron in aerated solutions.

The critical concentration of an anion such as benzoate required to protect a grit-blasted steel surface was about 100 times the amount required to protect an abraded steel (76). This type of behavior was not observed with chromate (77) and nitrite (78) inhibitors. The time of exposure of the iron sample to air before immersion in a solution of anionic inhibitor was found to be beneficial with respect to inhibition by the anion (65, 73).

The critical concentrations of inhibitive anions such as benzoate (76), chromate (77), and nitrite (78) needed for the corrosion inhibition of steel were found to increase with an increase in temperature.

5.11   PASSIVE OXIDE FILMS

Iron exposed to air becomes covered with an oxide consisting of an anhydrous cubic oxide, Fe_3O_4, γ-Fe_2O_3, or an intermediate compound. The thickness of the air-formed film depends on the metal treatment procedure shown below (81):

pickled iron 40 Å; pickled mild steel 36.5 Å; hydrogen-reduced iron 26 Å.

The value of 26 Å agrees well with the value of 29 Å obtained with pure iron, which was reduced with hydrogen, and the loss in weight determined gravimetrically. Other attempts by optical method and vacuum fusion method yielded values of 20–40 Å and 37 Å, respectively, showing the dependence of the values on the surface conditions.

The air-formed oxide film becomes reinforced and thickens to different extents on exposure to inhibitor solutions as shown by the data (82) given in Table 5.18.

The data show that the air-formed oxide film thickens on exposure to the inhibitor solutions to an extent of 20–120%, depending upon the inhibitor and the surface condition of the metal. It is also interesting to note that very little of the inhibitor is

TABLE 5.18   Film Thickness of Oxide Film

Sample	Inhibitor	pH	Thickness (Å)	Thickening (%)
Pickled	0.1 N sodium benzoate	7.1	56	40
iron	0.1 N sodium acetate	8.4	56	40
	0.1 N sodium borate	9.2	54	35
	0.1 N sodium carbonate	11.0	55	37
	0.1 N sodium hydroxide	12.5	63.5	59
Pickled	0.1 N sodium benzoate	7.1	52	40
mild steel	0.1 N sodium borate	9.2	43	18
	0.1 N sodium carbonate	11.0	45	23
	0.1 N sodium hydroxide	12.5	57	56
Hydrogen-reduced	0.1 N sodium benzoate	7.1	38	45
iron	0.1 N sodium borate	9.2	39.5	51
	0.1 N sodium carbonate	11.5	51	95
	0.1 N sodium hydroxide	12.5	58	120

incorporated in the oxide film; for example, about 1% of benzoate ions in sodium benzoate solutions.

Iron freed from the air-formed oxide film on immersion in 0.1 N sodium hydroxide solution containing dissolved oxygen becomes covered with a thin film of γ-Fe_2O_3, which is responsible for passivity. This protective film is formed by the reaction of the dissolved oxygen in solution and the iron. This mechanism is also confirmed by the observation that corrosion occurs in a deaerated sodium hydroxide solution. The film formed by the anodic discharge of hydroxyl ions is composed of species having the cubic Fe_3O_4 or γ-Fe_2O_3, which is indistinguishable from the air-formed film. The mechanism of inhibition in sodium hydroxide solution proceeds as follows.

Weak areas in the air-formed film are repaired by the electrochemically formed ferrous hydroxide, which later reacts with oxygen, giving rise to cubic Fe_3O_4 or γ-Fe_2O_3. Further repair may occur by the direct electrochemical formation of ferric oxide and thus lead to thickening of the protective film (83).

Protective films formed on iron in 0.1 N solutions of disodium hydrogen phosphate, sodium phosphate, sodium borate, and sodium carbonate in the presence of air but in the absence of air-formed film were studied by electron diffraction (84). All the films contained Fe_2O_3, γ-Fe_2O_3, or an intermediate compound. In the case of the inhibitor disodium hydrogen phosphate, large particles of $FePO_4 \cdot 2H_2O$ were present in addition to cubic Fe_3O_4, γ-Fe_2O_3. Trisodium phosphate gave rise to films containing a small amount of ferric phosphate. No second phase particles were present in the case of sodium carbonate- and sodium borate-treated samples. Sodium carbonate, sodium borate, and trisodium phosphate are salts of weak acid, and solutions of these salts are alkaline; the ferrous compounds formed at anodic sites hydrolyze to form ferrous hydroxide, which eventually is oxidized to an oxide of cubic structure. In the case of disodium hydrogen phosphate of pH 8.4, the ferrous phosphate formed at anodic sites will not hydrolyze but is oxidized to ferric phosphate and deposited on the ferric oxide.

The corrosion inhibition of iron by potassium chromate has been studied in detail and three theories of inhibition have been advanced. According to precipitation theory of Hoar and Evans, the inhibition of corrosion of iron is due to plugging of weak points in the primary air-formed film by a mixture of ferric and chromic oxides. According to Uhlig, the corrosion inhibition of iron by chromate is due to adsorption of a unimolecular layer of chromate ions in such a manner that they satisfy the secondary valency forces of iron ions without disruption of the metal lattice. The oxide film theory of Evans states that the main function of chromate is to repair the discontinuities in the oxide film along with the thickening of the oxide film. The oxide films formed on iron in the presence of chromic acid and potassium chromate were studied by electron diffraction and the view is advanced that iron ions are oxidized to γ-ferric oxide, which forms a thin continuous film on iron and prevents corrosion (85).

Detailed surface analysis of mild steel samples exposed to chromate, molybdate, and tungstate inhibitors using X-ray photoelectron spectroscopy (XPS), Auger electron spectroscopy (AES), and electron microprobe analysis showed (86, 87)

TABLE 5.19　Electron Microprobe Data on Inhibitor Films

Sample	Conditions	Element	Concentration (mg/m^2)
1	Deaerated molybdate solution for 2 h	Mo	0.75
2	Aerated molybdate solution		
	(a) 2 h	Mo	0.28
	(b) 18 h	Mo	0.32
3	Deaerated sodium tungstate solution for 2 h	W	1.07
4	Aerated sodium tungstate solution for 2 h	W	1.2
5	Deaerated potassium chromate solution for 2 h	Cr	5.2
6	Aerated potassium chromate solution for 2 h	Cr	5.0

some interesting features. Electron microprobe analysis of AISI 1010 steel coupons exposed to 0.01 M solutions of chromate, molybdate, and tungstate gave the data shown in Table 5.19.

The microprobe data show the elements present on the surface of the steel in the order: chromium > tungsten > molybdenum. It is possible to calculate the surface concentrations to be expected based on the simple assumption that the metal surface is covered by a single layer of chromate, molybdate, or tungstate inhibitors in a close-packed array. Based on such a calculation, close-packed oxygen atoms of the inhibitor cover an area of $\sim 23 \text{ Å}^2$. On this basis, the amounts of chromium, tungsten, and molybdenum in the films corresponded to several monolayers of chromium, a monolayer of tungsten, and a fraction of a monolayer of molybdenum. This work is unique since this is the first time that the electron microprobe has been shown to be capable of surface analysis.

The surface analysis data obtained by X-ray photoelectron spectroscopy and Auger electron spectroscopy are given in Table 5.20.

The elemental concentrations are in the order: chromium > tungsten > molybdenum, in complete agreement with electron microprobe data. The XPS

TABLE 5.20　Surface Analysis Data (wt.%)

Element	XPS ($\pm 10\%$)	AES ($\pm 20\%$)
Mo	1.7	6.7
	5.8	
	7.2	4.9
W	23.6	27.1
	21.7	22.3
Cr	31.0	26.7
	30.8	33.4

data showed the presence of molybdenum and tungsten in hexavalent state, while chromium is in trivalent state. The species detected were $FeO \cdot OH$ with MoO_3; Fe_2O_3 or Fe_3O_4 with WO_3; and iron oxides with $CrO \cdot OH$. The depth profiles showed chromium layers of ~ 12 nm, 10 nm layer of iron and tungsten, and 3 nm of molybdenum layer with 6 nm layer of iron oxide.

Passive film formation with the species being γ-Fe_2O_3 has been confirmed by electron diffraction studies of active, passive, and transpassive oxide films formed on iron samples and optical studies of passive film of 15–20 Å thickness (88, 89).

Electron diffraction studies (90) of passive films formed on iron in sodium nitrite showed the formation of γ-Fe_2O_3 with a small amount of γ-$Fe_2O_3 \cdot H_2O$.

The structure and mechanism of formation of oxide films formed on single crystals of iron in the presence of sodium nitrite gave evidence (91) for the presence of γ-Fe_2O_3 along with a small amount of magnetite, Fe_3O_4. Trace amounts of $FeO \cdot OH$ were also found in samples exposed to dilute solutions.

The anodic oxidation of iron in a borate–boric acid buffer solution of pH 8.4 resulted in the formation of an oxide film of 10–30 Å thickness. Electron diffraction studies (92) of the passive film showed it to contain an inner layer of "Fe_3O_4" and an outer layer of "γ-Fe_2O_3."

The kinetics of growth of oxide films on steel has been studied and documented in the literature (93–97). When mild steel is passivated in inhibitive solutions, the thickening of the oxide film on the surface follows a logarithmic rate law, as it is in dry air, and the constants of the logarithmic growth are of the same order of magnitude in all the inhibitive solutions, which suggests the operation of a similar oxide film growth mechanism.

The constants for oxide film growth in various inhibitive solutions show minor variations, and the values for growth in aerated solutions were found to be slightly greater than the values for growth in dry air. The data given in Table 5.21 show the ratio of oxide film growth rate constant in solution, k_s to that of the rate constant in air, k_a. The rate expression used to describe the film growth is $k = k_o \ln(t + 1) + c$.

TABLE 5.21 Data on the Ratio of Oxide Film Growth Rate Constants (97)

Medium	pH	k_s/k_a
Azelate	4.8	0.9
Nitrite	7.1	1.25
Benzoate	7.9	1.55
	7.1	1.30
Phosphate	9.1	1.35
Borate	9.2	1.30
Carbonate	11.0	1.30
Hydroxide	12.5	1.55
	13.0	1.30

The inhibition of corrosion of iron by inhibitive ions depends upon the effects of inhibitive ions on the oxide film present on iron and the nature of the inhibitive ion and the experimental conditions.

5.12   INTERACTION OF ANIONS WITH OXIDE FILMS

Corrosion inhibition of iron by oxyanions such as chromates and phosphates has been studied, and the results show the presence of chromic oxide (86) and ferric phosphate (84) on the surface of ferric oxide. Tracer studies have also confirmed the earlier work on the adsorption of inhibitive anions chromate (97) and phosphate (98) on the oxide film. Other inhibitive anions such as benzoate (99), pertechnate (100), and azelate (101) are less strongly adsorbed on the oxide film. Phosphate ions were firmly bound to oxide films on mild steel (102) after immersion in 0.1 M sodium phosphate over the pH range 7–13. The major part of the phosphate was found to be distributed uniformly over the steel surface with some locally increased uptake to a variable extent at some points and areas of the metal surface. At pH 8 approximately a monolayer of phosphate ions was present on the oxide. The uptake of phosphate ions increased linearly with the logarithm of time of immersion. Increase in pH resulted in decrease in the initial uptake of phosphate ions. The results were interpreted in terms of ion exchange of phosphate ions initially with the surface layer of air-formed oxide film and continuing during the thickening of the oxide film in the solution.

The uptake of inhibitive anions into the passive iron oxide film varies and some approximate amounts with various inhibitive anions are shown in Table 5.22.

The ion exchange type of mechanism proposed for the adsorption of inhibitive anions on the surface of passive films present on iron (103, 104) can be surmised by the presence of $FeO \cdot OH$ on iron, with the labile hydroxyl group being replaced by an inhibitive anion. The adsorption of anions such as phosphate (102) and azelate (101) may result in separate phases positioned in the oxide film, especially at low pH values and unfavorable conditions for inhibition, such as the presence of aggressive anions and thin oxide film.

The adsorption of inhibitive anions on oxide films is the fundamental basis of pore plugging theory enunciated by Evans (105). According to this theory the inhibitive

TABLE 5.22   Thickness of Inhibitor Layers

Inhibitive Anion[a]	Degree of Uptake into Passive Oxide
Phosphate	Monolayer
Benzoate	0.06–0.1 of monolayer
Pertechnate	0.05 of monolayer
Chromate	12.0 nm layer
Tungstate	10.0 nm layer of Fe and W
Molybdate	3 nm layer of Mo and 6 nm layer of FeO

[a] pH = 8.

anion takes part in repairing weak points or pores in the oxide film. Corrosion occurs at weak points in the film, giving rise to iron cations, and these ions react with the inhibitive anions to form insoluble precipitate or a separate phase, which plugs the gap or weak spots in the passive oxide film. The insoluble precipitate may contain an inhibitive anion such as phosphate (84) or azelate (101) or chromic oxide from chromate inhibitor (93). The pore plugging method of protection by inhibitive anions is quite effective when the pH in the pore region is high, as in the case of phosphate, borate, silicate, and carbonate ions, which prevent the drop in pH at anodic sites due to their buffering properties. The dissolved oxygen or oxidizing anions such as chromate and nitrite facilitate the oxidation of ferrous salts to insoluble ferric salts and can cause corrosion inhibition.

Pore plugging in oxide films does occur in the presence of some inhibitive anions, such as chromates and nitrite. The inhibitive properties of azide and pertechnate cannot be explained based on the formation of insoluble salts and buffering action, since these anions do not form insoluble salts and do not exert buffering action. Oxidizing power is not a necessary condition for the inhibitive action of anions since the well-known permanganate is a poor inhibitor, although it readily oxidizes ferrous salts into ferric salts (106). The extent of inhibition is not a direct function of the degree of incorporation of the inhibitive anions in the protective oxide film (102). It is useful to note that the inhibitive action of anions is operative as long as the metal sample covered with the oxide film is in contact with the solution containing the inhibitive anion. Studies on the uptake of inhibitive and aggressive anions on the surface of iron have shown that the adsorption of the ions is a competitive phenomenon.

The inhibitive effect of anions is operative as long as the metal sample with the oxide film is in contact with the solution containing the inhibitive anion. These observations suggest the transient nature of inhibition by anions due to their temporary presence in the oxide film on the iron surface. The transient nature of inhibition may be due to a transient adsorptive interaction to yield an unstable species:

$$OFe \cdot OH + An^- \rightarrow FeO \cdot OH - An^-$$

$$OFe - O - FeO + An^- \rightarrow OFe - O(An^-) - Fe - O.$$

These adsorptives are of limited lifetimes, yielding a product consisting of iron oxide, a stable product.

It is a well-recognized fact that inhibitive anions help in the repair of weak points, pores, or damaged parts of the oxide film on iron through promotion of the formation of a passive form of iron oxide at these areas. This mechanism of action of inhibitive anions has been advanced by Stern (107). According to this mechanism, a passivating inhibitor functions by producing local-action current, which anodically polarizes a metal into the passive potential region and thereby provides the means for obtaining a noble mixed potential. This mechanism is independent of the cause of the passivity, whether or not it is primarily caused by oxide in the presence of inhibitive anions due to an increase in the rate of cathodic process, which arises from either the reduction of the oxidizing anion or acceleration of oxygen reduction reaction as follows

$$CrO_4^{2-} \rightarrow Cr_2O_3$$
$$2H^+ + O_2 + 4e^- \rightarrow 2OH^-$$

A greater anodic current is available to exceed the required critical current density for passivation. The inhibitive anions might facilitate the passive oxide formation by reducing the magnitude of critical current density or by making the Flade potential more negative. Flade potential is the potential at which the onset of passive region occurs. Later work showed that the inhibitive anions affect primarily the anodic process. Thus, in the case of solutions of oxidizing inhibitive anions such as chromate (108), nitrite (108), and pertechnate (109), reduction of the dissolved oxygen is the main cathodic reaction. There is some evidence that some anions can increase the rate of reduction of oxygen, although this effect is not significant. Anodic polarization studies (107, 110) have shown that the critical current density for passivation is significantly less in the presence of inhibitive anions than aggressive anions. It has also been shown in the case of a number of inhibitive anions that the critical current densities for passivation increase in the same order as the inhibition efficiencies decrease. The critical passivation current densities in the deaerated 0.1 M solution increase in the order:

nitrite < hydroxide < chromate < borate < phosphate < carbonate < benzoate < bicarbonate < nitrate.

The critical passivation potential (NHE scale) in solutions of anions where passivation occurs relatively easily was found to vary with pH in the range of 6–12 according to the equation

$$E_p = +0.09 - 0.06 \, pH \, V.$$

Critical passivation potential, critical passivation current density, and leakage currents through the passive films were found to depend on the pH and the nature of the anion present. In solutions of inhibitive anions, the current density for passivation generally increases as the pH decreases. In the case of benzoate solutions, dissolved oxygen reduces the critical current density for passivation to a significant degree. This is borne out by the electrode potential shift of 0.67 V for iron in aerated solution to that in deaerated solution. This is also corroborated by the corrosion potential of -0.8 V at zero current in deaerated solutions and -0.15 V in aerated solutions, with the difference between the two values being 0.65 V in good accord with the noble potential shift of 0.67 V. The dissolved oxygen has very little effect on the passivation of iron in carbonate solutions (110). The anodic passivation of iron in carbonate solutions of pH 11.3 consists of two well-defined peaks. The first oxidation peak at -0.58 V is due to the probable formation of magnetite.

The iron surface is also covered by a film of ferrous carbonate. The second peak at -0.22 V is attributed to the formation of hydrated γ-ferric oxide from ferrous carbonate film. The degree of passivation has been found to be dependent on the extent of adsorption of the inhibitive anion on the iron surface. An example of this is provided by the studies on adsorption of phosphate and o-iodohippurate on an iron surface at different potentials during anodic passivation in near-neutral solutions. The degree of adsorption of the phosphate and iodohippurate is dependent on the potential in the active potential region (111). These studies indicate the greater the degree of adsorption of the anion on the iron surface the smaller the critical current density for passivation.

Brasher made an important observation on the inhibition of an actively corroding steel by immersion in a solution containing an inhibitive anion. The corroding steel sample in a solution of an aggressive anion can be inhibited by the addition of a nonoxidizing inhibitive anion as long as the potential has not become more negative than "critical potential" for inhibition. The critical potential for inhibition depends on (i) the nature of the anion and (ii) the relative concentrations of inhibitive and aggressive anions in solution. The critical potential will become more positive as the concentration of aggressive anion increases (113). The critical potential for inhibition is related to the effect of potential on the adsorption of anions and is thought to be a potential at which adsorption of the inhibitive anion on the active corroding region has the appropriate value necessary to reduce the metal dissolution rate to such a level as to enable formation of oxide film. The potentials for inhibition in benzoate and phosphate are -0.28 and -0.43 V, respectively, and these values are close to critical passivation potentials for iron, which supports the notion of formation of passive oxide at corrosion sites.

The passive oxide film on iron will be stable and protective when the rate of formation of the oxide film is greater than the rate of dissolution of the oxide film. At potentials more positive than the Flade potential, dissolution of the oxide film occurs by the passage of ferric ions from the oxide film on the surface into solution (114). The rate of passage of ferric ions from the oxide film into solution is smaller in solutions of chromate (115) than sulfate. The rates of dissolution of the oxide film are in the order: phthalate > sulfate > chromate. The rates of dissolution in solutions of these anions decrease with increase in pH of the solutions. The dissolution rate of the oxide film also decreases with increase in thickness of the oxide film. Under unfavorable conditions such as low pH, low inhibitive nature of anions, and lower thickness of the passive oxide film, considerable ferric ion dissolution current can lower the potential to the regime of the Flade potential. Thus, the rate of oxide dissolution increases due to the reductive dissolution, resulting in ferrous ions in the solution. The origin of formation of ferrous ions can be due to the reduction of ferric ion in the form of γ-ferric oxide:

$$\gamma - Fe_2O_3 \rightarrow 2Fe^{3+}$$
$$Fe \rightarrow Fe^{2+} + 2e^-$$
$$2Fe^{3+} + 2e^- \rightarrow 2Fe^{2+}.$$

The rate of reductive dissolution of the oxide film is greater in solutions of aggressive anions than in the presence of inhibitive anions. The rate of dissolution also increases with a decrease in the amount of dissolved oxygen. In the pH range 6–9, the oxidation by dissolved oxygen of the initial breakdown products such as ferrous ions to form insoluble species, which may delay film breakdown, was found to be more rapid in inhibitive than in corrosive solutions. The inhibitive anion promotes the oxidation of any ferrous ion on the surface of γ-Fe_2O_3 film and thus retards the dissolution. These effects of anions on the reduced reductive dissolution of the oxide film result in the displacement of the Flade potential to more negative values. The effects of many anions at different pH values on the Flade potential have been documented in the literature (116).

The passive oxide film consists of an inner magnetite layer covered by an outer γ-Fe_2O_3 layer. The reductive dissolution of the outer γ-Fe_2O_3 layer of the oxide film gives rise to the exposure of the inner magnetite layer. The inner magnetite layer dissolves in solutions of pH less than 4. The magnetite layer is stable in near-neutral solutions depending upon the concentration and nature of the anion. The magnetite layer has been found to be stable in solutions of inhibitive anions such as benzo-ate (117), carbonate (110), hydroxide (110), and borate (92). The nature and stability of the magnetite layer is a decisive factor in the corrosion inhibition of iron when it is in conjunction with other metals such as aluminum, zinc, or cadmium.

The inhibitive anions (i) retard the dissolution of protective γ-Fe_2O_3 and magnetite layers and (ii) facilitate the formation of passivating oxide film on the iron surface.

The inhibitive anions have to overcome the effects of aggressive anions, which accelerate dissolution and breakdown of the passive oxide film on the iron surface. The relationship between the highest tolerable molar concentration of aggressive anion in the presence of a given concentration of inhibitive anion is of the form (104):

$$\log C_{inh} = n \log C_{agg} + K,$$

where C_{agg} is the highest tolerated molar concentration of the aggressive anion in the presence of a molar concentration C_{inh} of the inhibitive anion. Some data on the order of tolerated concentrations of aggressive anions by some inhibitive anions and the values of n in the equation are given in Table 5.23.

TABLE 5.23 Order of Tolerated Concentrations of Inhibitive Anions (104)

Inhibitive Anion	Order of Tolerated Concentrations of Aggressive Anions	Values of n		
		Nitrate	Chloride	Sulfate
Benzoate	$NO_3^- > Cl^- > SO_4^{2-}$	1.04	0.82	0.59
Chromate	$NO_3^- > SO_4^{2-} > Cl^-$	1.20	0.68	0.67
Nitrite	$NO_3^- > Cl^- > SO_4^{2-}$	1.01	1.00	0.61

The values of n are related to the ratio of the valencies of the pairs of the anions. Thus, a monovalent inhibitive anion with a monovalent aggressive anion gives a value of $n \sim 0.8$–1.0, and a monovalent inhibitive anion with a divalent aggressive anion gives $n = 0.5$–0.6.

The oxide film on iron is considered to act as an anion exchanger. Some metal oxides such as alumina or silica behave as inorganic ion exchangers, a concept that may be applied to the anodic oxide film on iron. In the absence of inhibiting anions, aggressive ions penetrate the oxide film and cause the breakdown of the protective film. On the other hand, inhibitive anions occupy ion-exchange sites of the oxide film and thus prevent the aggressive ions from the disruption of the oxide film. Thus, one can envision competitive adsorption on the oxide film between inhibitive and aggressive anions. The effects of valencies of the competing inhibitive and aggressive anions are consistent with the total charge due to anion uptake remaining constant. It is also observed that the iron metal protected in inhibitive anion-bearing solution begins to corrode upon the addition of aggressive anion or on transfer into distilled water. All of these observations lead to the conclusion that the inhibitive anions compete with the aggressive anions in adsorption on the oxide surface and reduce the surface concentration of aggressive anions below a critical value.

The inhibitive anions are, in general, anions of weak acids, and aggressive anions are anions of strong acids. Solutions of inhibitive anions are generally alkaline and resist displacement to more acid values. These factors contribute to the stability and repair of the oxide film. The important difference between inhibitive anions and aggressive anions lies in the difference in the rates of oxide dissolution reaction on the iron surface. In the case of oxyanions MO_4^{n-} the difference between inhibitive and noninhibitive anions lies in the difference in the polarity of the M^+–O^- bond, leading to varied interactions in the Helmholtz double layer and affecting the passage of metal ions into solution (118). Anions of weak acids bond more strongly with the iron surface than anions of strong acids. The mechanism of dissolution of metal ions from the iron oxide surface is not clear and it probably occurs in the form of $Fe(OH)^{2+}$ ions into solution. It is thought that coordinate complexes formed by the inhibitive anions with the iron are less soluble than the ionically bonded complexes with aggressive anions. It is also possible that electron transfer to the ferric ion by inhibitive anions will confer stability to ferric ion from reduction to the ferrous state and consequent reductive dissolution of the oxide film.

Carboxylate anions were found to be inhibitors and the corrosion inhibition effectiveness was maximal, with the molecule containing 8–10 carbon atoms (119). This observation can be rationalized on the basis of the increased tendency for adsorption on the surface and decreased solubility of the ferric–carboxylate complex.

The mechanism of corrosion inhibition of iron in near-neutral solutions by inhibitive anions consists of (i) reduction of the dissolution rate of the passivating oxide film; (ii) repair of the oxide film by reformation of the oxide film; (iii) repair of the oxide film by plugging pores in the oxide film with insoluble compounds; and (iv) prevention of the adsorption of the aggressive anions on the iron surface. The important step consists of reduction of the dissolution rate of the passivating oxide film. The inhibitive anions probably form a complex with ferric ion in the passive oxide and render the complex less soluble than the aquo, hydroxo, or other anionic species of iron. Inhibitive anions retard the reductive dissolution of the ferric oxide film by reoxidation of ferrous ions formed in the oxide film. The plugging of pores in the oxide film helps to extend the range in which inhibition is achieved. The retarding of adsorption of aggressive anions is made possible by competitive adsorption of inhibitive anions.

Under favorable pH values, oxidizing power, and aggressive anion concentration, reduction of the dissolution rate of the oxide film will be predominant, and the oxide film will be stable. Under unfavorable conditions for corrosion inhibition, localized breakdown of the passive film might occur, and pore plugging and reformation of the passive oxide film play important roles.

REFERENCES

1. G Trabanelli, V Carassiti, *Advances in Corrosion Science and Technology*, Plenum Press, New York, London, 1970, p. 147.

2. (a) P Lacombe, *2nd European Symposium on Corrosion Inhibitors*, Ferrara, 1965, University of Ferrara, 1966, p. 517;. (b) E Gileadi, *2nd European Symposium on Corrosion Inhibitors*, Ferrara, 1965, University of Ferrara, 1966, p. 543.

3. F Wormwell, JGN Thomas, *Surface Phenomena of Metals*, Society of Chemical Industry, London, Monograph no. 28, 1968, p. 365.

4. JGN Thomas, *Werks Korros* **19**:957 (1968).

5. RR Annand, RM Hurd, N Hackerman, *J Electrochem Soc* **112**:138 (1965).

6. E Gileadi, *J Electroanal Chem* **11**:137 (1966).

7. Optical studies of adsorbed layers at interfaces, *Symposium of the Faraday Society*, Vol. 4, 1970.

8. GL Zuchini, F Zucchi, G Trabanelli, *3rd European Symposium on Corrosion Inhibitors*, Ferrara, 1970, University of Ferrara, 1971, p. 577.

9. A Makrides, N Hackerman, *Ind Eng Chem* **47**:773 (1955).

10. TP Hoar, RD Holliday, *J Appl Chem* **3**:502 (1953).

11. FM Donahue, A Akiyama, K Nobe, *J Electrochem Soc* **114**:1006 (1967).

12. BB Damaskin, OA Petrii, VV Batrakov, *Adsorption of Organic Compounds on Electrodes*, Plenum Press, New York, 1971.

13. AC Makrides, N Hackerman, *Ind Eng Chem* **46**:523 (1954).

14. Z Szklarska-Smialowska, B Dus, *Corrosion* **23**:130 (1967).

15. K Wippermann, JW Schultze, R Kessel, J Penninger, *Corros Sci* **32**:205 (1991).

16. RC Ayers, N Hackerman, *J Electrochem Soc* **110**:507 (1963).

17. PF Cox, RL Every, OL Riggs, *Corrosion* **20**:299t (1964).

18. FM Donahue, K Nobe, *J Electrochem Soc* **114**:1012 (1967).

19. J Vosta, J Eliasek, *Corros Sci* **11**:223 (1971).

20. VP Grigoryev, VV Ekilik, *Prot Met* **4**:23 (1968).

21. VP Grigoryev, V Gorbachev, *Prot Met* **6**:282 (1970).

22. VP Grigoryev, VV Kuznetsov, *Prot Met* **5**:356 (1969).

23. AA Kiyama, K Nobe, *J Electrochem Soc* **117**:999 (1970).

24. H Brandt, M Fischer, K Schwabe, *Corros Sci* **10**:631 (1970).

25. FM Donahue, K Nabe, *J Electrochem Soc* **112**:886 (1965).

26. V Carassiti, F Zucchi, G Trabanelli, *Proceedings of the 3rd European Symposium on Corrosion Inhibitors*, Sez. V(Suppl. 5), Ann. University Ferrara, NS, 1971, p. 525.

27. VS Sastri, JR Perumareddi, *Corrosion* **50**:432 (1994).

28. JO Bockris, M Gamboa-Aldeco, MS Sklarczk, *J Electroanal Chem* **339**:355 (1992).

29. DJ Barclay, *J Electroanal Chem* **19**:318 (1968).

30. G Horanyi, *Corros Sci* **46**:1741 (2004).

31. G Horanyi, E Kalman, *Corros Sci* **44**:899 (2002).

32. VS Sastri, JR Perumareddi, M Elboujdaini, *Corros Eng Sci Technol* **40**:270 (2005).

33. FM Donahue, K Nobe, *J Electrochem Soc* **114**:1012 (1967).

34. AI Altsybeeva, SZ Levin, AP Dorokhov, *3rd European Symposium on Corrosion Inhibitors*, Ferrara, University of Ferrara, 1971, p. 501.

35. G Gece, *Corros Sci* **50**:2981 (2008).

36. A Chakrabarti, *Br Corros J* **19**:124 (1984).

37. J Costa, JM Lluch, *Corros Sci* **24**:929 (1984).

38. R Sayos, M Gonzalez, J Vosta, *Corros Sci* **26**:927 (1986).

39. VS Sastri, JR Perumareddi, *Corrosion* **53**:617 (1997).

40. VS Sastri, JR Perumareddi, unpublished data.

41. RG Pearson, *Hard and Soft Acids and Bases*, Dowden, Hutchinson and Ross Inc., Stroudsberg, PA, 1973.

42. VS Sastri, JR Perumareddi, M Lashgari, M Elboujdaini, *Corrosion* **64** (8):283 (2008).

43. M Duprat, F Dabosi, *Corrosion* **37**:89 (1981).

44. OL Riggs Jr., KL Morrison, DA Brunsell, *Corrosion* **35**:356 (1979).

45. AA Al-Suhybani, IH Al-Hwaidi, *Corros Meth Mater* **41** (8):9–14 (1994).

46. KC Emzegul, E Düzgün, O Atakol, *Corros Sci* **48**:3243 (2005).

47. F Bentiss, M Bounais, B Mernari, M Traisnel, H Vezin, M Legrenee, *Appl Surf Sci* (2006).

48. VS Sastri, M Elboujdaini, JR Perumareddi, *Corrosion* **64**:657 (2008).

49. CA Mann, BE Lauer, CT Hultin, *Ind Eng Chem* **28**:159 (1936).

50. N Hackerman, ES Snavely, JS Payne, *J Electrochem Soc* **113**:677 (1966).

51. G Trabanelli, GL Zucchini, F Zucchi, V Caassiti, *Br Corros J* **4**:267 (1969).

52. H Ertel, L Horner, *2nd European Symposium on Corrosion Inhibitors*, Ferrara, University of Ferrara, 1966, p. 71.

53. GW Poling, *J Electrochem Soc* **114**:1209 (1967).

54. H Kaesche, N Hackerman, *J Electrochem Soc* **105**:191 (1958).

55. S Thibault, J Talbot, *3rd European Symposium on Corrosion Inhibitors*, Ferrara, University of Ferrara, 1971, p. 75.

56. H Kasehe, *Die Korrosion der Metalle*, Springer Verlag, Berlin, 1966, p. 159.

57. EJ Kelly, *J Electrochem Soc* **112**:124 (1965).

58. L Cavallaro, L Felloni, G Trabanelli, F Pulidori, *Electrochim Acta* **9**:485 (1964).

59. EJ Duwell, *J Electrochem Soc* **109**:1013 (1962).

60. VP Grigoryev, VV Ekilik, *Prot Met* **4**:517 (1968).

61. VP Grigoryev, OA Osipov, *3rd European Symposium on Corrosion Inhibitors*, Ferrara, 1970, University of Ferrara, 1971, p. 473.

62. LI Antropov, *Corros Sci* **I**:607 (1968).

63. ZA Iofa, *2nd European Symposium on Corrosion Inhibitors*, Ferrara, 1965, University of Ferrara, 1966, p. 93.

64. DM Brasher, *Br Corros J* **4**:122 (1969).

65. P Hersch, JB Hare, A Robertson, SM Sutherland, *J Appl Chem* **11**:251 (1965).

66. MJ Pryor, M Cohen, *J Electrochem Soc* **100**:203 (1953).

67. DM Brasher, JG Beynon, KS Rajagopalan, JGN Thomas, *Br Corros J* **5**:264 (1970).

68. JEO Mayne, EH Ramshaw, *J Appl Chem* **1**:419 (1960).

69. RA Legault, MS Walker, *Corrosion* **20**:282 (1964).

70. F Wormwell, AD Mercer, *J Appl Chem* **2**:150 (1952).

71. DE Davies, QJM Slaiman, *Corros Sci* **11**:671 (1971).

72. MJ Pryor, M Cohen, *J Electrochem Soc* **98**:263 (1951).

73. D Gilroy, JEO Mayne, *Br Corros J* **1**:102 (1965).

74. QJM Slaiman, DE Davies, *3rd European Symposium on Corrosion Inhibitors*, Ferrara, 1970, University of Ferrara, 1971, p. 739.

75. JEO Mayne, CL Page, *Br Corros J* **5**:93 (1970).

76. DM Brasher, AD Mercer, *Br Corros J* **3**:120 (1968).

77. AD Mercer, IR Jenkins, *Br Corros J* **3**:130 (1968).

78. AD Mercer, IR Jenkins, JE Rhoades-Brown, *Br Corros J* **3**:136 (1968).

79. DM Brasher, D Reichenberg, AD Mercer, *Br Corros J* **3**:136 (1968).

80. RA Legault, S Mori, HP Leckie, *Br Corros J* **3**:144 (1968).

81. P Hancock, JEO Mayne, *J Chem Soc* 4167 (1958).

82. P Hancock, JEO Mayne, *J Chem Soc* 4172 (1958).

83. JEO Mayne, JW Menter, *J Chem Soc* 99 (1954).

84. JEO Mayne, JW Menter, *J Chem Soc* 103 (1954).

85. JEO Mayne, MJ Pryor, *J Chem Soc.* 1831 (1949).

86. VS Sastri, RH Packwood, JR Brown, JS Bednar, LE Galbraith, VE Moore, *Br Corros J* **24**:30 (1989).

87. VS Sastri, RH Packwood, VE Moore, J Bednar, *Proceedings of the 6th European Symposium on Corrosion Inhibitors*, Vol. 8, University of Ferrara, 1985, p. 741.

88. CL Foley, J Kruger, CJ Bechtold, *J Electrochem Soc* **114**:994 (1967).

89. J Kruger, *J Electrochem Soc* **110**:653 (1963).

90. M Cohen, *J Phys Chem* **56**:451 (1952).

91. PHG Draper, *Corros Sci* **7**:91 (1967).

92. M Nagayama, M Cohen, *J Electrochem Soc* **109**:781 (1962).

93. DM Brasher, AH Kingsbury, *Trans Faraday Soc* **54**:1214 (1958).

94. O Kubaschewski, DM Brasher, *Trans Faraday Soc* **55**:1200 (1959).

95. DM Brasher, AD Mercer, *Trans Faraday Soc* **61**:803 (1965).

96. DM Brasher, *Br Corros J* **1**:184 (1966).

97. DM Brasher, CP De, AD Mercer, *Br Corros J* **1**:188 (1966).

98. MJ. Pryor, M Cohen, F Brown, *J Electrochem Soc* **99**:542 (1952).

99. DM Brasher, ER Stove, *Chem Ind*, 171 (1952).

100. GH Cartledge, *J Phys Chem* **59**:979 (1955).

101. JEO Mayne, CL Page, *Br Corros J* **7**:115 (1972).

102. JGN Thomas, *Br Corros J* **5**:41 (1970).

103. GH Cartledge, GH Spahrbier, *J Electrochem Soc* **110**:644 (1963).

104. DM Brasher, D Reichenberg, AD Mercer, *Br Corros J* **3**:144 (1968).

105. UR Evans, *J Chem Soc* 1020 (1927).

106. GH Cartledge, *J Electrochem Soc* **114**:39 (1967).

107. M Stern, *J Electrochem Soc* **105**:638 (1958).

108. JGN Thomas, TJ Nurse, *Br Corros J* **2**:13 (1967).

109. GH Cartledge, *J Phys Chem* **64**:1882 (1960).

110. JGN Thomas, TJ Nurse, R Walker, *Br Corros J* **5**:87 (1970).

111. JGN Thomas, *Br Corros J* **1**:156 (1966).

112. DM Brasher, *Nature, London* **185**:835 (1960).

113. LI Antropov, NF Kuleshova, *Prot Met* **3**:131 (1967).

114. KE Heusler, *Ber Bunsenges Phys Chem* **72**:1197 (1968).

115. D Gilroy, JEO Mayne, *Br Corros J* **1**:107 (1965).

116. LI Freiman, YM Kolotyrkin, *Prot Met* **5**:113 (1969).

117. AD Mercer, JGN Thomas, *3rd European Symposium on Corrosion Inhibitors*, Ferrara, 1970, University of Ferrara, 1971, p. 777.

118. GH Cartledge, *Br Corros J* **1**:293 (1966).

119. P Hersch, JB Hare, A Robertson, SM Sutherland, *J Appl Chem* **11**:251 (1961).

6

INDUSTRIAL APPLICATIONS OF CORROSION INHIBITION

6.1 CORROSION INHIBITION OF REINFORCING STEEL IN CONCRETE

Corrosion of reinforcing steel in concrete, also known as rebar corrosion, is a serious and significant problem from both structural integrity and economic points of view. The cost can be considerably reduced by mitigating the corrosion problems in existing bridges. It is estimated that $450–550 million could be saved per year by combating and solving the corrosion problems in bridges. Most of the reinforcement corrosion in North America is due to the use of deicing salts during the winter season to maintain ice-free highways and roads. Deicing salts cause potholes, spalls, and delamination of reinforced highway structures such as bridge decks. Parking garages and support pillars are also subject to deterioration since automobiles carry salt-bearing snow and ice through the garages, which upon melting allow aggressive chloride ions to permeate the concrete. Similarly, coastal structures are subject to corrosion when exposed to seawater containing high amounts of chloride.

Some of the methods of combating corrosion of reinforcing steel entail the use of (i) epoxy-coated reinforcement, (ii) protective coatings and membranes, (iii) cathodic protection, (iv) low-permeability concrete, and (v) corrosion inhibitors. The use of corrosion inhibitors is probably more attractive from the point of view of cost and ease of application. It is estimated that it costs US$730/m^3 to construct a new bridge. This cost increases by US$52/m^3 when corrosion inhibitors are used in conjunction with

Green Corrosion Inhibitors: Theory and Practice, First Edition. V. S. Sastri.
© 2011 John Wiley & Sons, Inc. Published 2011 by John Wiley & Sons, Inc.

highly impermeable concrete. This increase in cost is very small in comparison to estimated rehabilitation costs of over US$2600/m^3.

The factors that govern corrosion protection of reinforcing steel by concrete are (i) provision of a physical barrier to ingress of aggressive ions such as chloride, (ii) formation of passive oxide due to the high pH of the pore solution, (iii) limiting galvanic corrosion by the high electrical resistance of concrete, and (iv) formation of protective mineral scales on the rebar, which prevents the reaction of the steel with the environment. Failure of any of these factors results in the corrosion of steel in concrete (1). Virgin concrete has a pH in the range of 12–13.5. In this pH range, the potentials existing in concrete favor the formation of iron oxides on steel.

The protective nature of the passive oxide film is affected by the (i) dissolution of CO_2 and SO_2 gases from the atmosphere in the pore solution resulting in reduced pH and corrosion, (ii) ingress of chloride causing breakdown of the passive film and initiating pitting corrosion, and (iii) insufficient oxygen for maintaining passive oxide film resulting in corrosion. Depending upon availability of oxygen and the pH at the steel surface, the anodic reactions are as follows:

$$3Fe + 4H_2O \rightarrow Fe_3O_4 + 8H^+ + 8e^-$$
$$2Fe + 3H_2O \rightarrow Fe_2O_3 + 6H^+ + 6e^-$$
$$2Fe + 2H_2O \rightarrow HFeO_2 + 3H^+ + 2e^-$$
$$Fe \rightarrow Fe^{2+} + 2e^-.$$

The cathodic reactions are as follows:

$$2H_2O + O_2 + 4e^- \rightarrow 4OH^-$$
$$2H^+ + 2e^- \rightarrow H_2.$$

Atmospheric gases such as CO_2 and SO_2 present in rain react with the lime in concrete and decrease the pH of the pore solution:

$$Ca(OH)_2 + CO_2 \rightarrow CaCO_3 + H_2O$$
$$Ca(OH)_2 + SO_2 \rightarrow CaSO_3 + H_2O$$
$$Ca(OH)_2 + SO_3 \rightarrow CaSO_4 + H_2O.$$

Carbonation of the concrete results in lowering of the pH of the pore solution leading to an increase in corrosion rate of the steel. The chloride originating from the deicing salts and the added calcium chloride as the setting accelerator are responsible for pitting corrosion of the reinforcing steel in concrete.

As in the case of any system, the corrosion inhibitor should minimize the corrosion rate of reinforcing steel over an extended time period. The other requirements are that (i) the inhibitor must be soluble in mixing water and not readily leachable from concrete, (ii) the inhibitor must be compatible with the aqueous cement phase, and

(iii) the inhibitor should not affect the properties of concrete such as setting time, strength, and durability of the concrete.

Corrosion protection can be achieved by the interaction of the inhibitor with the metal surface and formation of a protective film. The most commonly used inhibitor in concrete is calcium nitrite, although organic inhibitors have become prominent in recent years. Calcium nitrite has been used as an inhibitor in concrete on a large scale. Calcium nitrite provides protection for the steel in the concrete in the presence of chlorides, does not affect the properties of concrete, and is readily available for commercial use in concrete.

The detailed mechanism of corrosion inhibition of steel in concrete by calcium nitrite is given in the literature 2, 3. The main anodic reaction taking place in concrete is

$$Fe \rightarrow Fe^{2+} + 2e^-,$$

which is followed by other reactions leading to the formation of $Fe(OH)_2$, $Fe_3O_4 \cdot nH_2O$, or γ-$FeO \cdot OH$ at the surface of the reinforcing steel. The anodic inhibitor, calcium nitrite, is responsible for the constant repair of the weak spots in the iron oxide film as well as the reaction with ferrous ions:

$$2Fe^{2+} + 2OH^- + 2NO_2^- \rightarrow 2NO + Fe_2O_3 + H_2O$$
$$Fe^{2+} + OH^- + NO_2^- \rightarrow NO + \gamma FeO \cdot OH.$$

The nitrite ion competes with chloride and hydroxyl ions in reaction with ferrous ions. Since nitrite helps to repair the flaws in the oxide film, the probability of attack at the flaws of oxide film and dissolution of the oxide film as the soluble chloride complexes is reduced. Since nitrite is primarily involved in repairing the oxide film and the oxide film is of the order of monolayers, the nitrite inhibitor is not consumed to a large extent.

The behavior of calcium nitrite inhibitor in simulated pore solutions, concrete, and mortar samples was studied by polarization resistance, AC impedance, cyclic polarization, and the macrocell corrosion technique. Calcium nitrite was found to delay corrosion initiation, reduce corrosion rate, and increase the pitting and protection potentials with no detrimental effects on the mechanical properties of the concrete. Calcium nitrite also acted as an accelerator.

Calcium nitrite was found to be an effective inhibitor under long-term accelerated testing conditions. It was also found to be more effective when the quality of the concrete was better.

Pozzolans are substances used in concrete to prevent corrosion by reducing the permeability of concrete to chlorides. Microsilica is one of the most effective pozzolans due to its small particle size. Calcium nitrite is compatible with microsilica and a mixture of calcium nitrite and microsilica should provide good corrosion protection. Calcium nitrite has been found to improve the performance of fly ash concrete and concretes containing slag.

TABLE 6.1   Corrosion Inhibition of Reinforcing Steel in Concrete

Inhibitor Systems	Reference
Sodium nitrite, potassium dichromate, and formaldehyde in concrete blocks exposed to admixed chlorides and partially immersed in seawater	18
Sodium nitrite, sodium borate, and sodium molybdate in mortar lollipop samples	19
Evaluation of nine inhibitors and sealants	20
Calcium nitrite and sodium molybdate. Complete protection in limewater was observed with calcium nitrite and with sodium molybdate mixture	15

Calcium nitrite protection was also found to be effective in cracked concrete samples (8). Several studies on the use of calcium nitrite as inhibitor may be found in the literature 9–15. There was no evidence of corrosion in concrete samples containing 15 L/m^3 of calcium nitrite in a test lasting 1.5 years (10). Calcium nitrite was found to inhibit corrosion over time rather than inhibit the initiation of corrosion.

A mixture of calcium nitrite and stannous chloride in saturated limewater in the corrosion inhibition of iron showed effective corrosion inhibition in solutions containing less than 2% of calcium chloride (9). Steel samples corroded by calcium chloride could be passivated by calcium nitrite. Protective and adherent oxide films were observed on samples exposed to nitrite solution.

Some of the inhibitor systems used in corrosion inhibition in concrete are given in Table 6.1.

The potential use of calcium nitrate as an inhibitor and its binding capacity to cement showed the following mechanism similar to nitrite to be operative:

$$6Fe(OH)_{2(s)} + 2NO_2^- + 4H_2O \rightarrow 6Fe(OH)_{3(s)} + N_2 + 2OH^-$$
$$2Fe(OH)_{2(s)} + NO_3^- + H_2O \rightarrow 2Fe(OH)_{3(s)} + N_2^-.$$

Increasing use of organic inhibitors is becoming an alternative to the most commonly used calcium nitrite inhibitor. The organic corrosion-inhibiting admixture, often referred to as OCIA, offers protection by adsorption on the metal surface, thus forming a protective film. The inhibitor is adsorbed on the metal surface through the polar functional group and the hydrophobic chain of the inhibitor oriented perpendicular to the metal surface.

The hydrophobic chains of the organic inhibitor not only repel the aqueous aggressive fluids but also interact with each other to form aggregates thereby forming a tight film on the metal surface. It is thought that the hydrophobic portions of hydrocarbon molecules in process fluids interact with the hydrophobic portions of the inhibitor molecules, which may result in increased thickness and effectiveness of the hydrophobic barrier to the corrosive environment. The film will also be impermeable to chloride ingress into the concrete.

Organic amines and esters were used as inhibitors in both uncracked and precracked concrete samples and compared with calcium nitrite-treated precracked

TABLE 6.2   Corrosion Inhibitors Used in Concrete

Inhibitors	Reference
Glycerophosphate and nitrite in steel mortars	22
Ethanolamine in reinforced concrete	23
Polyalcohols, polyphenols, and sugars	24
Alkanolamine-type inhibitors	25

samples (21). The organic inhibitors lowered the corrosion and the ingress of chloride into the samples was low in samples treated with organic inhibitors. The ingress of chloride into the concrete samples was greater in samples treated with calcium nitrite than in samples treated with OCIA. Some of the organic inhibitors used in concrete are given in Table 6.2.

Alkanolamine-type inhibitors migrate through concrete. This type of inhibitor can be injected into cavities that can lead to extension of the useful life of the structure. Field tests with these inhibitors showed good protection without adverse effect on the mechanical properties of the concrete.

6.2   CORROSION INHIBITION IN COAL-WATER SLURRIES

Sodium nitrite has been used as an inhibitor in combination with sodium molybdate, sodium tungstate, and potassium chromate in coal-water slurries. The synergistic effect observed (26) with the mixture of sodium molybdate and sodium nitrite is depicted in Fig. 6.1. Some data on the corrosion rates (27) are given in Table 6.3. Potassium chromate is unacceptable from the environmental point of view, but it can be substituted by sodium molybdate or sodium tungstate.

6.3   CORROSION INHIBITION IN COOLING WATER SYSTEMS

The factors that govern inhibition of cooling systems are as follows (28): (i) oxygen saturation at the cooling towers, (ii) ingress of gases such as sulfur dioxide due to the scrubbing action of the towers, (iii) pH, (iv) total dissolved solids of the makeup water, (v) algae spores, which may proliferate, (vi) contamination due to leakage, (vii) sulfate-reducing bacteria, which produce hydrogen sulfide and cause localized corrosion under slime deposits, and (viii) silt and solids, which present special problems.

Carbon steel is widely used in heat exchangers and in distribution piping of recirculating cooling systems because it is cost-effective. Other materials used are cast iron copper tubes in heat exchangers, admiralty brasses, copper–nickel alloys, austenitic stainless steels, Inconel 625, Hastealloy C, and titanium.

The criteria to be met by an inhibitor to be effective in cooling water are that (i) the entire metal must be protected from corrosion, (ii) low concentration of the inhibitor must be effective, (iii) the treatment must be effective under a variety of conditions

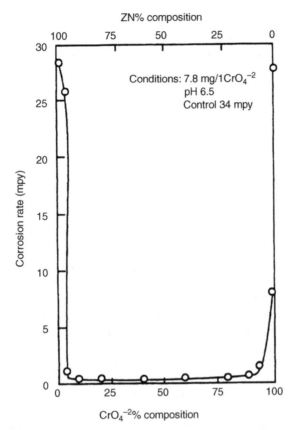

FIGURE 6.1 Influence of zinc-to-chromate ratio on steel corrosion.

such as pH, temperature, heat flux, and water quality, (iv) it should not produce deposits that impede heat transfer, (v) it must have acceptable toxicity for discharge, (vi) the treatment must prevent formation of carbonate and sulfate scales, and (vii) it must combat biological activity due to microorganisms.

Chromates are passivating inhibitors and render protection of both ferrous and nonferrous alloys when they are incorporated into the iron oxide layer. The chromate

TABLE 6.3   Data on Corrosion Inhibition by Oxyanions in Coal-Water Slurries

		Corrosion Rate (mpy)	
Inhibitors	Concentration (M)	15% Coal	30% Coal
No inhibitor		56	64
K_2CrO_4 + $NaNO_2$	0.005 + 0.005	2.4	1.3
Na_2MoO_4 + $NaNO_2$	0.005 + 0.005	0.6	1.9
Na_2WO_4 + $NaNO_2$	0.005 + 0.005	0.4	3.1

that is initially adsorbed on the metal produces a mixture of Fe_2O_3 and Cr_2O_3. Chromate, a powerful oxidizing agent, oxidizes ferrous hydroxide or ferrous oxide to ferric compound and self-reduces to chromic hydroxide and eventually to chromic oxide:

$$6FeO + 2CrO_4^{2-} + 2H_2O \rightarrow 3Fe_2O_3 + Cr_2O_3 + 4OH^-.$$

Chromate is effective in the pH range of 6–11 and temperature range of 38–66°C. The effective concentration is 500–1000 ppm of chromate. The toxicity of chromate is a limiting factor in its use as an inhibitor.

Nitrites have been used in cooling water systems at a concentration similar to that of chloride and in excess of sulfate (29) by 250–500 ppm. The oxidation of nitrite to nitrate by nitrobacteria is a problem that is countered by the use of biocides. The protective oxide is formed as a result of the reaction

$$4Fe + 3NO_2^- + 3H^+ \rightarrow 2\gamma\text{-}Fe_2O_3 + NH_3 + N_2.$$

Silicate inhibitors have been in use for over 60 years. The corrosion inhibition by silicates is affected by pH, temperature, and solution composition. Silicates are ineffective at high ionic strength and are effective in solutions containing salts at 500 ppm or less. The usual effective concentration range is 25–40 ppm of SiO_2. Pitting attack is less severe with silica than with chromate and nitrite as inhibitors. The inhibition by silicate is effective for low carbon steel when the ratio of SiO_2 to Na_2O is in the range of 2.5–3.0. The protective film is considered to be a hydrated gel of silica and metal oxide.

Sodium molybdate in the pH range of 5.5–8.5 is an effective inhibitor for carbon steel. Molybdate is an effective inhibitor in aerated solutions. The binary combination of sodium molybdate and sodium nitrite in the ratio of 0.6:0.4 exhibits synergism. Sodium molybdate at 1000 ppm is needed for inhibition when 200 ppm of chloride is present. About 200 ppm of sodium molybdate can offset corrosion due to 30 ppm of sodium chloride and 70 ppm of sodium sulfate. Competitive adsorption of chloride and sulfate on the iron may be written as

$$Fe(H_2O) + X^- \rightarrow Fe(X^-)_{ads} + H_2O$$
$$FeX_{ads} + OH^- \rightarrow FeOH^+ + X^- + 2e^-$$
$$FeOH^+ + H^+ \rightarrow Fe(H_2O)^{2+}.$$

The corrosion inhibition mechanism is not as well understood as in the case of chromate. Surface analysis of mild steel exposed to molybdate showed the presence of $FeO\cdot OH$ in combination with MoO_3. Auger electron spectra showed molybdenum oxide of 3 nm thickness in combination with a 6 nm layer of iron oxide. It is possible that the iron oxide layer is repaired by molybdate.

Molybdates were found to be useful inhibitors for localized corrosion. The relative values of potential (E_H) and pH, the low pH, and the reduction of potential to

0.3–0.4 V (SCE) after reduction of Mo(V) to Mo(IV) are considered to be responsible for the precipitation of MoO_2 in the crevice film. According to other schools of thought, the passivating film is thought to be ferric molybdate.

Phosphates are effective inhibitors in the presence of oxygen. Monosodium phosphate is not protective, while trisodium phosphate is a good inhibitor. The dissolved oxygen oxidizes iron to γ-Fe_2O_3 and the discontinuities in the oxide film are repaired by ferric phosphate:

$$Fe^{2+} + H_2PO_4^- \rightarrow FeH_2PO_4^+$$
$$FeH_2PO_4^+ + 2H_2O \rightarrow FePO_4 + 2H_2O + 2H^+ + e^-.$$

Chloride anions tend to attack the iron oxide film at voids and form soluble ferric chloride. The ferric salt can hydrolyze and generate hydrogen ion with consequent lowering of pH. This is followed by localized corrosion such as pitting at anodic sites.

High-hardness waters can cause solid deposit formation and the insoluble deposits promote underdeposit corrosion. In such cases, a dispersant agent should be used along with the inhibitor.

Polyphosphates have been used as cathodic inhibitors for the past six decades. The general formula of polyphosphates is

$$Na_4(P_2O_7)_n.$$

When $n = 0$, we have orthophosphate; when $n = 1$, pyrophosphate; $n = 2$, tripolyphosphate; and $n = 12$–14, polyphosphates. Polyphosphates are used in large amounts in highly corrosive media and are effective against galvanic corrosion (30).

The ratio of calcium ion to polyphosphate concentration should be in the range 0.2–0.5 for effective inhibition. A positively charged colloidal complex is thought to be present at cathodic sites as a protective film. The protective film contained (31) ferric pyrophosphates $Fe_4(P_2O_7)_3$ and iron calcium metaphosphate $NaHFeCa$ $(PO_4)_5 \cdot 8H_2O$. Polyphosphates also prevent the formation of calcium sulfate and calcium carbonate scales.

Polyphosphates with $n = 14$–16 at dose level of 15–20 mg/L have been found to be cost-effective. The useful pH range is 5–7 and 6.7–7.0 for ferrous metals and copper alloys, respectively. Polyphosphates suffer from the susceptibility to hydrolytic decomposition at higher pH and temperature in the presence of biological growth. Such hydrolytic decomposition yields orthophosphates. Polyphosphates contain P–O–P linkage, while phosphonates with P–C–P linkage are more stable with respect to hydrolytic decomposition. The structures of nitrilotris(methylenephosphonic acid) (AMP) and hydroxyethylidene diphosphonic acid (HEDP) are given below:

$C_3H_6N(H_2PO_3)_3$ (nitrilotris(methylenephosphonic acid) (AMP)

CH_3-$C(OH)(H_2PO_3)_2$ (1-hydroxyethylidine-1,1-diphosphonic acid) (HEDP)

The AMP molecules (i) control calcium carbonate and calcium sulfate scales, (ii) complex multivalent cations, and (iii) stabilize waters containing iron and manganese. These compounds are generally supplied as concentrated solutions and are stable beyond 100°C. The dosages vary between 1 g/m^3 as P$_2$O$_5$ and about 10 g/m^3 of commercial product.

The formation of phosphonate complexes can cause problems in dissimilar metal environments since the metal disposition and the galvanic attack cannot be ruled out should the pH value change. Both AMP and HEDP are marginally protective of steel at pH 8.0. The protection is due to mixed mode of protection. At a concentration of 100 mg/L, the cathodic corrosion control mode predominates.

Mixtures of inhibitors usually give better inhibition protection than similar concentrations of individual inhibitors. The mechanism of action of a multicomponent inhibitor system is synergistic in nature.

Zinc chromate is a heavy metal system that is effective as cathodic in nature and impedes the reduction of oxygen at the cathode. At 95% zinc and 5% chromate, the corrosion rate is nearly zero mpy (Fig. 6.2).

The concentration usually used in recirculating cooling water systems is 10 mg/L of zinc and chromate in the pH range of 5.5–7.0, with 7.0 as the optimum value. This is not an effective formulation since calcium carbonate and sulfate scales as well as insoluble zinc salts form at pH above 7.5. This may be overcome by the addition of phosphonates such as AMP. Biocidal treatment is not necessary since zinc chromate is not a biological nutrient.

Zinc polyphosphate protects both ferrous and nonferrous metals by maintaining insensitivity to electrolyte concentration, thereby preventing calcium carbonate and calcium sulfate scale formation and multivalent ions from forming positively charged colloidal complexes. Zinc increases the rate of formation of the protective film

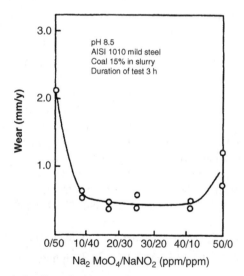

FIGURE 6.2 Synergistic effect of a binary mixture of sodium molybdate and sodium nitrite at pH 8.5.

TABLE 6.4 [a] **Corrosion Rates in Zinc-Polyphosphate System**

Amount of Zinc (mg/L)	Corrosion Rate (mpy)
0.2	6.8
0.3	5.7
0.4	4.1
0.5	2.7
0.6	1.6
0.7	1.3
1.0	1.1

[a] Conditions: 8 mg/L of polyphosphate, pH 6.8, 41°C.

and also the concentration of zinc needed for effective protection is less than the polyphosphate concentration. Some data on corrosion rates in zinc–polyphosphate are given in Table 6.4 (32).

Some data on corrosion rates in the presence of zinc and phosphonates show synergistic behavior (32) (Table 6.5).

The data clearly show the synergistic behavior of zinc–phosphonate system in the protection of steel from corrosion. The addition of zinc to phosphonates is particularly useful in the protection of copper base alloys because it prevents the formation of copper–phosphonate complexes. The mode of protection of zinc–phosphonate may be cathodic in nature, while phosphonates are anodic in nature. Zinc–phosphonate is effective over a wide pH range of 6.5–9.0 and the performance of the system is not affected by the electrolyte concentration, water quality, or chlorine and up to temperatures of 70–77°C.

Nonheavy metal systems in use are (i) mixture of AMP and HEDP, (ii) mixture of polyphosphate and HEDP, and (iii) mixture of polyphosphate and orthophosphate. These combination systems function by (i) formation of protective films consisting of complexes, (ii) prevention of the formation of calcium carbonate and calcium sulfate scales, and (iii) dispersion of surface deposits. Some data on corrosion rates with AMP–HEDP mixtures are given in Table 6.6 (32).

TABLE 6.5 Corrosion Rate as a Function of AMP Concentration

Zn (%)	AMP (%)	Corrosion Rate (mpy)
0	100	23
24	72	1.5
50	50	0
78	23	7
100	0	23

TABLE 6.6 Corrosion Rate as a Function of HEDP and AMP Concentration

HEDP (%)	100	80	50	35	0
AMP (%)	0	20	50	65	100
Corrosion rate (mpy)	27	22	5	10	17

TABLE 6.7 Corrosion Rate of Steel in Variable Concentration of Inhibitor

HEDP (%)	100	80	60	40	20	0
Polyphosphate (%)	0	20	40	60	80	100
Corrosion rate (mpy)	27	7	4	3.5	4.0	7.5

An AMP/HEDP ratio of 1.5–1.0 and a pH of 7.5 give a maximum inhibition. Polyphosphate–HEDP acts as a cathodic inhibitor and some corrosion rate data (32) are given in Table 6.7.

Complex corrosion inhibitor films are formed on iron under cooling water conditions. The intense iron and oxygen signals with small amounts of chromium and phosphorus and the absence of molybdenum and nitrogen signals indicate the formation of γ-Fe_2O_3 as the primary inhibition mechanism when nitrite, chromate, molybdate, and phosphate are used as inhibitors. The presence of calcium carbonate, calcium sulfate, calcium hydroxide, or calcium oxide in the film indicates inhibition by these species at cathodic sites.

The films formed by zinc and phosphonate consisted of two layers. The top layer consisted of zinc as hydroxide, oxide, carbonate, or phosphate and calcium as carbonate or phosphonate. The bottom layer consisted of iron oxides along with calcium carbonate and calcium phosphonate. Calcium was present in all the films with variable amounts, depending upon experimental conditions. Calcium phosphate and calcium carbonate were predominant when the phosphate was the inhibitor.

6.4 MOLYBDATE INHIBITOR IN CORROSION INHIBITION

Group VI oxyanions, chromate, molybdate, and tungstate have been used as corrosion inhibitors for the protection of metals. Environmental agencies have banned the use of chromate because of its toxicity. Thus, molybdate has been substituted in the place of chromate in a variety of applications (33–36). In neutral and alkaline solutions, corrosion of iron results in the formation of a nonprotective hydrated iron oxide film that allows passage of ferrous ions through the oxide film, leading to continued corrosion. Addition of molybdate to an aerated solution containing iron results in adsorption of molybdate and oxygen on the iron surface, leading to the formation of passive oxide (37). The adsorbed molybdate initially forms ferrous molybdate, which is later oxidized to ferric molybdate. The ferric molybdate prevents the outward diffusion of ferrous ions and inward diffusion of aggressive ions like chloride. Unlike chromate, molybdate is not reduced to a lower oxidation state. The surface analysis of passive films (38–40) on an iron surface exposed to a molybdate inhibitor showed the presence of hexavalent molybdenum, unlike chromium, which is in the trivalent state (Cr_2O_3; $CrO \cdot OH$). The passive film formed on mild steel (41) at pH 8.5 contained a monolayer of molybdate based on the calculation that three close-packed oxygen atoms cover 23 $\overset{\circ}{A}^2$.

The passivation of 304 stainless steel by molybdate in the absence of oxygen has been attributed to the redox potential and exchange current developed on the metal surface due to the adsorbed molybdate anions (62). Corrosion inhibition of mild steel in an acidic medium (pH 3– 5) was less effective than in neutral and alkaline solutions (pH 9.0). Films of MoO_2 are formed indicating reduction of molybdenum from the hexavalent to the tetravalent state.

Molybdates form isopolymolybdates in acidic solutions, which unlike molybdate are not good inhibitors. On the other hand, isopolymolybdates give good protection against pitting corrosion of stainless steels by repassivation of pits in iron (43) and formation of insoluble molybdates in the pits.

Corrosion inhibition of aluminum, copper, tin, zinc, cadmium, titanium, brass, and lead-based solders by molybdate has been reported (33). The inhibition of pitting corrosion of aluminum at pH 7 and 9 has been attributed to the adsorbed molybdate in the pits and the presence of hexavalent and tetravalent molybdenum in the film covering the pits (44). The possible existence of aluminum molybdate in the film at pH 9 has been reported (45).

Anodic passivation of zinc (46) in deaerated solutions of pH 7–9 by molybdate and tin (47) at pH 3–8.5 showed the presence of tetravalent molybdenum and $Sn(OH)_4$ and molybdate on zinc and tin, respectively. Some of the applications of molybdate as a green corrosion inhibitor are summarized in Table 6.8.

6.5 CORROSION INHIBITION IN ACID SOLUTIONS

Acid solutions are used extensively in industry such as acid pickling, industrial acid cleaning, acid descaling, and oil well acidizing. Inhibitors are needed to minimize corrosion. The selection of a suitable inhibitor depends upon the type of acid, concentration of acid, temperature, flow velocity, presence of dissolved inorganic or organic substances, and type of metallic material exposed to the acid solution. The most commonly used acids are hydrochloric, sulfuric, nitric, hydrofluoric, citric, formic, and acetic acid.

6.5.1 Acid Pickling

Metals or alloys subjected to painting, enameling, galvanizing, electroplating, phosphate coating, cold rolling, and other finishing processes must possess a clean surface free of salt or oxide scale. The sample is immersed in an acid pickling bath to remove undesirable scale. After the removal of the scale, the acid might attack the metal. Corrosion inhibitors are added to the pickling solution to minimize the attack of the metal by the acid.

TABLE 6.8 Applications of Molybdate as Corrosion Inhibitor

Application	System	Reference
Engine coolants	Molybdate	48
	Molybdate with nitrite; molybdate, arsenite, or arsenate and benzotriazole along with borate/phosphate/amine	49
	Nitrite, nitrate, phosphate, borate, silicate, benzoate, aminophosphonate, phosphinopolycarboxylate, polyacrylate, hydroxybenzoate, phthalate, adipate, benzotriazole, tolyltriazole, mercapto-benzothiazole, and triethanolamine are combined with molybdate. In glycol, 0.1–0.6 wt% of molybdate is used	50–58
Protection of solder and aluminum parts in cooling systems: open recirculating cooling water systems	Molybdophosphate and molybdosilicate	59
	Molybdate at hundreds of ppm level	61–63
	Molybdate with zinc salt	
	Nitrite, silicate, aminophosphonates, acrylates, benzotriazole, tolyl triazole, fatty acid carboxylates, ortho- and polyphosphates, phthalate, gluconate, polymers of maleic acid and amylene or allyl alcohol used in conjunction with molybdate	64–85
Closed recirculating cooling water systems	200 ppm sodium molybdate with 100 ppm of sodium nitrite	86
	50 ppm molybdate, 50 ppm phosphate, 2 ppm Zn^{2+}	87
	40 ppm sodium molybdate + 40 ppm sodium silicate	88
Pigments	Molybdate was proposed	89
	Use of zinc, calcium, and ferric molybdates	90
	For protection of steel and aluminum, zinc, Ca, Al, Fe, Ba, Ti, or Sr molybdate admixed with zinc oxide, phosphate, borate, or chromate or stannic oxide	91
	Extenders like alumina, zinc oxide, silica, talc, or calcium carbonate are used	33
	Solvent or water-base paints involving molybdate pigments	33
	Molybdenum disulfide in paints	92

Hydraulic and metal-working fluids	1 wt% sodium molybdate in polymer-thickened glycol–water protected Al, Cd-plated steel, mild steel, leaded bronze, yellow bronze, solder, copper- and zinc-plated steel	93, 94
	Sodium molybdate with sodium nitrite gave multimetal protection	102, 103
	Sodium molybdate with monoethanolamine	104–110
Conversion coatings	Molybdates incorporated in the oxide film during sealing process in anodized aluminum	
	$FeMoO_4$ coating on mild steel	111
	Conversion coatings on stainless steels using molybdate	112–117
	Molybdate coating on mild steel exposed to molybdate–phosphate and nitrite	118
	Mild steel exposed to molybdate–organic polymer followed by heating	119
	Molybdate rinse of zinc-phosphated steel	120, 121
	Molybdate used in phosphating solution	122
	Passivation of zinc and galvanizing involving molybdate	123–125
	Molybdate protective coating on titanium	126
	Molybdate coating on tin	127
	Black decorative coatings on Al, Zn, steel, Cu, brass, and tin	128
	Cd produced from ammonium heptamolybdate solutions	129
	Molybdate added to oxalate coatings used on ferrous stainless steel wires or heads	130
	Sintered steel parts dipped in ammonium heptamolybdate solution and heated	131
Brines: refrigeration, air conditioning, and humidity control	Molybdate used in lithium chloride and bromide brines protects mild steel, 90/10 and 70/30 cupronickel alloys, from galvanic and crevice corrosion	132–136
Oil wells containing calcium chloride and bromide	Sodium molybdate and sodium erythorbate	137
Cooling water of steam plant	Molybdate with an aluminum salt and thiourea	138
Aqueous sodium chloride	Molybdate with monoethanolamine	101, 102
Naval environments		139

(continued)

225

TABLE 6.8 (*Continued*)

Application	System	Reference
	Mixture of molybdate, chromate, nitrite, and borate inhibited stress-corrosion cracking of AISI 4340 steel and AA7075 Al	140
Air conditioning using lithium and zinc chlorides	Sodium molybdate	
Lubricants	Oil soluble Mo(VI) organic compounds in lubricant oils and greases. Mo–S and Mo–S–P compositions have anticorrosive, antifriction, and antioxidant properties	141–145
	Molybdate–alkaline earth sulfonate in motor oil gave antirust and antifriction properties	145
	Ammonium heptamolybdate in polymeric solid lubricants for cold rolling of metals gives corrosion protection	146
Boiler waters	Corrosion protection of steam-generating equipment, heating systems, and chemical reactor heating water by sodium molybdate and phthalate	147
	Mild steel corrosion inhibition in boilers by a mixture of sodium molybdate, sodium citrate, manganous sulfate, polymaleic acid, and morpholine	148
	Protection of mild steel in hard water boilers by sodium molybdate and sodium nitrite	149
Reinforcing steel in concrete	Molyorange pigment and strontium chromate in aqueous acrylic dispersion used to coat reinforcing steel bars	150
	Steel bars coated with polyurethane containing ammonium molybdate and ammonium dihydrogen phosphate	151
Metalworking lubricants	Mixture of sodium molybdate, borate, and bicarbonate in graphite lubricant thickened in glycol–water mixture	152
Batteries	Quicklime, soap, citric acid, and sodium molybdate mixture	153
		154

Volatile corrosion inhibitors	Alkali metal molybdate added to Li-MoO$_2$ battery electrolyte protected stainless steel cathode from pitting corrosion	155
	Wrapping paper coated with a mixture of ammonium molybdate and sodium nitrite protected ferrous parts from corrosion in highly humid atmospheres	156
	Triethanolamine molybdate is used in the protection of steel, nickel, brass, and zinc	
Beverage cans	Application of 10 mg/ft^2 of sodium molybdate on packaging paper prevented localized corrosion and staining of aluminum cans	157
Ore-grinding ball mills	Use of sodium molybdate along with sodium nitrite in grinding iron ore	158, 159
Coal-water slurries	Sodium molybdate with sodium nitrite or HEDP and with zinc sulfate, phosphate, and mercapto-benzothiazole protected AISI 1010 steel	160, 161
Brake linings	Two parts of MoO$_3$ per 100 parts of friction material gave corrosion protection	162
Temporary coatings	80 g/L potassium molybdate per 100 g vinyl alcohol	163
Acid systems	0.001 M of molybdate inhibited stress corrosion cracking of Ni–Mo alloy casing pipe in acidizing wells	164
	Molybdophosphate performed better than molybdate in inhibition of corrosion of stainless steel in acid medium	165
Metallic glasses	MoO$_3$ is used in metallic glasses to improve their corrosion resistance	166, 167
Coal-water slurries	20/30 Na$_2$MoO$_4$/NaNO$_2$, ppm/ppm, 0.5 mm/year; 0.005 M Na$_2$MoO$_4$ + 0.005 M NaNO$_2$ 0.6–1.9 mpy; 0.005 M Na$_2$WO$_4$ + 0.005 M NaNO$_2$ 0.4–3.1 mpy	160

Hydrochloric acid is usually used in the pickling bath. Large-scale continuous treatment, such as metal strip and wire pickling, and regeneration of pickling solutions are advantages of hydrochloric acid over sulfuric acid. Other acids such as nitric, phosphoric, sulfamic, oxalic, tartaric, citric, acetic, and formic acid are used for special applications.

The choice of inhibitor depends on the purpose of acid pickling. If the pickling is for the removal of mill scale from hot-rolled steel, the acid concentration, temperature, pickling time, and the type of steel will dictate the selection of the inhibitor. In general, pickling with HCl involves 200 g/L of HCl at 60°C for about 30 min, depending upon the type of steel and the steel mill operating conditions. Sulfuric acid at 200–300 g/L level at temperatures up to 90°C may be used. These are severe conditions that require effective inhibitors.

Inhibitors used in pickling solutions should satisfy the following conditions: (i) effective inhibition of metal dissolution, (ii) lack of overpickling process, (iii) effective pickling at low inhibitor concentration, (iv) effective at high temperatures, (v) thermal and chemical stability, (vi) effective inhibition of hydrogen entry into the metal, (vii) good surfactant properties, and (viii) good foaming properties. The main requirement is that the inhibitor must minimize metal dissolution even in the presence of dissolved salts such as ferrous sulfate that promotes metal dissolution. The corrosion rates of steel in 20% sulfuric acid at 90°C and in a mixture of 20% sulfuric acid containing 12% ferrous sulfate are 3100 and 5900 g/(m^2 h), respectively. The increased corrosion rate in the presence of ferrous sulfate can be countered by inhibitors that require longer pickling times. Pickling in sulfuric acid at 60–90°C to remove mill scale requires inhibitors with high thermal stability. Inhibitors must be resistant to hydrolysis, condensation, polymerization, and hydrogenation by nascent hydrogen.

Pickling inhibitors should also minimize hydrogen entry into the metal. Inhibitors used in acid pickling should inhibit both metal dissolution and hydrogen entry into the metal, although they are not tested for the latter. The inhibitor should also be effective at low concentrations for economic reasons. A good pickling inhibitor should also have good surfactant and foaming properties. Usually pickling inhibitors have poor foaming and surfactant properties and, therefore, wetting agents, detergents, and foaming agents are added to commercial inhibitor formulations.

Wetting agents help the acid penetrate the cracks and fissures in the scale and remove the scale. Wetting agents are also known as pickling accelerators and pickling degreasing agents. No single inhibitor can meet all the requirements for an effective inhibitor and hence multicomponent inhibitor formulations are used. An inhibitor formulation consists of an active inhibitor, wetting agent, detergent, foaming agent, solvent, and cosolvent. The components of detergents, wetting agents, and foaming agents consist of alkyl sulfonates (R–SO$_3$Na), sulfates (R–O–SO$_3$Na), nonionics, and ethoxylates. In general, 5–50 g of commercial inhibitor is used per 1 L of the pickling acid. Corrosion inhibitors used for various metals in HCl pickling media are given in Table 6.9.

Inhibitors used in sulfuric acid pickling medium are given in Table 6.10.

Iron and steel are passivated in concentrated nitric acid (65%), but dissolve in lower concentrations of nitric acid. Unlike hydrochloric and sulfuric acids, nitric

TABLE 6.9 Inhibitors Used for Various Metals in HCl Pickling Media

Medium	Metal or Alloy	Inhibitors	Reference
HCl	Iron, steel	Primary, secondary, and tertiary amines, oximes, nitriles, mercaptans, sulfoxides, thioureas, complexones	168–171
	Copper, brass	Thiourea, benzimidazoles, 2-mercapto-benzothiazole, phenyl thiourea	
1 M HCl + 2.5% NaCl	Al and its alloys	Aliphatic, aromatic aldehydes, carboxylic acids, ketones, sulfonamides, sulfones, dihydric phenols, tetrazolinium compounds, formazan compounds, sulfoxides, aromatic sulfonic acids, cupferron, N-allylthiourea, dibenzoyl sulfoxide	172–174
HCl + H$_3$PO$_4$ (IN)		p-Thiocresol and sodium diethyl dithiocarbamate	
10% HCl		Mixture of aromatic aldehyde, 1 M urea/triethylamine/ dodecylamine, >1 M thiourea, and 1-hexyn-3-ol together with alkyl sulfonates	175
Hydrochloric acid	Zn	Inhibitors used for Al are satisfactory	
5.6 M HCl	Nickel	Thiourea and its derivatives; o-tolyl thiourea and sym-diisopropyl thiourea gave 90% inhibition over a wide temperature range	176
HCl	Titanium and its alloys	Copper sulfate, nitric acid, and chromic acid at 0.5–1.0% act as inhibitors. Nitrobenzene, trinitrobenzoic acid, picric acid, benzene arsenic acid, at 0.03–0.003 M	177–180

TABLE 6.10 Inhibitors Used in Sulfuric Acid Pickling Medium

Metal	Inhibitors and Conditions	Reference
Steel	2-Mercaptobenzimidazole	171
	2-Mercaptobenzimidazole with propargyl alcohol or 1-hexyn-3-ol; sulfonium compounds at low temperatures	181–183
	Amines/quaternary ammonium salts with halides $Cl^-/Br^-/I^-$ show synergistic effect p-Alkylbenzyl pyridinium halides and p-alkylbenzyl quinolinium halides with 0.01–0.001 M KI in 5–30% H_2SO_4 give good corrosion inhibition	184
	Nitrogen-containing inhibitors with SH^- from H_2S show synergistic effect	185
	Acetylenic alcohols and acetylenic alcohols with halides	186
Copper	Thiourea	187–189
Brass	Quinolines, azoles	190–192
Titanium	2,4,6-Trinitrophenol, 5-nitroquinoline, 8-nitro-quinoline, o-nitrobenzene, and picric acid	193–196

acid is an oxidizing acid and it is difficult to inhibit high and medium concentrations effectively. Some of the inhibitors used in various pickling media are given in Tables 6.11–6.14.

6.6 OXYGEN SCAVENGERS

Corrosion involves the anodic oxidation of the metal and the cathodic reduction of oxygen:

$$\text{Anodic reaction: } M \rightarrow M^{n+} + ne^-$$
$$\text{Cathodic reaction: } O_2 + 2H_2O + 4e^- \rightarrow 4OH^-.$$

TABLE 6.11 Inhibitors Used in Nitric Acid Pickling Medium

Metal	Inhibitors and Conditions	Reference
Ferrous metals	Propargyl mercaptan, propargyl sulfide, ethane dithiol, thioacetic acid, mercaptoacetic acid, diethyl sulfide, and thiourea in low concentration	197–200
Copper	Aliphatic and aromatic amines and diamines	201–210
Brass	Aminophenols, aminobenzoic acids, heterocyclic amines, thiourea and derivatives; p-thiocresol, o-chloroaniline, o-nitroaniline, and diethyl thiourea	211–215
Titanium	Silicone oil in 20–70% HNO_3 reduced the corrosion rate from 1–10 mm/year to a negligible value	216

TABLE 6.12   Inhibitors Used in Hydrofluoric Acid Pickling Medium

Metal	Inhibitors and Conditions	Reference
Ferrous metals	In 0.5–10 wt% hydrofluoric acid, dibenzyl sulfoxide, C_{10}–C_{14} alkyl pyridinium salts, di-o-tolylthiourea, benzotriazole, mixture of thiourea and Mannich base derived from rosin amine	217–219
Ferritic steels	2–5% Thiourea in a mixture of nitric and hydrofluoric acids	220

TABLE 6.13   Inhibitors Used in Phosphoric Acid Pickling Medium

Metal	Inhibitors and Conditions	Reference
Steel	Benzylquinolinium thiocyanate	221
Steel vessels	Transport of 50–116% phosphoric acid; dodecylamine or 2-aminobicyclohexyl (0.03–0.1%) in 0.007–0.014%potassium iodide	222
Aluminum alloys	Urotropin and p-toluidine	223, 224
Titanium	Inhibition in concentrated phosphoric acid by ammonium molybdate and potassium dichromate	225

Depending upon the predominant reaction, the corrosion process is termed as anodically or cathodically controlled. When the cathodic reaction is rate-determining, the corrosion reaction depends on the oxygen concentration.

The inhibition by oxygen scavengers may occur by cathodic inhibition in which chemical removal of oxygen predominates over cathodic reduction of oxygen to hydroxyl ions and anodic inhibition by way of passivation. Thus, the rate of consumption of oxygen by the oxygen scavenger is important in the cathodic mode of inhibition. This mode of inhibition in the absence of passivation is possible in oilfield waters and industrial boilers when the feedwater is insufficiently deionized. In this case, an oxygen scavenger together with pH control will suffice to inhibit corrosion.

Hydrazine treatment in boiler waters results in the formation of magnetite:

$$12Fe(OH)_3 + N_2H_4 \rightarrow 4Fe_3O_4 + 20H_2O + N_2$$
$$12FeO \cdot OH + N_2H_4 \rightarrow 4Fe_3O_4 + 8H_2O + N_2$$

The rate of reaction between an oxygen scavenger and the dissolved oxygen is not fast and hence catalysts are used to accelerate the reaction. Some typical catalysts are transition metal salts, transition metal complexes, and organic compounds. Some of the oxygen scavengers generally used are given in Table 6.15.

6.7   INHIBITION OF CORROSION BY ORGANIC COATINGS

The main function of a coating on a metallic structure is to provide a protective barrier between the metal and the external corrosive environment. Corrosion inhibition is the

TABLE 6.14 Inhibitors Used in Scale Removal in Hydrochloric Acid Medium

Metals/Scales	Inhibitors and Conditions	Reference
Scale of calcium and iron salts	5–15 wt% HCl at 80°C mixture of imines, thioureas, and alkynols	226
Zinc	Mixtures of N,N^1-dicyclohexylthiourea, ethyl cyclohexanol, and plyamine-methylene phosphate	
Carbon steel	Sulfonium compounds	227, 228
Scale in refineries	14% HCl containing 33–37 g/L of hexamethylenetetramine	229
Copper with a scale of magnetite in high-pressure boilers	Mixture of thiourea or its derivatives and hexahydropyrimidine-2-thione	
Iron scales and silicate deposits	Hydrochloric acid mixed with hydrofluoric acid and ammonium bifluoride	
	Hydrofluoric acid can remove the scales	
Magnetite and hematite scales	Mannich bases, thiourea, and its derivatives and alkynols are used as inhibitors in hydrofluoric acid. Condensation product of diethyl thiourea with hexamethylene tetramine is used in hydrofluoric acid medium for scale removal	230
Oxide scales	5–15% Sulfuric acid is used to remove oxide scale	
Organics, coke	Mixture of sulfuric and nitric acid	
Calcium deposits	Sulfamic acid is used to remove the scale	
Oxide scale	Mixture of sulfamic acid and sodium chloride removes oxide scale	
Copper, brass, stainless steel and galvanized steel	Sulfamic acid is used in cleaning equipment	
Heat exchangers in nuclear plants	Citric acid is used in cleaning equipment. Inhibitors used in H_2SO_4 solutions may be used in citric acid	

Equipment made out of aluminum	Acetic acid is used to remove carbonate scales. Inhibitors used in sulfuric acid may be used	
Magnetite scale in boilers	2 wt% hydroxyacetic acid with 1% formic acid removes the scale. Inhibitors used in sulfuric acid can be used	231
Acidizing of oil and gas wells with HCl	Condensation products of amines with aldehydes, such as cyclohexylamine or aniline, and pyridine with formaldehyde	232
	Performance of amines in HCl improved by added cuprous iodide	233
	Propargyl alcohol or 1-heyn-3-ol gave improved performance	234
	Alkynoxymethyl amines give good protection	235
	Acetylenic inhibitors with ferrous ions give good protection due to the production of a surface film through iron-catalyzed condensation and polymerization	

TABLE 6.15 Some Oxygen Scavengers

Type	Reaction	Mechanism	Application	Reference
Na_2SO_3 $NaHSO_3$	$2SO_3{}^{2-} + O_2 \rightarrow 2SO_4{}^{2-}$ Co^{2+}, Mn^{2+} salts, peroxides, chlorine dioxide	Oxygen removal	Oilfields and boilers	236–239
Hydrazine	$N_2H_4 + O_2 \rightarrow N_2 + 2H_2O$. Catalysts: hydroquinone, cobalt–EDTA complex	Passivator	Boilers	240, 241
Carbohydrazide	$(H_2N-NH)_2CO + O_2 \rightarrow 2N_2 + 3H_2O + CO_2$. Transition metal salts (copper) used as catalysts	Passivator	Boilers	242
Diethyl hydroxylamine	$4(CH_3-CH_2)_2NOH + 9O_2 \rightarrow 8CH_3COOH + 2N_2 + 6H_2O$	Passivator	Boiler	243
Hydroquinone mixed with hydrazine	Hydroquinone $+ O_2 \rightarrow$ quinone $+ H_2O$		Boilers	244
D-Erythorbic acid with NH_3 or amine	Salts of Cu, Ni, and Fe used as catalysts	Passivator	Boilers	245
Methylethylketoxime	Methylethylketoxime $\rightarrow$ Methylethylketone $+$ $N_2O + H_2O$	Passivator		246

primary objective in the use of pigments. Paints also serve as cosmetic agents. Corrosion inhibitors are usually incorporated in primers.

Inhibitors bond to the metal surface by chemisorption or physisorption. The applied coatings succeed or fail, depending upon their ability to bond to the substrate to which they are applied. Inhibitive pigments consist of finely divided solids added to oils or resins, volatile solvents, and driers or catalysts. The coating containing polar molecules will bond to the metal surface better than one containing nonpolar molecules. This is due to the higher electron density at the polar functional group. Thus, the constituents of the coating determine the performance of the coating. For example, the coating will be unstable in alkaline solutions when it contains ester molecules because of saponification of esters and formation of soaps. The polar substituents in the coatings behave like oleates and petroleum sulfonates, in that they adsorb on the metal surface preferentially, displace the adsorbed water molecules, and orient in such a way that their hydrophobic tails extend to repel the aggressive aqueous phase. The functions of the various components of coatings are not well understood, and the science of coatings tends to be empirical rather than sound and scientific in nature. In spite of this fact, the role of inhibitors in coatings must be the same as in corrosion inhibition. Thus, the effective inhibitors in coatings must bond to the metal surface strongly and at the same time repel aggressive ions or neutralize the acidic corrosive environment, which otherwise would attack the coating–substrate interface.

Both anodic and cathodic inhibitors are used in coatings. Anodic inhibitors stifle the anodic reaction and cathodic inhibitors stifle the cathodic reaction. Some inhibitors can inhibit both cathodic and anodic reactions. Inorganic anodic inhibitors give about 80–95% inhibition. There is a vast array of inhibitors used in coatings and some chosen examples used in the protection of steel are given in Table 6.16.

Some of the components are used as fillers. Zinc and calcium molybdate are effective at low levels ($\sim$3%) and are nontoxic. These compounds may be used in place of toxic chromates. Red lead combines with the corrosion products of iron and gives protection. Red lead, zinc oxide, and calcium carbonate form soaps with linseed oil and then decompose in the presence of oxygen and water into lead salt of azelaic acid, which promotes cathodic reduction of oxygen. This helps to stabilize the iron oxide film by maintaining suitable potential. Calcium plumbate increases the pH and inhibits corrosion. The lead suboxide in oleoresinous coatings forms lead soaps and blocks the anodic sites through adsorption. Lead suboxide in linseed oil primer was found to penetrate rust film and move into crevices of rusted steel.

The performance of coatings that contain inhibitors showed that the onset of corrosion extended over long periods of time, as evidenced by the data given in Table 6.17.

The performance of polymer-based coatings containing inhibitors showed that the onset of corrosion extended over long periods, as evidenced by the data given in Table 6.17.

Corrosion inhibition by chromated pigments is similar to that by dissolved chromate inhibitor and this is supported by the slight solubility of the pigment. Chromate ions are adsorbed on the surface followed by the formation of the passive film containing Cr_2O_3 and Fe_2O_3.

TABLE 6.16   Some Inhibitors Used in Protective Coatings

Compounds	Reference
Calcium	
Carbonate	248
Molybdate	248
Plumbate	249
Chromium	
Barium potassium chromate	248
Basic lead silicochromate	250
Cadmium chromate	249, 251
Oxide	249
Strontium chromate	251
Zinc chromate	249–251
Lead	
Basic silicoplumbate	249, 250
Blue	249
Calcium plumbate	248
Carbonate (white)	249
Red	250
Titanate	249
Zinc	
Chromate	250, 251
Molybdate	248
Oxide	250
Tetroxychromate	248
Miscellaneous	
Antimony oxide	252
Carbon black, chalk	253
China clay	254
Iron oxide	250
Strontium chromate	251
Talcum	254
Titanium dioxide	250

TABLE 6.17   Performance of Coatings

Coating	Time for Onset of Corrosion (min)
Acrylic latex without inhibitor	0.5
Acrylic latex with inhibitor	14.0
Natural drying oil	0.5
Natural drying oil with inhibitor	12.0
Alkyd varnish	2.7
Alkyd varnish with inhibitor	18.2
Alkyd styrene varnish	0.5
Alkyd nitrocellulose varnish	0.5
Alkyd nitrocellulose varnish with inhibitor	9.0

While potassium chromate protects steel by the anodic dissolution mechanism, protection by zinc chromate involves the formation of zinc hydroxide film inhibiting the cathodic reaction. Zinc, cadmium, and strontium chromates have been used in the protection of aluminum involving crevices. These inhibitors are reduced at cathodic sites and hydrogen evolution is prevented. Disbonding of coatings is due to hydrogen evolution, and addition of cadmium phosphate yielded improved inhibition.

Chromates such as zinc chromate, potassium chromate, or barium chromate may be used as primers in vehicles such as linseed oil, long oil linseed penta alkyd, and double boiled linseed oil for corrosion inhibition. Satisfactory protection was obtained with modified phenolic stand oil and epoxy primers as confirmed by neutral pH and noble potentials and formation of soap, which covered the zinc chromate causing water repulsion.

Some pigments based on lead, silicate, and phosphate with some applications are given in Table 6.18.

The coatings discussed so far involve heavy metal compounds that are toxic with the exception of molybdates and phosphates. Attention has been directed toward development of environment-friendly coatings containing tannins. Tannins are used as corrosion inhibitors in water-based paints, and these paints are applied directly on rusted objects without pretreatment such as sandblasting. Some studies have focused on the mechanism of protection of iron from corrosion by tannins 259–265.

Tannins are known as "rust converters" that convert "active" rust into nonreactive oxide such as magnetite (266). X-ray diffraction studies of the effect of tannins on the phase transformations of rust in paint films showed the reduction of lepidocrocite into magnetite (267). The insoluble complex formed may act as a barrier for diffusion of

TABLE 6.18   Pigments Used in Corrosion Inhibition

Pigment	Description	Reference
Red lead	Red lead in linseed oil is used as a primer for steel structures. Red lead oxide forms soluble soaps with vehicle	255
Lead silicochromate	Consists of silicate case coated with its own pigment. Anticorrosive due to lead and passivating due to chromate. This coating is comparable to red lead in linseed oil in protection	256
Barium oxide–iron oxide	Developed for corrosion control of fertilizer plants	257
Zinc phosphate/barium phosphate primer	Nontoxic and yields excellent intercoating adhesion. The primer converts the metal into metal phosphate and reduces diffusion of ions through the paint film. Zinc phosphate–iron oxide in linseed stand oil performed better than red oxide–zinc chromate	258

oxygen (259). The action of tannins depends upon the properties of the paint–rust system and especially on the amount of rust present (263).

The formation of insoluble ferric–tannin complex at anodic sites acts as an insulator between anodic and cathodic sites due to the cross-linked structure of the chelate complex. Tannins convert "active" rust into nonreactive oxide such as magnetite (266). The insoluble complex formed may act as a barrier for the diffusion of oxygen. The action of tannins depends upon the properties of the paint–rust system and especially on the amount of rust. The ratio of ferrous to ferric states increased upon treatment of rust with tannins.

6.8   MECHANISM OF PROTECTION BY TANNINS

The atmospheric corrosion of iron can be described as

$$2Fe \rightarrow 2Fe^{2+} + 2e^-$$
$$8\gamma\text{-FeO} \cdot OH + Fe^{2+} + 2e^- \rightarrow 3Fe_3O_4 + 4H_2O.$$

The net reaction is

$$8\gamma\text{-FeO} \cdot OH + Fe \rightarrow 3Fe_3O_4 + 4H_2O.$$

Upon drying, magnetite is reoxidized to trivalent γ-FeO·OH:

$$2Fe_3O_4 + \frac{1}{2}O_2 + 3H_2O \rightarrow 6\gamma\text{-FeO} \cdot OH.$$

The corrosion inhibition by tannins in paints is due to (i) gallic acid forming a complex with ferrous ion, which prevents the formation of magnetite, and (ii) the oxidation of ferrous gallic acid complex and precipitates like Fe(III) complex that blocks the access of ferrous ion and oxygen to cathodic sites and increases the ohmic resistance of local corrosion cells between the metal and magnetite (266).

6.9   CORROSION INHIBITION OF TITANIUM AND ZIRCONIUM IN ACID MEDIA

Titanium and zirconium as well as their alloys are extensively used in aircraft and nuclear plants, respectively. Titanium and zirconium are highly resistant to attack by chloride and acids such as nitric acid, hydrochloric acid, or sulfuric acid and hence are used in chemical process equipment. The ease with which passive films of TiO_2 and ZrO_2 are formed is responsible for their corrosion resistance. Thus, corrosion of titanium and zirconium is possible in media that dissolve the passive oxide film or prevent its formation. Thus, corrosion inhibition of these metals consists of improving the passivity of these metals.

TABLE 6.19 Corrosion Rates of Titanium in Acids with Added Salts (268)

Additive	Concentration (mol/L)	Corrosion Rate (mdd)	
		1% H_2SO_4	3% HCl
None	–	1130	672
Copper sulfate	0.01	1.92	–
Silver sulfate	0.01	0.0	–
Mercuric sulfate	0.01	0.0	–
Platinum chloride	0.01	0.96	2.88
Gold chloride	0.01	0.96	2.88
Mercuric chloride	0.01	–	0.96
Cupric chloride	0.01	–	0.96

One of the simple approaches to improving the passivity of titanium is the addition of salts of electropositive metals such as Cu, Hg, Pd, Ir, Pt, Au, Rh, and Re. Some data on the corrosion rates of titanium in the presence of added noble metal salts are given in Table 6.19.

The concentration of the noble metal salt must exceed the critical value in order for the added inhibitor to be effective. The effect of the concentration of noble metal ion on the corrosion rate of titanium in 15% HCl is illustrated by the data given in Table 6.20. The data show the increase in corrosion rates when the concentration of the noble metal ion is below the critical value for inhibition (269).

Nickel ions also reduce the corrosion rate of titanium. Even 0.2 ppm of nickel ions renders passivity to titanium in 3.5% NaCl. Nickel ions are thought to enhance the kinetics of the cathodic process. The nickel ions may also lower overvoltage of hydrogen liberation on titanium.

Ferric ions have been found to be a very effective inhibitor in acid media. The concentration of ferric ions must exceed the critical value for effective protection.

TABLE 6.20 Corrosion Rates of Titanium in 15% HCl

Additive	Concentration (g·ion/L) $\times 10^{-5}$	Corrosion Rate (g/(m^2 h))
None	–	0.23
Cu^{2+}	0.05	0.25
	0.10	0.32
	0.20	0.39
	0.30	0.38
	0.40	0.00
Pt^{4+}	0.05	0.55
	0.075	0.66
	0.10	0.59
	0.20	0.00

TABLE 6.21   Passivation Current and Critical Fe^{3+} Oxidizer Required to Passivate Titanium in Boiling H_2SO_4 Solution Containing 5% Na_2SO_4

H_2SO_4 (%)	Passivation Current Density (mA/cm^2)	Critical Concentration of (Fe^{3+}, mg/L)
1	0.05	0.040
2	0.20	0.085
3	0.40	0.195
5	1.80	0.250

TABLE 6.22   Effect of Oxidizing Inhibitors on the Corrosion Rate of Titanium (268)

Additive	Concentration	Corrosion Rate (g/(m^2 h)) H_2SO_4 (1%)	HCl (3%)
None	–	4.710	2.800
Ferric sulfate	0.01	0.00	–
Cerium sulfate	0.01	0.00	0.025
Sodium nitrite	0.01	0.004	0.008
Potassium permanganate	0.01	0.000	0.004
Sodium dichromate	0.01	0.004	0.000
Sodium molybdate	0.01	0.0	0.0
Sodium tungstate	0.01	0.0	0.0
Sodium iodate	0.01	0.0	0.004
Sodium bromate	0.01	0.004	0.016
Hydrogen peroxide	0.10	0.00	0.05

TABLE 6.23   Corrosion Rates of Titanium in Acids (270)

Corrosive Medium	Inhibitor	Corrosion Rate (mpy)	Reference
65% H_2SO_4	None	1430	270
70% H_2SO_4	5% HNO_3	<1.0	270
20% HCl	None	137	270
20% HCl	1% HNO_3	<1.0	270
5% HCl	None	2800	270
5% HCl	1% CrO_3	1.0	270
10 N H_2SO_4	$K_2Cr_2O_7$	–	271
5 N H_2SO_4	$K_2H_2Sb_2O_7$	–	272
H_2SO_4	Chloride	–	273
H_2SO_4	$SbCl_3$	–	274
HCl	SbO	–	275
H_2SO_4	Sulfite	–	276
H_2SO_4, HCl, HBr, HI	Chlorine	–	277

TABLE 6.24   Effect of Some Organic Inhibitors on the Corrosion of Titanium (278)

Inhibitors	Corrosion Inhibition (%)	Reference
10 N H_2SO_4	0	278
Picric acid	98.4	
α-Nitroso β-naphthol	92.4	
p-Nitroaniline	87.5	
Cupferron	85	
m-Dinitrobenzene	–	
Hexamethylenetetramine	–	
8-Nitroquinoline	>90	279
Azo compounds (neutral red, methyl orange)	~90	280
p-Benzoquinone	~90	281
Condensation products of formaldehyde with p-anisidine, p-toluidine, and nitroanilines at 3 mM/L in HCl	~90	282
Quinhydrone, 2,6-dinitrotoluene at 3 mM/L	~90	
0.04–0.08 g/L katapin A and ferric sulfate, p-nitroaniline or 0.1–0.2 mM/L neutral red	~90	283
Complexing agents as inhibitors		
3 mM xylenol orange with 30 mM of p-nitroaniline	~90	284
Phenyl arsonic acid	~90	
Mixture of diantipyryl methane and potassium thiocyanate inhibitor in HCl and H_2SO_4 solutions	~90	285
Cupferron is an effective inhibitor at 5 mM	~90	280
Hexamethylene tetramine, tartaric acid, EDTA	~70	
Hexamethylenediamine chromate	~90	
Phthalocyanines are effective when air-formed film is present on the metal	~80	
Cupferron at >5 mM is an effective inhibitor	>85	280

The data given in Table 6.21 show increased amounts of ferric ion with increase in acid concentration.

From polarization studies it has been shown that 0.03 M ferric ion acts as an effective inhibitor for the titanium alloy in 20% H_2SO_4 at 65°C. The inhibition of corrosion of titanium by chromium ions and sodium vanadate in sulfuric, hydrochloric, and phosphoric acids has been reported (270). Strong oxidizing agents such as nitric acid, potassium dichromate, potassium permanganate, potassium iodate, sodium chlorate, chlorine, and hydrogen peroxide have been found to be effective inhibitors. Some data on the effect of oxidizing agents in the corrosion inhibition of titanium are given in Table 6.22.

Corrosion rates of unalloyed titanium in acids with inhibitors (270) are given in Table 6.23

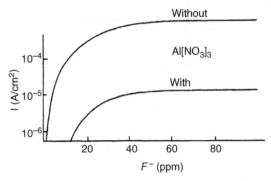

FIGURE 6.3 Polarization curve of zirconium alloy in 10% HCl containing 15 ppm fluoride with and without aluminum nitrate.

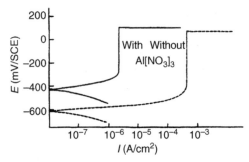

FIGURE 6.4 Effect of fluoride ion on passive current density of Zr alloy with and without added aluminum nitrate.

TABLE 6.25 Corrosion Resistance of Magnesium and Its Alloys in Various Environments

Resistant to	Not Resistant to
HF (>2%)	Inorganic acids except HF
H_2CrO_4 free from Cl^- and SO_4^{2-} ions	Organic acids
NaOH, KOH at <60°C	NaOH, KOH at >60°C
NaClO, concentrated NH_4OH	Chlorides, bromides, and iodides
Chromates, fluorides, and nitrates	Sulfates, persulfates, and chlorates
Phosphates of Na, K, Ca, Ba, Mg, and Al	Hypochlorites; salts of heavy metals displaced by Mg
Dry gases, F_2, Br_2, S_2, H_2S, and SO_2	Cl_2, I_2, NO, and NO_2; alkyl halides
Dry Freons	Moist Freons
Passivity at pH > 11.5	
Inhibition by porphyrins and phthalocyanines	

TABLE 6.26 Corrosion Resistance of Aluminum and Its Alloys in Various Environments

Resistant to	Not Resistant to
Acetic, citric, tartaric, and boric acid at room temperature	HCl, HBr, HF, H_2SO_4, $HClO_4$, H_3PO_4, formic acid, and trichloroacetic acid
80% HNO_3 at <50°C, fatty acids	Oxalic acid
$Ca(OH)_2$, >10% xNH_4OH (at <50°C), $(NH_4)_2S$, and Na_2SiO_3	LiOH, NaOH, KOH, NH_4OH, $Ba(OH)_2$, Na_2S, NaCN
Sulfates, nitrates, phosphates, and acetates of NH_4, Na, K, Ca, Ba, Mg, Mn, Zn, and Cd	Salts of metals such as Hg, Sn, Cu, Ag, Pb, Co, and Ni
Sodium hypochlorite inhibited with sodium sulfite, $NaClO_4$ without Cl^-; sodium sulfite, $NaClO_4$ without Cl^-, and 1–10% $KMnO_4$ at room temperature	$NaClO_4$-containing chloride, NaClO, and $CaClO_2$
Dry gases: Br_2, Cl_2, F_2, HCl, HBr, ozone, SO_2, SO_3, H_2S, CO_2, NO, NO_2, and NH_3	Moist SO_2, SO_3, Cl_2, HCl, NH_3, CCl_4, CH_3Cl, and CH_3Br
Passivity in the pH range 4–9	
Passivating inhibitors like potassium chromate, complexing agents such as EDTA, and hydroxy quinoline	

TABLE 6.27 Corrosion Resistance of Titanium in Various Environments

Resistant to	Not Resistant to
All concentrations of nitric acid up to boiling point, aqua regia at room temperature	
<10% HCl, H_2SO_4 at room temperature	>10% HCl, H_2SO_4, HF, fuming HNO_3
<30% H_3PO_4 at 35°C; <5% boiling H_3PO_4, chromic, acetic, oxalic, lactic, and formic acid at room temperature	>30% H_3PO_4 at 35°C, boiling >5% H_3PO_4, boiling formic acid, oxalic acid, and trichloroacetic acid
Dilute solutions of NaClO at room temperature	Hot concentrated alkali
Most salt solutions including chlorides such as $FeCl_3$ and $CuCl_2$	ALF_3, boiling $AlCl_3$, $MgCl_2$, $CaCL_2$, and F_2
Moist Cl_2, ClO_2; air, O_2 at <425°, N_2, H_2 at <750°C	Dry Cl_2, air, O_2 at >500°C, N_2 > 800°C, H_2 > 750°C
Passivating inhibitors may be used, organic inhibitors such as nitrophenols and nitroanilines may also be used	

TABLE 6.28 Corrosion Resistance of Chromium in Various Environments

Resistant to	Not Resistant to
$<50\%$ HNO$_3$ at $<75°C$; H$_2$SO$_4$ + CuSO$_4$; H$_2$SO$_4$ + Fe$_2$(SO$_4$)$_3$	HCl, HBr, HI, conc. HNO$_3$ at high temperatures
$<5\%$ H$_2$SO$_4$ aerated at room temperature	$>5\%$ H$_2$SO$_4$ at $>50°C$
SO$_2$ solutions	HF, H$_2$SiF$_6$, HClO$_3$, $>60\%$ H$_3$PO$_4$ at $>100°C$; H$_2$CrO$_4$
Aerated H$_3$PO$_4$ solutions at room temperature	Aerated alkalis at high temperatures
Nonhalide salts	FeCl$_3$, CuCl$_2$, HgCl$_2$, NaClO, thiosulfates, and dithionites
O$_2$ $<1100°C$; H$_2$O $<850°C$; SO$_2$ $<650°C$	O$_2$ $>1100°C$; H$_2$O $>850°C$; SO$_2$ $>650°C$
H$_2$S, S$_2$ $<500°C$	H$_2$S, S$_2$ $>500°C$
NH$_3$ $<500°C$; Cl$_2$, HCl $<300°C$	NH$_3$ $>500°C$; Cl$_2$, HCl $>300°C$
F$_2$, HF $<250°C$	F$_2$, HF $>250°C$
Corrosion inhibitors consist of passivating inhibitors and organic ligands containing nitrogen and oxygen donor atoms	

The performance of some organic compounds on the corrosion of titanium in acid solutions is shown in Table 6.24.

The effective inhibitors for titanium and zirconium are passivating inhibitors. Oxidizing inhibitors act as cathodic depolarizers. It is necessary to maintain a critical concentration of the oxidizing inhibitor for effective inhibition. In fluoride media, inhibition is achieved by adding an aluminum salt that forms aluminum fluoride complex in preference to titanium and zirconium. The polarization curve of a zirconium alloy in HCl solution containing fluoride and the effect of fluoride on passive current density of Zr are shown in Figs. 6.3 and 6.4, respectively.

TABLE 6.29 Corrosion Resistance of Iron in Various Environments

Resistant to	Not Resistant to
H$_2$CrO$_4$, conc. HNO$_3$; 70% H$_2$SO$_4$; 70% HF	All acids not listed in resistant column
Most alkaline solutions	Hot concentrated alkalis in stressed condition
>1 g/L KMnO$_4$	<1 g/L KMnO$_4$
>3 g/L H$_2$O$_2$	<3 g/L H$_2$O$_2$
K$_2$CrO$_4$	Oxidizing salts FeCl$_3$, CuCl$_2$, and NaNO$_3$; hydrolyzing salts: AlCl$_3$, Al$_2$(SO$_4$)$_3$, ZnCl$_2$, and MgCl$_2$
Air $<500°C$; Cl$_2$ $<200°C$	Air $>450°C$; Cl$_2$ $>200°C$
Dry SO$_2$, $<300°C$; NH$_3$ $<500°C$	F$_2$, SO$_2$ moist, NH$_3$ $>500°C$
H$_2$O(g) $<500°C$; H$_2$S $<300°C$	H$_2$O(g) $>500°C$; H$_2$S $>300°C$
Passivating inhibitors such as chromate, molybdate, and tungstate	
Primary amines and aromatic amines	

TABLE 6.30   Corrosion Resistance of Nickel in Various Environments

Resistance to	Not Resistant to
Dilute nonoxidizing acids	Oxidizing acids
Deaerated H_2SO_4 (<80%) at room temperature; <1% aerated HCl at room temperature	HNO_3
HF at room temperature	HF at high temperatures; hot concentrated H_3PO_4
Deaerated dilute organic acids	Aerated organic acids
Deaerated H_3PO_4 at room temperature; LiOH, NaOH, KOH at bp	>1% NH_4OH
Most nonoxidizing salts; $NaClO_4$, $KMnO_4$ at room temperature	Most oxidizing salts ($FeCl_3$, $CuCl_2$, and $K_2Cr_2O_7$) NaClO
Dry halogens <200°C; dry hydrogen halides, <200°C	Moist halogens and hydrogen halides
$H_2O(g)$ <500°C	$H_2O(g)$ >500°C
H_2 <550°C	Cl_2 >450°C
SO_2 <400°C	H_2S >65°C; NH_3 at high temperatures
S_2 <300°C	S_2 >300°C
Passivating inhibitors and organic inhibitors containing nitrogen and oxygen donor atoms such as dimethyl glyoxime	

TABLE 6.31   Corrosion Resistance of Copper in Various Environments

Resistant to	Not Resistant to
Deaerated, nonoxidizing acids	HNO_3, H_2SO_4, hot conc. aerated acids
<10% HCl at <75°C	>10% HCl
<70% HF at <100°C	Strong NaOH, KOH, NH_4OH, NaCN, KCN, and NaClO solutions
<60% H_2SO_4 at <100°C	Most oxidizing salts: $FeCl_3$, $Fe_2(SO_4)_3$, $CuCl_2$, and $Hg(NO_3)_2$
H_3PO_4 and acetic acid at room temperature	$AgNO_3$
Dilute solutions of NaOH, KOH, Na_2CO_3, K_2CO_3, $KMnO_4$, K_2CrO_4, and $NaClO_3$; deaerated solutions of sulfates, nitrates and chlorides; seawater	Aerated and agitated salt solutions
Dry gases: CO, CO_2, F_2, Cl_2, Br_2, SO_2, and H_2	Moist gases: SO_2, H_2S, CS_2, CO_2, F_2, Cl_2, and Br_2; H_2 containing O_2
O_2 <200°C, OF_2, ClF_3, and ClO_3F	Dry O_2 >200°C
Corrosion inhibitors such as organic compounds containing nitrogen and sulfur donor atoms; passivity in the pH range 8–12	

TABLE 6.32   Corrosion Resistance of Silver in Various Environments

Resistant to	Not Resistant to
Dilute HCl at room temperature	Concentrated HCl at high temperatures
HF at low temperatures, H_3PO_4 at room temperature	Aerated HF at high temperatures
H_2SO_4 at room temperature, organic acids; LiOH, NaOH, and KOH	Dilute HNO_3 at room temperature
Most nonoxidizing salts	Hot concentrated H_3PO_4
$KMnO_4$ at room temperature	Hot conc. H_2SO_4
F_2, Cl_2, and Br_2 at room temperature	NH_4OH, Na_2S, and NaCN
HCl at $<200°C$	Oxidizing agents—$K_2S_2O_8$, $FeCl_3$
SO_2 at room temperature	$CuCl_2$, $HgCl_2$
Air, O_2	Complexing salts: cyanides, thiosulfates, and polysulfides
Corrosion inhibitors used contain nitrogen as donor atom	HCl and Cl_2 at $>200°C$
	H_2S and S_2 at room temperature
	SO_2 at high temperatures

6.10   CORROSION RESISTANCE OF SEVERAL METALS AND ALLOYS

Corrosion resistance of several metals and their alloys in various environments are given in Tables 6.25–6.38.

TABLE 6.33   Corrosion Resistance of Zinc in Various Environments

Resistant to	Not Resistant to
pH <12	pH >12
Na_2CrO_4, $Na_4B_2O_7$, Na_2SiO_3, $(NaPO_3)_6$ (inhibitors)	All common inorganic and organic salts
1 g/L at room temperature	
Na, CO_2, CO, N_2O	Aerated salt solutions
Dry Cl_2, NH_3	Moist Cl_2, C_2H_2
Passivity in the pH range 8–12	

TABLE 6.34   Corrosion Resistance of Cadmium in Various Environments

Resistant to	Not Resistant to
Dilute LiOH, NaOH, KOH, NH_4OH; Na_2CrO_4, $Na_4B_2O_7$, Na_2SiO_3	All common inorganic and organic acids; concentrated LiOH, NaOH, KOH
$(NaPO_3)_6$ act as inhibitors at 1 g/L and 25°C	Aerated salt solutions
Dry NH_3; air, O_2 at $<250°C$	Cl_2, Br_2
Dry SO_2 at room temperature	Moist SO_2
Passivity in the pH range 6–13.5	

TABLE 6.35   Corrosion Resistance of Tin in Various Environments

Resistant to	Not Resistant to
Dilute, deaerated nonoxidizing inorganic and organic acids	Oxidizing acids, aerated mineral and organic acids
pH < 12; higher pH values in the presence of silicates, phosphates, and chromates	pH > 12; chlorides, sulfates, and nitrates
Phosphates, chromates, borates	Salts of metals nobler than tin; Cl_2, Br_2, I_2 at room temperature
Fluorine at <100°C	F_2, O_2, H_2S at above 100°C

TABLE 6.36   Corrosion Resistance of Lead in Various Environments

Resistant to	Not Resistant to
H_2SO_4 <96%, room temperature	H_2SO_4 >96%, room temperature
H_2SO_4 <80%, 100°C	H_2SO_4 >70% boiling
Commercial H_3PO_4 containing some H_2SO_4	Pure H_3PO_4
H_2CrO_4 <60% HF at room temperature	HNO_3 < 80%
H_2SO_3	HCl
pH <11.0	Organic acids; LiOH, NaOH, KOH, pH >12
<1% NH_4OH	>1% NH_4OH
Na_2CO_3	
Concrete; sulfates, carbonates, and bicarbonates	$FeCl_3$
Pure $NaClO_4$	$NaClO_4$ with NaCl
	Nitrates, acetates
Cl_2, moist or dry, <100°C	Cl_2 >100°C
Br_2, dry, room temperature	Moist Br_2 or at high temperatures
SO_2, SO_3, H_2S	HF
Passivity in the pH range 5–11	

TABLE 6.37   Corrosion Resistance of Zirconium in Various Environments

Resistant to	Not Resistant to
Acids such as HCl, H_2SO_4, acetic acid	HF and Cl_2 at high temperatures
Passivity due to ZrO_2 in the pH range 4–13	

TABLE 6.38   Corrosion Resistance of Tantalum in Various Environments

Resistant to	Not Resistant to
Most acids HCl, HNO_3, and H_2SO_4	Mixture of HNO_3 and HF
	Fused alkalis and nonmetals at high temperatures

REFERENCES

1. DS Leek, AB Poole, *Corrosion of Reinforcement in Concrete*, Elsevier Science Publishers Ltd., Wishaw, Warwickshire, 1990, pp. 65–73.

2. NS Berke, TG Weil, *Advances in Concrete Technology*, CANMET, Montreal, 1992.

3. B El-Jazairi, NS Berke, *Corrosion of Reinforcement in Concrete*, Elsevier Science Publishers Ltd., Wishaw, Warwickshire, 1990, pp. 571–585.

4. NS Berke, *ASTM STP 906*, American Society for Testing and Materials, Philadelphia, PA, 1986, pp. 78–91.

5. NS Berke, DF Shen, KF Sundberg, *ASTM STP 1065*, American Society for Testing and Materials, Philadelphia, PA, 1990, pp. 38–51.

6. NS Berke, M Hicks, BI Abdelrazig, TP Lees, A belt and braces approach to corrosion protection. In: RN Swamy (ed.), *Corrosion and Corrosion Protection of Steel in Concrete,* Sheffield, Academic Press, Sheffield, 1994, pp. 893–904.

7. NS Berke, MC Hicks, RJ Hoopes, PJ Tourney, Use of laboratory techniques to evaluate long-term durability of steel reinforced concrete exposed to chloride ingress. *ACI SP 145*, Nice, France, pp. 299–329, 1994.

8. NS Berke, *Mater Perform* **27**(10):41–44 (1989).

9. BB Hope, AKC Ip, *Corrosion Inhibitors for Use in New Concrete Construction,* Research and Development Branch, Ontario Ministry of Transportation, 1987.

10. R Cigna, G Familiari, F Gianetti, E Proverbio, Influence of calcium nitrite on the reinforcement corrosion in concrete mixtures containing different cements. In: RN Swamy (ed.), *Corrosion and Corrosion Protection of Steel in Concrete*, Sheffield Academic Press, Sheffield, 1994, pp. 878–892.

11. WD Collins, RE Weyers, IL Al-Qadi, *Corrosion* **49**:74–88 (1993).

12. BB Hope, AKC Ip, *ACI Mater J* **86**:602–608 (1989).

13. CA Loto, *Corrosion* **48**:759–763 (1992).

14. KK Sagoe-Crentsil, VT Yilmaz, FP Glasser, *Adv Cement Res* **415**: 91-96 (1991–1992).

15. SV Thompson, The use of chemical inhibitors to prevent corrosion of reinforcing steel in reinforced concrete structures, thesis, Queen's University, Kingston, ON, 1991.

16. F Tomosawa, Y Masuda, I Fukushi, M Takajura, T Hori, Experimental study on the effectiveness of corrosion inhibitor in reinforced concrete, *RILEM Symposium on Concrete Durability*, Barcelona, Spain, 1991, pp. 382–391.

17. VT Yilmaz, KK Sagoe-Crentsil, FP Glasser, *Adv Cement Res* **415**:97–102 (1991–1992).

18. CA Loto, *Corrosion* **48**:759 (1992).

19. KK Sagoe-Crentsil, VT Yilmaz, FP Glasser, *Adv Cement Res* **415**: 91–96 (1991–1992).

20. WD Collins, RE Weyers, IL Al-Qadi, *Corrosion* **49**(1):74–88 (1993).

21. CK Nmai, SA Farrington, GS Bobrowski, *Concrete Int* **45**:51 (1992).

22. C Monticelli, A Frignani, G Trabanelli, G Brunoro, *Proceedings of the 8th European Symposium on Corrosion Inhibitors*, Vol.1, University of Ferrara, 1995, p. 609.

23. IP Anoschenko, EV Puzey, AN Tulenev, *Proceedings of the 8th European Symposium on Corrosion Inhibitors*, Vol.1, University of Ferrara, 1995, p. 671.

24. G Wieczorek, J Gust, *Proceedings of the 8th European Symposium on Corrosion Inhibitors*, Vol. 1, University of Ferrara, 1995, p. 599.

25. B Miksic, L Gelner, D Bjegovic, L Sipos, *Proceedings of the 8th European Symposium on Corrosion Inhibitors*, Vol. 1, University of Ferrara, 1995, p. 569.

26. VS Sastri, R Beauprie, M Desgagne, *Mater Perform* **25**:45 (1986).

27. VS Sastri, J Bednar, *Mater Perform* **29**:41 (1990).

28. GB Hatch, In: CC Nathan (ed.), *Corrosion Inhibitors*, NACE, Houston, TX, 1973, p. 126.

29. JF Conoby, TM Swain, *Mater Prot* **6**(4):55 (1967).

30. GB Hatch, *Ind Eng Chem* **44**(8):1780 (1952).

31. G Butler, *Proceedings of the 3rd European Symposium on Corrosion Inhibitors*, University of Ferrara, 1970, p. 753.

32. BP Bofardi, In: A Raman, P Labine (eds), *Reviews on Corrosion Inhibitor Science and Technology*, NACE, Houston, TX, 1993.

33. MS Vukasovich, In: A Raman, P Labine (eds), *Reviews on Corrosion Inhibitor Science and Technology*, Paper No. II-12-1, NACE, Houston, TX, 1993.

34. GD Wilcox, DR Gabe, ME Warwick, *Corrosion* **6**:327 (1986).

35. MS Vukasovich, JPG Farr, *Mater Perform* **25**:9 (1986).

36. MS Vukasovich, *Mater Perform* **29**:53 (1990).

37. MJ Pryor, M Cohen, *J Electrochem Soc* **100**:203 (1953).

38. MA Stranick, *Corrosion* **40**:296 (1984).

39. JPG Farr, M Saremi, *Surf Technol* **19**:137 (1983).

40. VS Sastri, RH Packwood, *Werkst Korros* **38**:77 (1987).

41. VS Sastri, RH Packwood, JR Brown, JS Bednar, LE Galbraith, VE Moore, *Br Corros J* **24**:30 (1989).

42. M Stern, *J Electrochem Soc* **105**:638 (1958).

43. K Sugomoto, Y Sawada, *Corrosion* **32**:347 (1976).

44. MA Stranick, *Corrosion 85*, Paper No. 380, NACE, Houston, TX.

45. RC McCune, R Shilts, SM Ferguson, *Corros Sci* **22**:1049 (1982).

46. D Bijimi, DR Gabe, *Br Corros J* **18**:138 (1983).

47. D Bijimi, DR Gabe, *Br Corros J* **18**, 88 (1983).

48. H Lamprey, US Patent 2,147,395 (1939).

49. AD Meighen, US Patent 2,803,604 (1957).

50. ST Hirozawa, JC Wilson, US Patent 4,210,549 (1980).

51. ST Hirozawa, European Patent Appl 0042937AI (1982).

52. T Yoshioka, Japanese Patent JP57174472 (1982).

53. LC Rowe, RL Chance, MS Walker, *Mater Perform* **22**:17 (1983).

54. PR Engelhardt, EM Ventura, UK Patent Appl 8409522 (1984).

55. PR Engelhardt, EM Ventura, German Offen DE3413416 (1984).

56. JC Wilson, ST Hirozawa, JJ Conville, US Patent 4,440,721 (1984).

57. JW Darden, US Patent 4,561,990 (1985).

58. T Kamei, T Fuji, M Sootome, Japanese Patent JP60240778, 60243185 (1985).

59. ST Hirozawa, European Patent Appl 248346 (1987).

60. WD Robertson, *Chem Eng* **57**:290 (1950).

61. A Weisstuch, CE Schell, *Corrosion* **28**:299 (1972).

62. DR Robitaille, *Mater Perform* **15**:40 (1976).

63. RD Hogue, TM King, RS Mitchell, US Patent 3,989,637 (1976).

64. MS Vukasovich, DR Robitaille, *J Less-Common Metals* **54**:437 (1977).

65. TC Breske, *Mater Perform* **16**:17 (1977).

66. DR Robitaille, MS Vukasovich, US Patent 4,149,969 (1979).

67. KM Verma, MP Gupta, RN Ghosh, BB Simka, *Fet Technol* **16**:135 (1979).

68. F Suzuki, US Patent 4,176,059 (1979).

69. RJ Lipinski, US Patent 4,138,353 (1979).

70. DR Robitaille, *Ind Water Eng* **16**(6):14 (1979).

71. RD Hogue, TM King, RS Mitchell, US Patent 4,209,487 (1980).

72. A Marshall, N Richardson, US Patent 4,239,648 (1980).

73. SK Saksenberg, *Corrosion 81*, Paper No. 189, NACE, Houston, TX, 1981.

74. DR Robitaille, *Chem Eng* **89**:139 (1982).

75. P Labine, T Wells, J Minalga, S Roberts, B Ritts, *Corrosion 82*, Paper No. 227, NACE, Houston, TX, 1982.

76. Kurita Water Industries, Japanese Patent JP58171577 (1983).

77. DM Drazic, CS Hao, *Corros Sci* **23**:683 (1963).

78. M Khobaib, L Quakenbush, CT Lynch, Corrosion **39**:253 (1983).

79. EJ Latos, JC Payne, US Patent 4,409,121 (1983).

80. Kurita Water Industries, Japanese Patent JP58164790, 1979.

81. TF Koitwii, KJ Kozelski, *Mater Perform* **23**:43 (1984).

82. MR Palmer, US Patent 675,115 (1985).

83. QJM Slaiman, AF Al-Shammary, *J Pet Res* **4**:75 (1985).

84. CA Jones, *J Cooling Tower Inst* **6**:9 (1985).

85. RJ Franco, N Dineilli, RJ Nowicki, BT Corrosion 85, Paper No. 133, NACE, Houston, TX, 1985.

86. TR Weber, MA Stranick, MS Vukasovich, *Corrosion* **42**:542 (1986).

87. A Cepero, *Rev Cienc Quim* **16**:55 (1985).

88. T Takahasi, T Imari, S Ano, T Kenya, Japanese Patent JP62280381 (1987).

89. WD Robertson, *Chem Eng* **57**:290 (1950).

90. DH Killefer, *Paint Oil Chem Rev* **117**:24 (1954).

91. AK Choudhury, SC Shome, *J Sci Ind Res* **17A**:30 (1958).

92. Sumitomo Metal Industries Ltd., Japanese Patent JP59211580 (1984).

93. JE Brophy, VG Fitzsimmons, JG O'Rear, TR Price, WA Zisman, *Ind Eng Chem* **43**:884 (1954).

94. NS Dempster, *Corrosion* **15**:395 (1959).

95. M Noda, H Nakai, M Sasaki, Z Kanno, Japanese Patent JP79116338 (1979).

96. MS Vukasovich, *Lubric Eng* **36**:708 (1980).

97. KW Koh, US Patent 4,218,329 (1980).

98. A Marshall, N Richardson, US Patent 4,239,648 (1980).

99. MS Vukasovich, DR Robitaille, US Patent 4,313,837 (1982).

100. S Kimura, T Sato, Japanese Patent JP60177099 (1985).

101. MS Vukasovich, *Lubric Eng* **40**:456 (1984).

102. IL Rozenfeld, S Ch Verdiev, AM Kyazimov, A Kh Kyazimov, *Zasch Met* **18**:866 (1982).

103. S Ch Verdiev, LP Kazanski, A Kh Bairomov, AM Kyazimov, *Zasch Met* **23**:264 (1987).

104. RC Spooner, US Patent 2,899,368 (1959).

105. EF Barkman, US Patent 3,257,244 (1966).

106. HB Romans, US Patent 3,272,665 (1966).

107. W Fassell, US Patent 3,852,124 (1974).

108. ER Reinhold, US Patent 4,146,410 (1979).

109. AK Bairamov, S Zakipor, C Leygrof, *Corros Sci* **25**:69 (1985).

110. Yu N Mikhailovskii, GA Berdzenishvili, *Zasch Met* **22**:699 (1986).

111. DE Miller, US Patent 2,557,509 (1951).

112. Asahichem Industry Co. Ltd., Japanese Patent JP59200769 (1984).

113. Sumitomo Metal Industries Ltd., Japanese Patent JP59211580 (1984).

114. CR Clayton, YC Lu, *J Electrochem Soc* **133**:2465 (1986).

115. E Baron, GB Freeman, J Gluszek, J Kubicki, J Masalaski, *Zasch Met* **23**:82 (1987).

116. S Maeda, M Yamamoto, Japanese Patent JP6237377 (1987).

117. Y Sone, K Wada, S Narutani, S Suzuki, Japanese Patent JP62158898 (1987).

118. K Kurosawa, T Fukushima, *Nippon Kagaku Kaishi* **10**:1822 (1987).

119. H Lijima, Y Goto, M Teresaka, US Patent 3,586,543 (1971).

120. HR Charles, DL Miles, US Patent 3,819,423 (1974).

121. RD Wyvill, *Finish Indus* **3**:52 (1979).

122. GS Selyukova, RI Gudkova, N Satova, USSR Patent SU1339162 (1987).

123. D Bijimi, DR Gabe, *Br Corros J* **19**:196 (1984).

124. V Buettner, JL Josten, German Offen DE3407095 (1985).

125. YS Shindo, W Hotta, Japanese Patent JP62238399 (1987).

126. RS Glass, *Corrosion* **41**:89 (1985).

127. D Bijimi, DR Gabe, *Br Corros J* **18**:88, 93 (1983).

128. RA Hoffman, RO Hull, *Proc Am Electroplater Soc* **1**:45 (1939).

129. EW Schweikher, US Patent 2,351,639 (1944); 2,417,133 (1947).

130. Y Nakamura, TR Arakai, Japanese Patent JP6270584 (1987).

131. SM Gugel, *Poroshk Metl* **9**:101 (1987).

132. Showa Denko, Japanese Patent JP58224185, 58224186,58224187 (1983).

133. M Itoh, M Aizawa, K Tanno, *Bashoku Gijitsu* **36**:142 (1987).

134. Sanyo Electric Co. Ltd., Japanese Patent JP5993778 (1984).

135. M Ito, AM Aizawa, K Tanno, *Boshoku Gijutsu* **36**:142 (1987).

136. T Asano, K Nishimura, H Nonaka, A Ohayashi, Japanese Patent JP62129127, 1986.

137. AJ Son, MS Kuzlik, European Patent Appl EP15319228 (1985).

138. M Iochev, A Lilova, *Proceedings of the 6th European Symposium on Corrosion Inhibitors*, Sez. V, suppl. No. 8, Ferrara, Italy, 1985.

139. VS Agarwala, *New Mater New Process* **3**:178 (1985).

140. T Asano, K Nishimura, H Nonaka, A Manako, A Ohayashi, Japanese Patent JP62129127 (1987).

141. Chiyoda Kagaku Kenkyusho, Japanese Patent JP6018590 (1985).

142. S Katsumata, J Kuno, K Yanakihara, A Fukushima, T Kamakura, Japanese Patent JP61106587 (1986).

143. T Handa, N Tanaka, K Yanagihara, K Fukushima, T Kamakura, Japanese Patent JP6243491 (1987).

144. VM Shkolnikov, Yu N Shektev, MV Pospelov, A Ya Furman, NV Kardash, *Zasch Met* **23**:774 (1987).

145. JJ Valcho, SE Lindberg, MW Hunt, AR Sabol, US Patent 4,601,837 (1986).

146. X Xu, J Dun, G Zhou, J Zhao, Chinese Patent CN85102380 (1987).

147. Katayama Chemical Works Co. Ltd., Japanese Patent JP5916984 (1984).

148. Katayama Chemical Works Co. Ltd., Japanese Patent JP5992097 (1984).

149. R Rizzi, S Bonato, L Forte, *Proceedings of the 6th European Symposium on Corrosion Inhibitors*, Sez V, Suppl. No. 8, Ferrara, Italy, 1985.

150. F Drobny, J Pavlovic, Czechoslovakian Patent CS206067 (1983).

151. L Skupin, M Horyna, Czechoslovakian Patent CS222144 (1985).

152. SC Jain, CA Morris, US Patent 4,287,073 (1981).

153. H Nakagawa, E Nishi, Japanese Patent JP60161492 (1985).

154. Matsushita Electric Industrial Co. Ltd., Japanese Patent JP5060970, 5060865 (1984).

155. Honshu Paper Co. Ltd., Japanese Patent JP6071800 (1985).

156. SA Ginsberg, AV Schreider, *J Appl Chem USSR* **33**:2334 (1960).

157. NMR Castro, Use of inhibitors for control of crevice corrosion, MSc thesis, Colorado School of Mines, Golden, CO, 1986.

158. AW Lui, GR Hoey, *Mater Perform* **15**:13 (1976).

159. AW Lui, VS Sastri, J McGoey, *Br Corros J* **29**:140 (1994).

160. VS Sastri, R Beauprie, M Desgagne, *Mater Perform* **25**:45 (1986).

161. VS Sastri, J Bednar, *Mater Perform* **29**:42 (1990).

162. Aisin Kako Ltd., Japanese Patent JP5980542 (1984).

163. W Ullrich, D Lange, R Rosert, T Neubert, German Patent DD235880 (1986).

164. Sumitamo Metal Industries Ltd., Japanese Patent JP59228591 (1984).

165. M Mokhosoev, LV Tumurova, EV Kvashnina, *Zasch Met* **21**:825 (1985).

166. M Naka, K Hashimoto, T Masimoto, *J Non-Cryst Solids* **28**:403 (1978).

167. VS Raja, S Ranganathan, *Corrosion* **44**:263 (1988).

168. TI Kurilovich, NG Klyuchnikov, *Inhibit Korroz Met* **168** (1972); *Chem Abstr* **83**:210657j (1975).

169. MN Desai, HM Sheth, *Vidya* **B19**:118 (1976); *Chem Abstr* **88**:40661q (1978)

170. MN Desai, ST Desai, *Chem Concept* **5**(7):15 (1978); *Chem Abstr* **89**:219453 (1978).

171. MN Patel, NK Patel, JC Vora, *Chem Era* **11**(5):4 (1975); *Chem Abstr* **87**:9998 (1977).

172. L Homer, K Meisel, *Werkst Korros* **29**:654 (1978).

173. SC Makwana, NK Patel, JC Vora, *J Ind Chem Soc* **50**(10):664 (1973); *Chem Abstr* **80**:127351X (1974); *Werkst Korros* **24**:1036 (1974).

174. CM Makwana, AC Bhavsar, DC Gandhi, *Chem Era* **12**(4):145 (1976); *Chem Abstr* **86**:125613j (1977)

175. SC Makwana, NK Patel, JC Vora, *Trans Soc Adv Electrochem-Sci Technol* **12**(1):15 (1977); *Chem Abstr* **88**:64933f (1978)

176. PN Clark, E Jackson, M Robinson, *Br Corros J* **14**:33 (1979).

177. F Mansfeld, JV Kenkel, *Corros Sci* **15**:767 (1975).

178. EI Tupikin, NG Klyuchnikov, MK Verzilina, *Izv Vyssh Ucheb Zaved Khim Tekhnol* **17**(4):**514** (1974); *Chem Abstr* **81**:32513 (1974).

179. LI Gerasyutina, FM Tulyupa, LG Karvaka, *Zasch Met* **14**: **6** (1978); *Chem Abstr* **90**:158952e (1979)

180. EJ Tupikin, NG Klyuchnikov, GL Nemchaninova, *Zasch Met* **11**(3):351 (1975); *Chem Abstr* **831**:055287 (1975).

181. L Heiss, M Hille, US Patent 3,773,675 (Nov 20, 1973); *Chem Abstr* **80**:148130 (1974).

182. WW Frenier, WJ Settineri, US Patent 3,764,543 (1974); *Chem Abstr* **85**:86674 (1976); US Patent 3,969,414 (July 13, 1976); *Chem Abstr* **85**:123665 (1976).

183. WJ Settineri, WW Frenier, JR Oswald, US Patent 3,996,147 (Dec 7, 1976); *Chem Abstr* **86**:58948 (1977).

184. AN Frumkin, *Z Elektrochem* **59**:807 (1955).

185. ZA Iofa, *Zasch Met* **6**(5):491 (1970); *Chem Abstr* **73**:136607 (1973); *Zasch Met* **8**(2):139 (1972); *Chem Abstr* **77**:55465a (1972)

186. KD Allabergenov, FK Kurbanov, *Zasch Met* **15**(4):472 (1979); *Chem Abstr* **91**:179255r (1979)

187. VI Sorokin, VP Romasenko, *Ukr Khim Zh* **40**(8):804 (1974); *Chem Abstr* **83**:67624p (1975)

188. VS Kolevatova, *Zh Prikl Khim* **48**(10):2216 (1975); *Chem Abstr* **84**:10323t (1976)

189. VS Kolevatova, VI Korobkov, *Zh Prikl Khim* **49**(1):86 (1976); *Chem Abstr* **84**:127942y (1976)

190. Z Ahmad, M Ghafelchbashi, S Nategh, S Jahanfar, *Met Corros Ind* (1975).

191. NK Patel, MM Patel, KC Patel, *Chem Era* **10**(12):24 (1975); *Chem Abstr* **84**:77811 (1976).

192. NK Patel, MM Patel, LN Patel, SH Mehta, *Chem Era* **12**(2):46 (1976); *Chem Abstr* **86**:32773 (1977).

193. EI Tupikin, NG Klyuchnikov, MK Verzilina, *Zasch Met* **10**(1):65 (1974); *Chem Abstr* **81**:53488 (1974).

194. EI Tupikin, NG Klyuchnikov, *Izv Vyssh Uchebn Zaved Khim Tekhnol* **20**(5):790 (1977); *Chem Abstr* **87**:92514 (1977).

195. MM Gleizer, Kh Tseitlim, Yu L Sorokin, GI Isaenko, SM Babitskara, *Zasch Met* **12**(5):629 (1976); *Chem Abstr* **86**:94505 (1977).

196. LI Gerasyutina, FM Tulyupa, NL Gromova, LG Koryaka, *Zasch Met* **12**(2):195 (1976); *Chem Abstr* **85**:98142a (1976)

197. JF Eberhard, BD Oakes, US Patent 3,037,934 (1962); *Chem Abstr* **57**:6970a (1962).

198. JF Eberhard, US Patent 2,963,439 (Dec 6, 1960); *Chem Abstr* **55**:10294 (1961).

199. MV Uzlyuk, Yu V Federov, AM Pinos, VF Tolstykh, ZV Panfilova, LI Shatukhina, S Miskidzh'van, *Zasch Met* **13**(2):212 (1977); *Chem Abstr* **86**:162689j (1977)

200. JE Mahan, RA Stahl, US Patent 2,769,690 (Nov 6, 1956); *Chem Abstr* **51**:3430c (1957).

201. MN Desai, *Werkst Korros* **23**:483 (1972).

202. MN Desai, VK Shah, *Corros Sci* **11**:725 (1972).

203. MN Desai, GH Thanki, *J Electrochem Soc (India)* **21**: 13 (1972); *Chem Abstr* **76**:161550p (1972)

204. MN Desai, JS Joshi, *J Inst Chem (Calcutta)* **44**: 138 (1972); *Chem Abstr* **78**:516035 (1972).

205. MN Desai, GH Thanki, *Labdev* **10**(Part A): 73 (1972); *Chem Abstr* **79**:26416w (1973)

206. MN Desai, SS Rana, *Anti-Corros Method Mater* **20**(6):16 (1973); *Chem Abstr* **79**:11103m (1973)

207. MN Desai, DK Shah, HM Gandhi, *Br Corros J* **10**(1):39 (1975).

208. MN Desai, DK Shah, *Vidya B* **19**(2):226 (1976); *Chem Abstr* **88**:77679c (1978)

209. MN Desai, BC Thaker, PM Chhaya, *J Ind Chem Soc* **52**(10):950 (1975); *Chem Abstr* **84**:97026 (1976).

210. MN Desai, SS Rana, *Anti-Corros Method Mater* **20**(5):**16** (1973); *Chem Abstr* **79**:86712z (1973).

211. MN Desai, GH Thanki, DK Shah, *Anti-Corros Method Mater* **18**:8 (1972); *Chem Abstr* **76**:67228t (1972).

212. MN Desai, BC Thaker, BM Patel, *J Electrochem Soc (India)* **24**:84 (1974); *Chem Abstr* **84**:139675u (1976).

213. MN Desai, SS Rana, *Anti-Corros Method Mater* **20**:16 (1973); *Chem Abstr* **79**:86712z (1973).

214. M Patel, NK Patel, JC Vora, *India J Technol* **12**:469 (1974); *Chem Abstr* **82**:78400t (1975).

215. MM Patel, NK Patel, JC Vora, *J Electrochem Soc (India)* **27**:171 (1978); *Chem Abstr* **90**:129380m (1979).

216. H Keller, K Risch, *Werkst Korros* **15**:741 (1964).

217. J Vosta, J Pelikan, M Smrz, *Werkst Korros* **25**:750 (1974).

218. F Pearlstein, RF Weightman, *Chem Abstr* **84**:183167 (1976).

219. JD Anderson, ES Hayman, EA Rodzewich, US Patent 3,992,313 (Nov 16, 1976); *Chem Abstr* **87**:139496 (1977).

220. MN Maksimenko, NI Podobaev, *Inhibit J Corros Met* **2**:78 (1972); *Chem Abstr* **84**:48406 (1976).

221. MV Uzlyuk, Yu Federov, VF Voloshin, ZV Parfilova, LI Shatukina, LG Aleinikova, *Zasch Met* **9**:446 (1973); *Chem Abstr* **80**:51152 (1974).

222. JE Malowan, US Patent 2,567,156 (1951); *Chem Abstr* **46**:876e (1952).

223. JD Talati, M Pandya, *Anti-Corros Method Mater* **21**:7 (1974); *Chem Abstr* **83**:17561 (1975).

224. JD Talati, M Pandya, *Corros Sci* **16**:603 (1976).

225. BP Ignatov, IP Anoshenko, VV Pyaterikov, GI Luk'vanova, *EA Tr Novocherk Politek Inst* **41**:285 (1975); *Chem Abstr* **82**:130921 (1975).

226. HD Clark, US Patent 3,969,260 (July 13, 1976); *Chem Abstr* **85**:165428 (1976).

227. WW Frenier, WJ Settineri, US Patent 3,764,543 (1973); *Chem Abstr* **80**:86674 (1974).

228. WJ Settineri, WW Frenier, JR Oswald, US Patent 3,996,147 (Dec 7, 1976); *Chem Abstr* **86**:58948 (1977).

229. WW Frenier, US Patent 4,310,435 (Jan 12, 1982).

230. J Jones, J Geldner, German Offen 2616144 (Nov 3, 1977); *Chem Abstr* **89**:10126 (1978).

231. A Constantinescu, DC Craciun, *Rev Coroz* **1**:206 (1971); *Chem Abstr* **77**:37989 (1972).

232. BR Keeney, JW Johnson Jr., US Patent 3,773,465 (1973); *Chem Abstr* **80**:136319 (1974).

233. BR Keeney, *Mater Prot Perform* **12**:13 (1973); *Chem Abstr* **80**:17141 (1974).

234. Oude-Alink, BA, US Patent 3,984,203 (1976); *Chem Abstr* **87**:104314 (1977).

235. D Redmore, US Patent 3,997,293 (1976); *Chem Abstr* **88**:10770 (1978); US Patent 3,998,883 (1976); *Chem Abstr* **87**:84823 (1977); US Patent 4,089,650 (1976); *Chem Abstr* **87**:10912 (1977).

236. BETZ, *Betz Handbook of Industrial Water Conditioning*, 8th edition, Betz Laboratories, Inc., Trevose, PA, 1980, p. 79.

237. AK Dunlop, US Patent 3,634,232 (1972).

238. JM Martin, JR Stanford, GD Chappell, US Patent 3,996,135 (1976).

239. AK Dunlop, *Corrosion 86*, Paper No. 176, NACE, Houston, TX, 1986.

240. H Kaufrass, US Patent 3,551,349 (1970).

241. MG Noack, US Patent 4,079,018 (1978).

242. CA Batton, DG Wiltsey, JA Kelly, *Proc Am Power Conf* **47**:1034 (1985).

243. DG Cuisia, *Corrosion 81*, Paper No. 268, NACE, Houston, TX, 1981.

244. G Donath, G Heitmann, J Messer, M Schott, *Vom Wasser* **49**:221 (1977).

245. JA Kelly, CA Soderquist, US Patent 4,419,327 (1983).

246. CO Weiss, DE Emerich, US Patent 4,487,745 (1984).

247. CM Allen, LB Silbey, S Cosgrove, *14th Annual Meeting of ASLE*, Paper No. 59, Buffalo, NY, April 1959.

248. T Rassel, *Werkst Korros* **20**:854 (1969).

249. RM Burns, WW Bradley, *Protective Coatings for Metals*, 3rd edition, Rheinhold, New York.

250. JD Keane (Ed.), *Steel Structures Painting Manual*, Vol. 2: *Systems and Specifications*, Steel Structures Painting Council, Pittsburgh, PA, 2000.

251. DB Boies, BJ Northan, WP McDonald, *Proceedings of the 25th NACE Conference*, NACE, Houston, TX, p. 180.

252. SL Chisolm, *Mater Prot* **3**:52 (1964).

253. W Von Fischer, EG Bobalek, *Organic Protective Coatings*, Reinhold, New York, 1953.

254. S Cristea, P Marcu, *Proceedings of the 3rd International Congress on Metallic Corrosion*, Amsterdam, 1966.

255. S Guraviah, KS Rajagopalan, *Paint India* **16**:28 (1968).

256. M Selvaraj, S Guraviah, *Paint India* **37**:27 (1987).

257. BR Chakrabarty, KM Verma, *Corros Prev Control* **34**:127 (1987).

258. H Gomathi, V Chandrasekeran, S Guraviah, KC Narasimhan, HVK Udupa, *Proceedings of the 2nd National Conference on Corrosion and Its Control*, Calcutta, 1979, p. 205

259. AJ Seavell, *J Oil Colour Chem Assoc* **61**:439 (1978).

260. E Knowles, T White, *J Oil Colour Chem Assoc* **41**:10 (1958).

261. R Franiau, *Congr Colloq Univ Liege* **74**:129 (1973).

262. TK Ross, RA Francis, *Corros Sci* **18**:351 (1978).

263. PT Deslauriers, *Mater Perform* **11**:35 (1987).

264. JR Gancedo, M Gracia, *Hyperfine Interact* **46**:461 (1989).

265. D Vacchini, *Anti-Corrosion* **9**:9 (1985).

266. D Landolt, M Favre, *Progress in the Understanding and Prevention of Corrosion*, Institute of Materials, London, 1993. pp. 374–386.

267. TK Ross, RA Francis, *Corros Sci* **18**:351 (1978).

268. M Stern, *J Electrochem Soc* **105**:638 (1958).

269. ND Tomashov, GP Chernova, *Passivity and Protection Against Corrosion*, Plenum Press, New York, 1967.

270. D Schlain, *Corrosion Properties of Titanium and Its Alloys,* Bulletin 619, US Department of Mines, 1964.

271. Abdel Hady, J Pagetti, *J Appl Electrochem* **6**:333 (1976).

272. R Anoshchenko, AP Zorchenko, *Proceedings of the 3rd European Symposium on Corrosion Inhibitors*, Ferrara, Italy, 1970, p. 293.

273. M Levy, *Corrosion* **23**:236 (1967).

274. W Roger, H Leidheiser, *J Electrochem* **62**:619 (1958).

275. VI Kazarin, TM Sigalovskaya, VV Andreeva, *Zasch Met* **6**:43 (1970).

276. MM Gleizer, KL Tseitlin, YI Sorokin, SM Babitskaya, KV Potapova, *Zasch Met* **13**:684 (1977).

277. RS Sheppard, DR Hise, PJ Gegner, WL Wilson, *Corrosion* **18**:211 (1962).

278. JA Petit, unpublished data.

279. OL Riggs, KL Morrison, DA Brunsell, *Corrosion* **35**:356 (1979).

280. Z Abdal Hady, Inhibitors for the Corrosion of Titanium and Zirconium, PhD thesis, Paris, 1977.

281. MM Gleizer, KL Tseitlin, YI Sorokin, SM Babitskaya, KV Potapova, *Zasch Met* **13**:684 (1977).

282. EI Tupikin, NG Klyuchnikov, GL Nemchaninova, *Zasch Met* **11**:351 (1975).

283. AP Bryzna, SN Lobanova, VI Sobnikova, NI Globa, *Zasch Met* **10**:405 (1974).

284. LI Gerasyutina, FM Tulyupa, *Zasch Met* **10**:742 (1974).

285. LI Gerasyutina, FM Tulyupa, LG Karyaka, *Zasch Met* **13**:324 (1977).

7

ENVIRONMENTALLY FRIENDLY CORROSION INHIBITORS

7.1 STANDARDIZED ENVIRONMENTAL TESTING

There is increasing concern about the toxicity, biodegradability, and bioaccumulation of corrosion inhibitors discharged into the environment. A noteworthy example consists of oilfield chemicals used as inhibitors discharged into the environment from offshore production platforms. Corrosion inhibitors in the aqueous phase are discharged into the ocean; these become an environmental hazard to marine life.

Efforts are now underway for industry to come to terms with the environmental impact of corrosion inhibitors in discharge water. The conventional inhibitors used may be satisfactory with respect to corrosion mitigation, but their environmental implications are not fully understood. Yet, it is well known that the chemical components of commercial inhibitors are certainly harmful to marine life.

In spite of the need for regulation, universal and clear guidelines on the use and discharge of corrosion inhibitors are not yet available. Different countries are directing their efforts toward preparing and introducing their own legislation. Environmental problems should be of primary concern due to the importance of protecting marine life and the preservation of the ecosystem. The European Economic Community (EEC) has delegated the Paris Commission (PARCOM) with the task of preparing and providing environmental guidelines. PARCOM is assigned the task of developing a harmonized approach to environmental testing consisting of (i) standardized environmental testing, and (ii) developing a model to use the data derived from such testing in industrial practice.

Green Corrosion Inhibitors: Theory and Practice, First Edition. V. S. Sastri.
© 2011 John Wiley & Sons, Inc. Published 2011 by John Wiley & Sons, Inc.

7.2 SUMMARY OF PARCOM GUIDELINES

PARCOM protocol consists of testing for three things: (i) toxicity, (ii) biodegradation, and (iii) bioaccumulation, as determined by the partition coefficient. We look at each of these in what follows.

7.2.1 Toxicity: As Measured on Full Formulation

Table 7.1 is a succinct overview of toxicity tests used for detecting certain species.

7.2.2 Biodegradation

In testing for biodegradation, all the components of the formulation must be tested using the following model:

$$\text{OECD Marine BOD} \quad \text{28-day test} \quad \text{Limit} > 60\%$$
$$\text{(modified OECD 301 D)}.$$

7.2.3 Partition Coefficient

The partition coefficient is to be determined for all the components of the formulation via the following model:

$$\text{OECD 117} \log P_{o/w} \quad \text{Limit} < 3.0.$$

7.2.4 Toxicity Testing

The present PARCOM guidelines require toxicity testing on organisms belonging to different trophic levels, including primary producers such as algae, consumers such as fish, crustacean, or sedimentary re-workers such as seabed worms.

Toxicity must be measured as both LC_{50} and EC_{50}. LC_{50} refers to the "lethal concentration" to affect (actually kill) 50% of the population, and EC_{50} refers to the "effective concentration" of the chemical required to adversely affect 50% of the population, for example, to diminish by 50% the intensity of luminescent bacteria or

TABLE 7.1 Toxicity Tests

Group	Preferred Species	Test
Primary producer	*Skeletonema costatum*	72-h EC_{50}
Consumer	*Acartia tonsa*	48-h LC_{50}
Decomposer	*Corophium volutator*	10-day LC_{50}

to stifle the body growth by 50% of seabed worms. In general, EC_{50} values that represent concentrations required to stunt growth are lower than the LC_{50} values representing the 50% of lethal concentrations.

Biodegradation is a measure of the duration over which a chemical will persist in the environment. All the constituents of the inhibitor formulation are tested for biodegradation. Biodegradation is a measure of the duration over which the chemical will persist in the environment. The modified OECD 301D test provides data on biodegradation after 28 days.

Bioaccumulation is a test involving the measurement of the partition coefficient $P_{o/w}$ of the chemical compound between its octanol and water phases. Test data are expressed as the logarithm of partition coefficient values (log $P_{o/w}$):

$$P_{o/w} = \frac{\text{Concentration of inhibitor in octanol}}{\text{Concentration of inhibitor in water}}.$$

Higher values of partition coefficients indicate greater partitioning of the chemical compound (inhibitor) from water across cell membranes and consequent bioaccumulation. All the compounds of the inhibitor are tested by standard high-pressure liquid chromatography (HPLC) OECD 117 method for bioaccumulation.

PARCOM developed a model known as Chemical Hazard Assessment and Risk Management (CHARM), by which the test data for different products can be compared and evaluated for their environmental impact. The CHARM model developed by Aquateam in Norway and TNO in the Netherlands consists of four modules: prescreening, hazard assessment, risk analysis, and risk management.

Prescreening consists of identification or flagging "bad actors" before one uses the model. Criteria for bad actors are $<20\%$ biodegradation or log $P_{o/w} > 0.5$ and molecular weight > 700.

Hazard assessment involves determination of the potential of a chemical compound to cause harm to the environment. The potential of a species to cause harm to the environment is obtained from the ratio of predicted environmental concentration (PEC) to no effect concentration (NEC), and when the value of the ratio (Q) is equal to or less than unity, the ecosystem is not affected. Three separate ratios, Q_{water}, $Q_{sediment}$, and $Q_{foodchain}$, are calculated and the largest value represents $Q_{ecosystem}$. The NEC values are obtained from PARCOM toxicity data and the PEC values are based on the release and subsequent dilution of the chemical compound from a realistic, worst case, oil and gas platform model. The PEC_{water} is determined at a distance of 500 m from the platform and the $PEC_{sediment}$ is determined within a 10-km^2 sphere around the platform.

Risk analysis involves an estimate of the probability of occurrence of harmful effects. This is similar to a hazard assessment module including specific details about a real platform for the purpose of a more representative evaluation.

Risk management involves identification of measures undertaken to reduce the risk of harmful effects. The measures involve setting targets such as best available technology (BAT) and best environmental practice (BEP). It also includes alternative chemicals, treatment rate, and cost-effectiveness.

The UK government developed and introduced its own guidelines to present PARCOM data. This is known as the UK Notification Scheme and is administered by the Ministry of Agriculture, Fisheries and Food (MAFF). The scheme presents the data in a single category, A–E, corresponding to 0–4 in the old scheme. A limit on the amount of chemical used by one platform or complex of platforms is placed on each category. In the old scheme, the limit placed is based on the predicted discharge of each chemical. In the new scheme, the limit is based on the cumulative use of all the chemicals in the same category.

Based on toxicity data, each chemical is assigned between A and E, with A representing the highest and E representing the lowest environmental rating. Based on biodegradability values and partitioning ratios, the category rating is moved one category upward or downward. When the biodegradability is $\geq 60\%$, the category is moved one category lower. When the biodegradability is $<20\%$ or the log $P_{o/w}$ value is greater than 3 and the molecular weight is less than 600, it is moved one category higher. In a case where the movement is in both directions, the worst-case scenario holds good. The classification in the United Kingdom is as follows:

$$A < 1, B\ 1 - 10, C\ 10 - 100, D\ 100 - 1000, E > 1000 (LC_{50}, EC_{50}, ppm).$$

The UK notification scheme reporting limits are given in Table 7.2.

There is no agency similar to PARCOM in North America, but there are governmental agencies dealing with environmental guidelines on the discharge of effluents and toxicity toward marine life. The companies dealing with the supply of corrosion inhibitors are well aware of the environmental concerns of the users and are directing efforts in developing environment-friendly nontoxic green inhibitors.

The toxicity of corrosion inhibitors used in oil production systems containing sour gas (H_2S) toward marine crustaceans is given in Table 7.3.

TABLE 7.2   Notification Scheme Limits

Category	Cumulative Use Limits (T^a/Point Source/Year)
A	40
B	70
C	150
D	375
E	1000

Old Scheme Category	Individual Discharge Limits (T/Point Source/Year)
4	All
3	1
2	10
1	100
0	None

^{a}T = tonne (metric ton).

TABLE 7.3   Toxicity of Some Corrosion Inhibitors

Product	Chemistry	Exposure Time (h)	LC$_{50}$ (ppm)			
			<10	10–100	100–1000	>1000
A	Imidazoline	24	X			
		48	X			
B	Imidazoline salt	24	X			
		48	X			
C	Quaternary ammonium	24	X			
		48	X			
D	Tertiary amine	24		X		
		48	X			
E	Amphoteric	24				X
		48			X	

Products D and E are less toxic than A–C. New range of molecules were synthesized to produce inhibitors F and G, which are less toxic than A–C as shown by the data in Table 7.4.

Two environmentally acceptable inhibitors X and Y were developed that may be used as substitutes for imidazoline (Z), and the toxicity data for these compounds

TABLE 7.4   Toxicity Data for Corrosion Inhibitors

Chemistry	Exposure Time (h)	LC$_{50}$ (ppm)			
		<10	10–100	100–1000	>1000
Diamine	24	X			
	48	X			
Intermediate 1	24	X			
	48	X			
Intermediate 2	24		X		
	48	X			
Intermediate 3	24			X	
	48		X		
Product F	24				X
	48			X	
Product G	24			X	
	48			X	
Product G (route 1)	24			X	
	48			X	
Product G (route 2)	24				X
	48				X

TABLE 7.5   Toxicity Data for Some Inhibitors

Toxicity	Product X	Product Y	Product Z
Skeletonema	4.5 ppm	20.7 ppm	0.72 ppm
Acartia tonsa	1.6 ppm	18.4 ppm	1.7 ppm
Abra alba	86 ppm	22 ppm	–
Biodegradation	73%	88%	51%
UK Scheme	B	C	A
CHARM	1.7442	0.4197	2.1970
Q pelagic	1.7442	0.4197	2.1970
Q benthic	0.0011	0.0013	0.0680
Q food chain	0.0009	0.0000	0.0440
Corrosion rate (mpy)	34, 36	3, 28	–

TABLE 7.6   Mammal Toxicity LD_{50} Values

Compound	LD_{50} (mg/kg)
Propargyl alcohol	55
Hexynol	34
Cinnamaldehyde	2200
Formaldehyde	800
Dodecylpyridinium bromide	320
Naphthylmethyl quinolinium chloride	644
Nonylphenolethylene oxide surfactants	1310

are given in Table 7.5. The data clearly show that product Y is acceptable from both the performance and environmental points of view.

Corrosion inhibitors in acid solutions to remove mill scale or for some other purpose consist of a Mannich base, which is obtained by the condensation of formaldehyde with an amine and a ketone. Inhibitors used in acid solutions may also contain acetylenic alcohols such as propargyl alcohol, hexynol, or ethyloctanol that are toxic. Some toxicity data of inhibitor components are given in Table 7.6. It is clear from the data that formaldehyde is quite toxic. An inhibitor composed of cinnamaldehyde, aromatic quaternary nitrogen compound, and nonylphenolethylene oxide was developed and found to be suitable both from the performance and toxicity points of view.

7.3   MACROCYCLIC COMPOUNDS IN CORROSION INHIBITION

Nontoxic macrocyclic compounds such as porphyrins and phthalocyanines form chelate metal complexes with high stability constants. Phthalocyanines were used in lubricants as early as 1950. Both porphyrins and phthalocyanines have planar structure and four pyrrole subunits with the nitrogen atoms as the bonding sites. The metal is at the center bonded to the four nitrogen atoms of the pyrrole subunits.

TABLE 7.7   Corrosion Rates Using TMPyP and Various Counteranions in 1% NaCl, pH 6 (Conc. 0.05 mM)

Inhibitor Counterion	Corrosion Rate		Inhibition Efficiency (%)
	12 h	48 h	
Control	380	370	0
Iodide	53	74	80
Chloride	174	182	50
Benzoate	84	106	70
Sulfate	114	160	59
Toluene sulfonate	96	87	76

The corrosion rates of Armco iron in the presence of tetramethyl-4-methylporphyrin (TMPyP) are given in Table 7.7. The data show the degree of inhibition depends upon the counteranion present in the inhibitor.

Corrosion inhibition data obtained on Armco iron in 1% NaCl by the addition of metal complexes of TMPyP toluene sulfonate at a concentration of 0.3 mM are given in Table 7.8.

The effect of pH and porphyrin concentration on the corrosion rate of Armco iron is illustrated by the data given in Table 7.9.

TABLE 7.8   Corrosion Rate Data in the Presence of Metallo-Derivatives of TMPyP Sulfonate in 1% NaCl, pH 6 (Conc. 0.3 mM)

Incorporated Metal Ion	Corrosion Rate (mdd)		Inhibition (%)[a]
	12 h	48 h	
None	380	370	0
VO(II)	71	46	87
Co(II)	53	68	81
Cu(II)	62	49	87
Fe(III)	27	15	96
Ni(II)	59	45	87
Mn(III)	32	28	92
Rh(II)	48	27	93

[a] Based on 48 h data.

TABLE 7.9   Effect of pH and Porphyrin Concentration on the Corrosion Rate of Iron

Inhibitor TMPyP Iodide (mM)	Corrosion Rate (mdd)		
	Distilled Water	1% NaCl	0.1 mM HCl
0.001	79	97	91
0.01	39	97	80
0.10	21	97	80
1.0	18	103	80

TABLE 7.10 Data on Mass-Loss Studies on Iron Samples Prepared by Vapor Deposition of Porphyrin in 1% NaCl Solution

Porphyrin	Corrosion Rate (mdd)	Inhibition (%)
None	382 ± 37	0
TPhP	28 ± 10	93
TPyP	36 ± 20	91
T(o-chloro Ph)P	38 ± 8	90
5-Nitroporphyrin	62 ± 5	84
NiTPhP	202 ± 31	47
CoTPhP	182 ± 21	52
ZnTPhP	89 ± 17	77
VOTPhP	173 ± 40	55
Fe(III)TPhCl	312 ± 37	18
Mn(III)TPhCl	283 ± 18	26
RhTPhP	271 ± 26	29
ZnTPyP	103 ± 13	73
Fe(III)TPyPCL	327 ± 27	14
CoTPyP	120 ± 18	68

Corrosion rates of iron samples subjected to vapor deposition of porphyrin in 1% NaCl solution are given in Table 7.10.

The electrochemical data extracted from polarization curves along with the percent inhibition data are noted in Table 7.11. The Fe(III)TCPC system appears to be a promising inhibitor system.

The resistive and capacitative behavior of the interface of steel coated with Fe(III) TCA UpC, Co(II) TCA UpC, Fe(III) TCA CpC, and Fe(III) TCA UpC, where TCA UpC and TCA CpC refer to tetrakis(N-carboxy-12 amino undecanoic acid) and tetrakis(N-carboxy-6-amine caproic acid), respectively.

AC impedance measurements on mild steel treated with monomeric and polymeric phthalocyanines such as TCA UpC (tetrakis(N-carboxy-12 amino undecanoic acid) and TCA CpC [tetrakis (N-carboxy-6 amino caproic acid)] yielded the data given in Table 7.12.

TABLE 7.11 Electrochemical Data on Mild Steel Treated with Phthalocyanines in 1% NaCl (pH 2.0)

Inhibitor System	I_{corr} (mA/cm^2)	Inhibition (%)
None	0.48	–
Co(II)TCpC	0.20	55
Fe(III)TCpC	0.08	83

TABLE 7.12   Corrosion Inhibition Data with Phthalocyanine Inhibitors

Inhibitor System	Rp^a (ohm/cm^2)	Inhibition (%)
Blank	42	–
Co(II) TCpC	137	68
Fe(III) TCpC	240	83
Fe(III) TCpC	159	74
Fe(III) TCpC	174	77
Fe(III) TCA UpC	323	88
Co(II) TCA UpC	243	83
Fe(III) TCA CpC	255	84
Co(II) TCA CpC	114	65

^{a}Rp = polarization resistance.

7.4   ENVIRONMENTALLY GREEN INHIBITORS

Environmental concerns require corrosion inhibitors to be nontoxic and environment-friendly and acceptable. Green chemistry serves as a source of environment-friendly green corrosion inhibitors. Corrosion inhibitors are extensively used in corrosion protection of metals and equipment. Organic compounds with functional groups containing nitrogen, sulfur, and oxygen atoms are generally used as corrosion inhibitors. Most of these organic compounds are not only expensive but also harmful to the environment. Thus, efforts have been directed toward the development of cost-effective and nontoxic corrosion inhibitors. Plant products and some other sources of organic compounds are rich sources of environmentally acceptable corrosion inhibitors. An example of such a system is the corrosion inhibition of carbon steel by caffeine in the presence and absence of zinc (2). Plant products are the main sources of environment-friendly green inhibitors such as phthalocyanines. Some of the studies on the use of environmentally acceptable inhibitors are summarized in Table 7.13.

7.5   ROLE OF RARE EARTH COMPOUNDS IN REPLACING CHROMATE INHIBITORS

In spite of the toxicity, chromates are still used in the manufacture and maintenance of aircraft. Some applications are the use of chromate in deoxidizers, conversion coatings, anodizing, chromate-inhibited primers, wash primers, and repair processes. The health problems associated with the use of chromate such as respiratory cancer in aircraft painters and the increased costs associated with the safe use and disposal of chromates has led to efforts in finding suitable alternate inhibitors such as rare earth salts.

Hinton and coworkers discovered the inhibitive properties of rare earth compounds on a range of metals (149–151). The protective properties of rare earth as inhibitors have been documented for ferrous metals (152) and aluminum alloys (153) as conversion coatings (154, 155) and deoxidizers (156) and incorporated into paint

TABLE 7.13   Environmentally Green Inhibitors

Metal or Alloy	Inhibitor and Conditions	Reference
Carbon steel	Sodium molybdate and calcium gluconate synergism was observed with this binary system for cooling water systems	4
Mild steel	Leaves and roots of medicinal plants A–D in H_2S medium were used as inhibitors. Aqueous extracts of C and D performed better than A and B products. All the inhibitors showed low bioaccumulation	5
Aluminum alloys, low-carbon steels	Environmentally safe corrosion inhibitors with conversion coatings	6
Carbon steel	Zinc phosphate base, molybdate base inhibitors in cooling waters	7
Low-carbon steels	Five commercial green inhibitors were tested using electrochemical impedance and rotating disk electrode	8
Mild steel	Corrosion protection of mild steel by polyaniline as a substitute for chromate	9
Mild steel in H_2SO_4 solutions	Ethanol extracts of *Garcinia kola* and a mixture with KI gave improved inhibition. This inhibitor may be used in metal surface anodizing and surface coatings	10
Zinc	Synergistic effect of tolutriazole and sodium carboxylates on zinc corrosion in atmospheric conditions	11
Cold rolled steel	Volatile corrosion inhibitors; environment-friendly metal cleaners and metal protectors	12
Mild steel	Effect of acid extracts of *Calotropis giganta* latex on the corrosion of mild steel in HCl	13
Mild steel	Inhibitive action of *Carica papaya* extracts on the corrosion of mild steel in acidic media	14
Steels, aluminum base alloys	Inhibitors—chelated sol-gel coatings for metal finishing	15
Aluminum, zinc, mild steel	Influence of wild strain *Bacillus mycoides* on corrosion and protection	16
Carbon steels	Synergistic role of ascorbate in corrosion inhibition	17
Aluminum alloys	Use of inhibitors to improve the corrosion protection of E-coat systems on aluminum alloys. Chromate-free coatings were developed	18
Steels	Corrosion inhibition in cooling waters using calcium gluconate as the inhibitor	19
Steel	Anticorrosive effect of epoxy-allyl menthols on steel in molar hydrochloric acid	20

TABLE 7.13 (*Continued*)

Metal or Alloy	Inhibitor and Conditions	Reference
Steel, copper, aluminum	Corrosion inhibition by naturally occurring substances: molasses as inhibitor for copper, aluminum, and steel	21
High shear environments	Development of green corrosion inhibitors for high shear applications	22
Mild steel	Environment-friendly ascorbic acid formulations for industrial cooling systems	23
Steel weapons	Environment-friendly protection methods used by the armed forces (vapor-phase corrosion inhibitors)	24
Carbon steel	Ternary inhibitor formulation consisting of Zn^{2+}-N, N-bis-(phosphonomethyl) glycine and ascorbate	25
Steels, carbon, and stainless	Environment-friendly approaches to inhibit biocorrosion	26
Steels	Ecological compounds as corrosion inhibitors	27
Mild steel	A green FDA-approved corrosion inhibitor for boiler condensate systems	28
Nickel base alloys, super alloys, Al brasses	Development of environment-friendly biocides for marine applications	29
Reinforcing steels in concrete	Migrating corrosion inhibitors for reinforced concrete	30
Steel in concrete	Migrating corrosion inhibitors in concrete	31
Galvanized steel	Use of scanning reference electrode technique for the evaluation of environment-friendly nonchromate inhibitors	32
Carbon steel	Phosphonated glycine, a new corrosion inhibitor for carbon steel in an aqueous environment	33
Carbon steel	Corrosion in aqueous potassium chloride solutions and scale inhibition by sodium lignosulfonate	34
Steels, aluminum base alloys	Corrosion prevention and chromates, the end of an era	35
Stainless steels	Preventing biocorrosion without damaging the environment, four innovative strategies	36
Copper alloys	Corrosion of copper alloys in a marine environment containing plant-based alkaloids	37
Mild steel and copper	Field studies with a new environmentally acceptable scale and corrosion inhibitor	38
Steels and bronzes	Evaluation of inhibitor efficiencies in hydraulic fluids	39

(continued)

TABLE 7.13 (*Continued*)

Metal or Alloy	Inhibitor and Conditions	Reference
Carbon steel	Corrosion inhibition of carbon steel by blends of gluconate/benzoate	40
Low-carbon steel boiler tubes	A new green corrosion inhibitor for boiler condensate	41
Steel	Rosemary oil as a green corrosion inhibitor for steel in phosphoric acid	42
Steels	A contribution to the study of nontoxic inhibitor behavior	43
Carbon steels	Development of nontoxic corrosion inhibitor for MEA plants	44
Steel	Effect of nontoxic corrosion inhibitors on steel in chloride solutions	45
Carbon steel	Synergistic inhibition of corrosion of carbon steel using a new phosphonate, zinc, and heterocyclic compound	46
Steel	Evaluation of corrosion inhibition by ascorbic acid	47
Carbon steel	Laboratory and field development of a novel environmentally acceptable scale and corrosion inhibitor	48
Structural steels	Combined use of dehumidification and vapor corrosion inhibitor for corrosion protection	49
Aluminum base alloys, brasses	The development of multifunctional corrosion inhibitors for aerospace applications	50
Steels	Corrosion inhibition in drilling muds	51
Carbon steel	Corrosion and scale control of carbon steel in near neutral media by a nontoxic inhibitor	52
Steel	Inhibition of corrosion in electrolyte solutions	53
Carbon steel	Environmentally acceptable corrosion inhibitors for carbon steel	54
Steels	Advanced corrosion technology—the watery revolution (corrosion of automotives, atmospheric corrosion)	55
Mild steel	Effect of acid extracts of powdered seeds of *Eugenia jambolans* on corrosion of mild steel in HCl	56
Steel	Study of low cost, eco-friendly corrosion inhibitors for cooling systems	57
Steels	CAHMT—a new eco-friendly acidizing corrosion inhibitor	58
Mild steel	Eco-friendly inhibitor for corrosion inhibition of mild steel in phosphoric acid medium	59

TABLE 7.13 (*Continued*)

Metal or Alloy	Inhibitor and Conditions	Reference
Carbon steel	Eco-friendly corrosion inhibitors: inhibitive action of quinine for corrosion of low-carbon steel	60
Mild steel, zinc-coated steel	Effect of metallic substrate composition on the protective power of a waterproof epoxy primer in marine environment	61
Carbon steels	Predicting the performance of oil field corrosion inhibitors from their molecular properties	62
Carbon steels	Phosphonocarboxylate inhibitor for highly cycled cooling water systems	63
Galvanized steels	Use of scanning reference electrode for the evaluation of environment-friendly nonchromate corrosion inhibitors	64
Al alloys, galvanized steel	Environment-friendly inhibitors for aluminum alloys and galvanized steel	65
Carbon steel	Evaluation of environment-friendly corrosion inhibitors for sour service	66
Carbon steels	Evaluation of environment-friendly corrosion inhibitors for sour service	67
Steels	Industry update—molybdate corrosion inhibitors	68
Carbon steels	Environmental challenges facing corrosion inhibition in North Sea crude oil export pipelines	69
Low-carbon steels	Environment-friendly volatile corrosion inhibitors	70
Steels	A packaging concept for the new millennium	71
Carbon steels	Effect of different polymers on the efficiency of waterborne methylamine adduct as corrosion inhibitor for surface coatings	72
Carbon steels	Sugar beets against corrosion	73
Galvanized steel	Growth mechanism of cerium layers on galvanized steel	74
High-strength steels	Trials and tribulations of inhibitors	75
Carbon steel	A novel environment-friendly scale and corrosion inhibitor	76
Zinc coatings	Corrosion inhibitor-doped protein films for protection of metallic surfaces	77
Mild steel	Some environment-friendly inhibitors for mild steel corrosion in sulfuric acid solution	78
Carbon steel	Synergistic inhibition of carbon steel by tertiary butyl phosphonate, zinc ions, and citrate	79

(continued)

TABLE 7.13 (*Continued*)

Metal or Alloy	Inhibitor and Conditions	Reference
Galvanized steel	Environment-friendly anticorrosion primers for hot-dip galvanized steel	80
Steels	Corrosion inhibitors in conversion coatings. Magnetite coatings	81
Steels	New possibilities of corrosion inhibition: improving the conversion coatings	82
Steels	Mechanisms of nontoxic anticorrosive pigments in organic waterborne coatings	83
Carbon steels	Mannich base derivatives—a novel class of corrosion inhibitors for cooling water systems	84
Mild steel	Encapsulated cerium nitrate inhibitors to provide anticorrosion sol-gel coatings on mild steel	85
Carbon steel	Variation of carbon steel corrosion rate with flow conditions in presence of inhibitive formulation	86
Zn, Zn–Al steels	Nicotinic acid as a nontoxic corrosion inhibitor for hot-dipped zinc and Zn–Al alloy coatings on steels in HCl	87
Steel	Acid cleaning of steel using Arghel eco-friendly corrosion inhibitor	88
Pipes (steel)	Corrosion problem solutions for pipe equipment during manufacture, fabrication, and storage using vapor-phase corrosion inhibitors	89
Carbon steel	Environment-friendly corrosion inhibitors of natural resources	90
Carbon steel–zinc alloys	Corrosion of metals in petrol	91
Carbon steels aluminum	Effect of metallic substrate composition on the protective power of a waterproof epoxy primer in marine environment	92
Steels	Stretch packaging: this is a wrap	93
Steel	Influence of additives on the corrosion protection of waterborne epoxy-coated steel	94
Carbon steel	Polyphosphate–silicate–zinc: a new corrosion inhibitor blend for freshwater cooling systems	95
Steels	Waterborne corrosion inhibitive primers evolve to be competitive	96
Steel	Corrosion inhibition of steel by triazolidines in saline water	97
Steels	Mitigating steel corrosion in cooling water by molybdate-based inhibitor	98

TABLE 7.13 (*Continued*)

Metal or Alloy	Inhibitor and Conditions	Reference
Mild steel	Evaluation of *Foenum graecum* as inhibitor for acid corrosion of steel	99
Steels	Mangrove tannins as corrosion inhibitors in acidic medium: study of flavonoid monomers	100
Carbon steels	Influence of flow on the synergistic effect of an inhibitive mixture used for water treatment in cooling circuits	101
Carbon steel	Influence of flow on the corrosion inhibition of carbon steel by fatty amines in association with phosphonocarboxylic acid salts	102
Steel	Electrochemical study of corrosion inhibition of steel reinforcement in alkaline solutions containing phosphate-based compounds	103
Carbon steel	A SERS investigation of carbon steel in contact with aqueous solutions containing benzyldimethylphenyl ammonium chloride	104
Carbon steel	Effect of inhibitor concentration and hydrodynamic conditions on inhibitive behavior of combinations of sodium silicate and HEDP for corrosion control in carbon steel water transmission pipes	105
Mild steel	Inhibitors and inhibition mechanisms for mild steel in cooling water systems	106
Carbon steels	Ecologically acceptable N-based compounds in corrosion inhibition in acidic media	107
Carbon steels	Corrosion inhibitors in conversion coatings, Part III	108
Mild steel	Carboxymenthylchitosan as an eco-friendly inhibitor for mild steel in 1 M HCl	109
Structural steels	Corrosion inhibitive performance of some commercial water-reducible nontoxic primers	110
Low-carbon steel	The inhibition of low-carbon steel corrosion in hydrochloric acid solutions by succinic acid	111
Steels	The role of organophosphonates in cooling water treatment	112
Cast iron, steels	The testing of corrosion inhibitors for central heating systems	113
Carbon steels	Control of ferric ion corrosion during chemical cleaning	114
Carbon steels	Inhibition of steel corrosion in brine coolants	115

(*continued*)

TABLE 7.13 (*Continued*)

Metal or Alloy	Inhibitor and Conditions	Reference
Steel	Anticorrosive compositions based on industrial wastes	116
Steel	Corrosion inhibition of steel by tannins	117
Mild steel	Effect of oxidants on protection of mild steel by zinc hydroxyethylidene diphosphonate	118
Mild steel	Natural compounds as corrosion inhibitors for highly cycled systems	119
Carbon steels	Approach to corrosion inhibition by some green inhibitors based on oleochemicals	120
Mild steel	Studies on the inhibitive effect of *Datura stramonium* extract on the acid corrosion of mild steel	121
Steels	Corrosion inhibition of steels in alkaline chloride solutions	122
Carbon steels	Nontoxic anticorrosive pigments for paints	123
Steels	Waterborne nontoxic high-performance inorganic silicate coatings	124
Carbon steels	Influence of environmental conditions on microbiologically induced corrosion	125
Carbon steels	An ecologically acceptable method of preventing corrosion scale problems by control of water conditions	126
Steels	Corrosion inhibitors in rust converters containing mimosa and oak tannins	127
Steel	Triethylene oxide polymers and copolymers as corrosion inhibitors for steel in acid media	128
Steels	Influence of sodium sulfite on the protective properties of a zinc phosphonate inhibitor	129
Iron, steels	The phosphonate inhibitors of metal corrosion	130
Steels, Cu, and Zn plating	The mechanism of action and principles for choice of organic inhibitors of metalselectrodeposition	131
Steels	Corrosion and biocorrosion protection of hot water supply systems of heat power stations	132
Carbon steels	Evaluation of corrosion inhibitors for cooling water	133
Carbon steels	Time dependence of inhibiting effect of amidoamines in CO_2-saturated chloride solutions	134
Steel	Corrosion resistance of nitrided layers on steel impregnated with inhibiting formulations containing tribological additives	135
Steels	Influence of time on the protection of steels by combined inhibitors in low-mineralized water	136

TABLE 7.13 (*Continued*)

Metal or Alloy	Inhibitor and Conditions	Reference
Steels	Continued studies on the use of mixed chromate–phosphate inhibitor in desalination service	137
Al-base alloys	High-performance multifunctional corrosion inhibitor for aircraft	138
Iron	The effect of complexing on the inhibiting properties of high molecular weight inhibitors on iron corrosion in neutral media	139
Carbon steels	Estimated and actual inhibitor availabilities for multiphase lines and wet gas lines	140
Steel	Citric acid cleaning of steel	141
Steels	Critical factors in predicting CO_2–H_2S corrosion in multiphase systems	142
Stainless steels	Surfactants as green inhibitors for pitting corrosion of stainless steels	143
Mild steel	Corrosion inhibition of mild steel by *Datura metel*	144
Steels	Adsorption and corrosion inhibitive properties of *Azadirachta indica* in acid solutions	145
Mild steel	Inhibitive effect of black pepper extract on the sulfuric acid corrosion of mild steel	146
Mild steel	New green inhibitors incorporating a rare earth and an organic compound for corrosion mitigation	147
Tin plate	Cerium: a suitable green corrosion inhibitor for tin plate	148
Carbon steel	Inhibition by tannins in cooling waters	171
Steel	Corrosion inhibition by *A. indica*	172
Steel	Corrosion inhibition by *Calotropis procera*	173
Steel	Corrosion inhibition by xanthates	174
Steel	Corrosion inhibition by lignosulfonates	175
Steel	Corrosion inhibition by pomegranate fruit shells	176
Steel	Corrosion inhibition by strychnine and quinine	177
Steel	Corrosion inhibition by *Cardia latifolia*	178
Steel	Corrosion inhibition by eucalyptus extract	179
Steel	Corrosion inhibition by *Jasminum auriculatum*	180
Steel	Corrosion inhibition by rosemary leaves extract	181

TABLE 7.14　Coating Times for Aluminum Alloys

Al Alloy	Immersion (Days)	H_2O_2 Accelerated (min)	Activated (min)
2024-T3	20	2–4	–
3004	–	18	5
5005	50	> 60	6
6016	–	50	15
7075-T6	20	10	NT

systems (157). Rare earth salts can be used in place of chromates in inhibitor formulations, deoxidizing solutions, conversion coatings, and anodizing and inhibiting pigments in paints.

The deposition times of coatings were longer than a few days and the times were shortened by a process involving hydrogen peroxide exposure (Table 7.14).

The data in Table 7.14 show the drastic reduction in coating times from several days to a few minutes by the hydrogen peroxide treatment. This acceleration of the coating process by hydrogen peroxide is termed the cerate process. The coatings resulting from the hydrogen peroxide (cerate process) resembled chromium conversion coatings in appearance.

The mechanism of deposition is proposed as increased pH due to the cathodic reduction of hydrogen peroxide:

$$H_2O_2 + 2e^- \rightarrow 2OH^-.$$

The increase in pH facilitates the precipitation of hydrated cerium oxide.

The effect of some inhibitors on the corrosion rate of AA 7075 in the presence of 0.1 M of sodium chloride is shown in Table 7.15.

The presence of 100 ppm of cerium chloride in 0.1 M sodium chloride solution reduced the corrosion rate of an Al–Zn–Mg–Cu alloy by about two orders of magnitude. The rare earth salts act as cathodic inhibitors. Rare earth salts appear to form a film of rare earth metal oxide at cathodic sites due to the increased local alkalinity at these sites as a result of oxygen reduction reaction and exceeding the solubility product of rare earth oxide.

TABLE 7.15　Corrosion Rates in the Presence of Rare Earth Salts

Inhibitor	Relative Corrosion Rate
None	10.0
Cerium salt	0.10
Lanthanum salt	0.09
Yttrium salt	0.08
Rare earth mixture	0.03
Chromate	0.02

TABLE 7.16   Corrosion Inhibition of Aluminum

Inhibitor	Duration (Weeks)	Factor of Improvement over Control
Cerium dibutyl phosphate	3	70
	6	40
Mischmetal dibutyl phosphate	3	65
	6	>100

Aqueous immersion involving cerium salts in corrosion of aluminum resulted in the data given in Table 7.16.

A novel approach to corrosion inhibition pioneered by Forsyth and Deacon (158–161) involves combining rare earths with organic inhibitors to produce multifunctional inhibitor compounds that exhibit synergistic behavior. The corrosion inhibition by cerium dibutyl phosphate and Mischmetal dibutyl phosphate of AA2024-T3 alloy at an inhibitor level of 200 ppm gave satisfactory inhibition. The dibutyl phosphate performed better than the carboxylic inhibitor. Surface analysis data showed the presence of a 200–500 nm film comprised of cerium and phosphorus.

Long-term exposure of AA2024-T3 to cerium salicylate, $Ce(Sal)_3$, and $Ce(dibutyl phosphate)_3$ showed similar performance in the shorter term, while $Ce(dbp)_3$ performed better in the longer term.

The performance tests carried out with chromate replacement inhibitors consist of (i) corrosion testing the conversion coating without paint, and (ii) corrosion and adhesion testing of the conversion coating and paint system.

Chiefly, rare earth salts such as cerium salts are used in place of chromates in conversion coatings. Hydrogen peroxide is in the application of cerium salt coating. Hydrated cerium oxide is deposited on copper alloys, but not on aluminum alloys devoid of copper. The cerium-deposited coatings require a sealing step in order to be effective.

The times to failure of cerate and chromate coatings in neutral salt spray (NSS) are given in Table 7.17 (162).

The adhesion performance of both the sealed and unsealed cerate conversion coatings is similar to chromate conversion, as evidenced by the data in Table 7.18.

Cerate is found to perform just like chromate in a range of underpaint corrosion tests.

In filiform corrosion testing, Mol (163) found cerate and chromate conversion coatings on aluminum alloys, including 2024-T3 and 7075-T6. There was no difference in performance of chromate and cerate under polyurethane topcoats.

TABLE 7.17   Times to Failure of Cerate and Chromate in Neutral Salt Spray (NSS)

Al Alloy	Chromate	Cerate
2024-T3	700	980
7075-T6	1300	360
7475-T7651	120	72

TABLE 7.18 Typical Adhesion as a Percentage of a Chromate Control

Al Alloy	Chromate	Cerate Unsealed	Cerate Sealed
2024-T3	100	119	124
7075-T6	100	89	80
7475-T7651	100	85	97

Recently, cerium salicylate and cerium dibutylphosphate, which are quite effective in neutral aqueous solutions, have been incorporated into epoxy primers. AA2024-T3 panels were coated and a cross-scribe cut into epoxy primer. The panels were exposed to constant immersion or alternate immersion cycles in chloride solutions at neutral pH. After 500 alternate immersion cycles (3 weeks testing), both cerium salicylate and cerium dibutyl phosphate suppressed filiform corrosion effectively. The data on the corroded area after removal of epoxy primer show good corrosion protection by cerium salicylate and cerium dibutylphosphate (Table 7.19).

In summary, the data in the literature show the cerium ions in solution and cerium compounds can be used to effectively inhibit corrosion of many metals, including aluminum. The inhibitive effects of cerium compounds on the alloy AA2024-T3 show the promise for replacement of toxic chromate inhibitors by rare earth-based inhibitors.

Cerium salts in aqueous solutions containing sodium chloride lower corrosion rates of many metals, including aluminum alloys used widely in the aerospace industry. Cerium-based conversion coatings have been found to be as effective as chromate-based coatings when tested under paint in a range of tests comprising both adhesion and corrosion tests. The corrosion inhibition data obtained with cerium salts revealed that a rare earth/organic inhibitor in a coating system for corrosion protection of AA2024-T3 is promising from both the corrosion inhibition and environmental protection points of view. The cerium dibutyl phosphate inhibitor was successfully incorporated into the basic epoxy-amide primer system, and the resulting system performed well, showing a high percentage of inhibition. The inhibition mechanism probably involves leaching of the inhibitor from the primer coating into the surroundings as well as protection. This mechanism is similar to the behavior of strontium chromate-based epoxy primers. The experimental data now available on cerium passivation in some important areas of metal corrosion point to the possibility of the use of a combined cerium passivation system.

TABLE 7.19 Corroded Area After Removal of Epoxy Primer

Inhibitor	Time Until Appearance of Corroded Area (cm^2)	
	21 Days	63 Days
No inhibitor	2.8	5.0
Cerium salicylate	0.5	0.8
Cerium dibutyl phosphate	0.3	0.3

Some inhibitors tested for suitable replacement of chromates include molybdates (164), vanadium-based compounds (165), boron-based compounds (166, 167), and rare earth salts (168). Many of these compounds have been studied for their performance of corrosion inhibition of aerospace Al alloys (168–170).

Single-component inhibitor systems are generally less effective than binary or ternary systems. This is true in corrosion inhibition of aerospace alloys. Synergistic combinations of rare earth salts and transition metal salts have been used in corrosion inhibition of aerospace alloys (182). Synergy occurs when inhibition by the combination exceeds the arithmetic sum of the inhibition by individual components. Synergistic combinations of rare earth and transition metal salts were studied with a ratio of 1:1 at a chosen single concentration. Synergy is a nonlinear process, and it is difficult to predict; hence a large number of experiments are required for discovery and characterization of synergy phenomenon.

A synergistic combination of inhibitors has been studied for steel in acidified (183–190) and neutral (191–194) aqueous environments, as well as copper in neutral aqueous environments (195, 196). All these studies examined the effect of a binary mixture of inhibitors in 1:1 ratio. Since synergy is a nonlinear process and difficult to predict, a large number of experiments involving various proportions of inhibitors have to be carried out in order to assess synergistic behavior of inhibitors.

A combinatorial process has been found to increase the rate of material discovery and experimental throughput (197–199). In the combinatorial process, a large number of variables are screened to identify the optimum process or condition of interest. Corrosion protection properties of the inhibitors can be determined by electrochemical methods, but the available methods are not rapid enough (being on the order of minutes to complete a determination) to predict long-term (i.e., years) corrosion protection.

The problem facing the corrosion scientist is to screen thousands of corrosion inhibitors in different combinations under a variety of conditions. Once the large numbers of tests to screen the inhibitors are done to identify target inhibitor compounds, more detailed testing can be done to arrive at the optimum inhibitor system.

Aluminum alloy 2024 wire with a diameter of 1.50 mm was used as an electrode for testing. The inhibitors studied were cerium chloride, yttrium chloride, sodium metatungstate, sodium metavanadate, lanthanum chloride, sodium molybdate, sodium metasilicate, potassium phosphate, sodium phosphate, and barium metaborate. With the use of these 10 individual compounds, 44 binary combinations were characterized for synergies.

High-throughput test methods for inhibitor performance consisted of a multichannel microelectrode analyzer (MMA) in combination with an 8×12 array of electrochemical cells. Cells were established through the use of a conventional 2 mL reaction frame and fabricated top to contain AA2024 wire electrodes, as shown in Fig. 7.1. The multichannel microelectrode analyzer is a group of 10 modules of 10 zero-resistance ammeters (ZRA) that can be used for measurement of current or potential of electrodes.

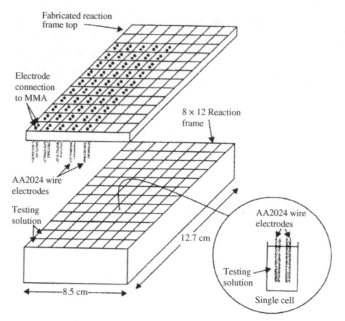

FIGURE 7.1 Array of electrochemical cells.

Fifty cells of conventional 8 × 12 reaction frame were used to house 50 independent chemistries. Two AA2024 wire electrodes were inserted into electrical contacts contained in the fabricated top for each of the 50 cells, totaling 100 wire electrodes connected to the MMA. The fabricated top was then placed on the reaction frame containing the chemistries of interest. Each module on the MMA was set to establish a potential of one wire electrode versus the other.

The corrosion resistance (polarization resistance minus diffusional impedance) indicated the performance of the inhibitor. Samples of AA2024-T3 coupons were placed in 1.8 mL of 3.4 mM inhibitor in 0.6 M NaCl solution of pH 7. One electrode in each cell was polarized at 100 mV with respect to the other electrode at $-525\,\mathrm{mV_{SHE}}$, and the resulting current was measured. The MMA measured the current between a pair of electrodes using a zero-resistance ammeter.

Based on the dealloying mechanism of copper-bearing aluminum alloys and the redeposition of copper, the amount of copper on the surface was determined (197–199) by cyclic voltammetry to assess the corrosion damage of AA2024-T3.

A third high-throughput test method consisted of a fluorometric technique based on the detection of metal cations released during corrosion by fluorescence measurement.

The best inhibiting efficiencies found in all the inhibitors and binary combinations (200) are shown in Fig. 7.2. The ratio with the best inhibitor efficiency was used for each combination. The observed inhibition is promising, although none of the inhibitor combinations can match sodium chromate in performance.

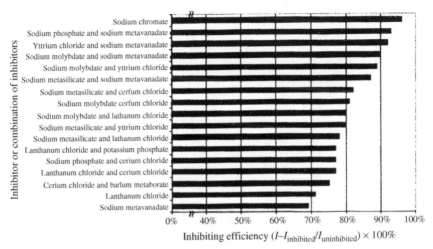

FIGURE 7.2 Inhibition efficiencies of various inhibitors.

The main advantage of the high-throughput screening approach is its ability to survey inhibitor performance over a range of test conditions within a single experiment. An ideal inhibitor should be effective in a wide pH and temperature range. An experiment involving a 50-cell array used to test seven inhibitor ratios and two inhibitors alone at pH of 2, 4, 7, 10, and 12 showed synergistic behavior at 50% of each inhibitor at pH 12; it showed the same synergistic behavior at pH 7.0 for 3.2 mM sodium metavanadate and 0.2 mM sodium metasilicate.

7.6 OLEOCHEMICALS AS CORROSION INHIBITORS

Oleochemicals are of vegetable, animal, and marine origin. These compounds consist of fatty acids, fatty alcohols, amines, and amides as well as fatty acid esters. Alkali and heavy metal soaps, polyoxyethylated compounds, and quaternary ammonium compounds are also used as inhibitors. Some of the applications of oleochemicals as inhibitors are given in Table 7.20.

Major concerns such as safety, environmental pollution, and economics require that the corrosion inhibitors be nonpolluting, safe, and economical. Cleaner production involves reconciling increased production with minimizing waste production at the source. Both clean production methods and sustainability are the subject of the report "Our Common Future" by G. H. Brundtland (301). According to this report, the concept of sustainable development is "Development that meets the needs of the present without compromising the ability of future generations to meet their own needs." It is useful to note that efforts may be directed toward beneficial use of resources in that less waste is generated. Sustainable development requires use of limited resources in the best possible manner, devoid of waste accumulation (302, 303). Efforts must be made to reduce waste at the source, failing which a

TABLE 7.20 Oleochemical Inhibitors

Metal/Alloy	Inhibitor	Reference
Al, Zn	Ammonium stearate spray	201
Al, nonferrous metals	Ethyl stearate	202
Al	Glyceryl monooleate coating	203
Al	Stearyl maleate	204
Al	Diethylene glycol monostearate	205
Brass	Dicyclohexylamine salt of oleoyl sarcosine	206
Brass	Potassium ricinoleate	207
Brass	Stearyl maleate	204
Bronze	Maleic anhydride adducts of butyloleate	208
Copper	Cobalt binoleate	208
Copper	Dicyclohexylamine salts of oleoyl sarcosine	206
Copper	Dilinoleic acid with sodium dodecylbenzene sulfonate	209
Copper	Monoethanolamine oleate with benzotriazole	210
Copper	Monoethanolamine stearamide	211
Copper	Potassium ricinoleate	207
Copper	Tristearyl borate	212
Magnesium	Caproic acid	213
Magnesium	Cetyl thiopropionate; distearyl thiodipropionate	214
Silver	Diacetyl sulfide; laurylamine acetate	215
Silver	Lauryl mercaptan	216
Silver	Palmitylamine acetate	217
Silver	Stearyl amine acetate	214
Ferrous metals	Ammonium ricinoleate; glyceryl monostearate; lauryl acid phosphate	218
Ferrous metals	5% lead stearate with cetylpyridinium stearate and sorbitan monooleate	219
Ferrous metals	Sodium and laurylamine salts of oleoyl sarcosine	220
Ferrous metals	Pentaerythritol monooleate monoaldyl carbonate	221
Ferrous metals	Sorbitan monooleate with propylene diamine	222
Ferrous metals	Zinc stearate	223
Ferrous metals	Lauryl acid phosphate with oxazolines and petroleum sulfonates	224
Metal pipes	Oleic acid hydrazide	225
Ferrous metals	Sorbitan monopalmitate	226
Ferrous metal	Palmitate helps phosphate coatings	227
Cold-rolled steel	Barium palmitate	228
Steel	Cetyl trimethylammonium bromide	229
	Lead linoleate; 0.1–2.0% lead stearate in perchloroethylene	230
	Lauryl diethanolamide	231
Steel	Maleic anhydride adducts of butyl oleates	232
	Morpholine stearate	233
	Octanolamine laurate stearate	234
	Piperidyl methylamine caproamide	235

TABLE 7.20 (*Continued*)

Metal/Alloy	Inhibitor	Reference
Iron	Stearamide/zinc stearate helps benzoyl peroxide	236
	Stearyl maleate with benzoyl peroxide	237
Steel	Stearyl methacrylate in mineral oil	238
	Triethanol amine monooleate ester with aminoethyl morpholine oleate	239
Tin plate	Glyceryl monooleate	240
Metal rubber	Calcium ricinoleate/sodium ricinoleate	241
Solder-covered metal	Ethylene glycol distearate	242
Steel	Squalene with oleyl methyl glycine	243
Razor blades	Squalene/sorbitan sesquioleate blend	243
Nuts and bolts	Isopropyl oleate	244
Zinc	Heptyloleate inhibits zinc in 9% HNO_3	245
Zinc	Reaction product of palmitonitrile and SO_2	246
Zinc	Ricinoleic acid hydrazide in methanol	247
Zinc	Stearyl maleate; stearyl mercaptan	248
Galvanized steel	Potassium stearate coating	249
Steel surfaces	Dicyclohexylamine oleate	250
Dewatering and water-displacing processes	Diethylaminoethanol oleate	251
	Ethylenediamine oleoamide ethoxylate	252
	Glyceryl monooleate	253
	Morpholine oleate	254
	Oleonitrile	255
	Triethanolamine dilaurate	256
Lubricants	Dicyclohexylamine oleate/dicyclohexyl amine stearate	257
	Sorbitan monostearate	258
	Reaction products of stearylamine and isoamyl-octyl acid phosphate	259
Rust preventives	Erucamide as a stripping agent for ethylene/vinyl copolymer coatings; glyceryl monooleate/octyl epoxy stearate	260
	Pentaerythritol dioleate	261
	Stearylamine	262
	Glyceryl monolaurate as a stripping agent; oleoyl sarcosine used in strippable coatings; linoleoyl sarcosine as a stripping auxiliary	263
Steel	Ammonium stearate in vapor phase formulation cyclohexylamine laurate/cyclohexylamine oleate	264
	Dicyclohexylamine oleate/dicyclohexylamine palmitate	265
	Dicyclohexylamine stearate	266
	Lithium caprate/sodium acid stearate	267
Wrapping iron/steel/copper	Capric acid with methylamine caprate	268
Petroleum refineries	Hydroxyethyl ethylene distearamide	269

(*continued*)

TABLE 7.20 (*Continued*)

Metal/Alloy	Inhibitor	Reference
Food cans	Sodium palmitate	270
Metal cleaners	Polyglycololeates	271
Detergents	Aluminum stearate	272
Electrical parts of conductors	Ethylene dipalmitamide with oleic acid in castor oil	273
Electrical conductors made from Al alloy	Sodium stearate	274
Petroleum processing	Reaction product of methyl oleate and H_2S (methyl mercapto stearate)	275
Jet fuel storage equipment	Dilinoleic acid with dimethyl dialkyl; ammonium chloride	276
Steel pipes in gas wells	Copper stearate, hexadecylamine esters of linoleic acid	277
Stainless steel	Nickel stearate	278
Water cooling	Glyceryl oleate/lauryl sarcosine/dodecyl; morpholine oleate and coco-morpholinium oleate	279
Water and steam lines	Morpholine oleate/salts of benzylamine	280
Protective wax coatings for cars	Morpholine stearate	281
	Alkylamino propylamine dioleate ester	282
	Aluminum stearate, calcium stearate, glyceryl stearate, glyceryl palmitate, zinc stearate	283
Engines	Dilauryl phosphate	284
	Ricinoleic acid in oils	285
	Triethanolamine oleate in mineral oils	286
Acid pickling	Stearylamine caprate	287
Antifreeze	Glyceryl monooleate	288
Metals	Ethyl tetrahydropyridine oleoamide in corrosive oil media	289
Steel	Piperidinyl butyral palmitate in H_2S medium	290
Electrodeposition	Ethyl oleate sulfate	291
Rusting	Caprylamine/hydroxyethyl ethylene dioleoamide	292
Iron, steel	Amine or ammonium salts of sulfostearic acid in hydrocarbon media	281
Steel, tin plate	Caproic acid in plum juice	293
Steel	Ethylstearylamine in sea water	294
	Propylene glycol monostearate together with cyclohexylamine carbonate	295
Carbon steel	Capric acid and caprylamine together with stearylaminehydrochloride in NaCl and H_2S medium	296
Stainless steel	Stearylamine benzoate in NaCl, $MgCl_2$ medium	297
Al alloys	Sodium stearate in NaCl medium	298
Steel	Laurylamine with capric acid in H_2SO_4 medium; stearyl trimethylammonium bromide	299
Iron	Mercaptostearic acid in acid media	300

safe process must be in place for safe disposal of the generated waste. Sustainable development involves manufacturing taking utmost care and adopting safe chemical practices. Thus, tenets of sustainability, eco-efficiency, and green chemical processes are very much involved in the selection and application of corrosion inhibitors (304).

Green chemistry is thus the foundation of selection and application of environment-friendly corrosion inhibitors (305). Synthesis of organic inhibitors involves adopting green chemical principles from synthesis to application in practice. Safety and environmental concerns are difficult to translate into economic terms. According to a study sponsored by the U.S. Federal Highway Administration (FHWA), the corrosion of metals costs in the United States in excess of $276 billion per annum in 1998. Loss for 1975 was estimated at $82 billion or about 4.9% of the gross national product.

Corrosion can impact the national economy as well as cause accidents involving loss of lives. Thus, it is difficult to estimate corrosion losses true economic value.There are many methods available for combating corrosion and the use of corrosion inhibitors is one of the methods to inhibit corrosion. Green chemistry is a readily available source of green inhibitors. The use of corrosion inhibitors for corrosion protection of metals is a well-established practice, as seen in the literature (306–308). A large number of organic compounds with N, S, and O as heteroatoms function as corrosion inhibitors but they fail the toxicity test (309). Plant extracts and some organic compounds provide corrosion inhibitors that are environmentally agreeable (310–319). Plant products containing natural organic corrosion inhibitors are nonhazardous and eco-friendly. These natural products form the basis of green inhibitors.

Adsorption mechanisms play an important role in corrosion inhibition. Elayyachi synthesized and tested two new telechelic compounds as inhibitors for the corrosion of steel in 1.0 M HCl solution (320). The two inhibitors are methyl 4-{2-[(2 hydroxyethyl) thio]ethyl}benzoate (T2) and 11-[(2-hydroxyethyl)thio]undecan-1-ol (T3). The corrosion inhibition increased with an increase in inhibitor concentration and gave greater than 90%inhibition. Adsorption of the inhibitor conformed to the Langmuir model of adsorption. The role of some surfactants in the corrosion of aluminum in 1 M HCl was studied (321) using weight loss and galvanostatic polarization techniques. The inhibition was found to be due to adsorption on the metal surface, and the degree of inhibition increased with inhibitor concentration and decreased with increasing temperature. The studies revealed that the inhibition conformed to the Freundlich adsorption isotherm. The percent inhibition increased upon the addition of potassium iodide, which may be due to increased surface coverage by quaternary ammonium iodide ($R - NH_3^+ \ I^-$). The thermodynamic parameters for adsorption and activation processes were determined and the galvanostatic polarization data indicate the inhibitors function as mixed-type inhibitors.

Fuchs-Godec studied (322) electrochemical aspects of cationic surfactants such as myristyltrimethyl ammonium chloride (MTAC), cetyldimethylbenzylammonium (CDBAC), and trioctyl methylammonium chloride (TOMAC) in the context of corrosion of ferritic stainless steel ion 2 M H_2SO_4. The potentiodynamic polarization data showed the hindering of both anodic and cathodic processes, which may be construed as mixed inhibitive action, and the adsorption of N-alkyl ammonium ion in 2 M H_2SO_4 corresponded to Langmuir adsorption isotherm. The calculated values of the free energy of adsorption were relatively high and characteristic of chemisorption when the charge on the metal surface is negative with respect to the point of zero charge. In the case of positive metal surfaces, the sulfate anions were adsorbed and the cationic species limited by the surface concentration of anions.

The corrosion inhibition of mild steel in perchloric acid solution by substituted pyridyl aminotriazoles was studied by both gravimetric and electrochemical imped-ance techniques. Respective protection percent of 95, 92 were obtained with 10^{-4} M of pyridylamino triazoles (323) (PAT), 3-PAT and 4-PAT, while 2-PAT gave only 65% inhibition. The degree of inhibition appears to depend on the position of the nitrogen atom in the pyridinium substituent. The adsorption of the inhibitors was in conformity with the Langmuir isotherm and the corrosion inhibition involving both physio- and chemisorption.

Schiff bases such as 2-[2-aza-2-(5-methyl-2-pyridyl)vinyl]phenol and the corre-sponding bromo- and chlorophenol were synthesized and their effect on the corrosion behavior of aluminum in 0.1 M HCl was studied by both electrochemical impedance of polarization techniques (324). All the data obtained supported the notion that the inhibitors function as mixed type of inhibitors. The degree of inhibition increased with increase in inhibitor concentration. The corrosion inhibition is due to adsorption on the aluminum surface in accordance with Langmuir model.

Corrosion inhibition of copper in aerated static 3% sodium chloride solutions in the temperature range of 15–65°C with sodium oleate as an anionic surfactant inhibitor resulted in a sigmoidal adsorption isotherm, confirming the application of Frumkin's equation for describing the adsorption process. The presence of chloride or bromide substitution in the Schiff bases resulted in greater strength of bonding between the metal and Schiff base. This improved inhibition due to substituents such as chloride or bromide in Schiff bases is in keeping with the earlier observation of Elewady (321) involving the synergism observed by the addition of potassium iodide to telechelic compounds. Quantum chemical calculations using the modified neglect of differential overlap (MNDO) semiempirical self-consistent field-molecular orbital (SCF-MO) methods showed the dependence of adsorption on the electronic charge density of the donor atom of the inhibitor molecule and the dipole moment. Along similar lines, quantum chemical calculations on benzimidazoles were performed (325) and the inhibition efficiency was correlated with E_{HOMO} energy (energy of highest occupied molecular orbital), dipole moment, linear solvation energy terms, total negative charge on the inhibitor, molecular volume, and linear solvation energy terms. Increase in inhibition efficiency with increase in E_{HOMO} was clearly established by El-Ashry et al. (325).

7.7   HYBRID COATINGS AND CORROSION INHIBITORS

Corrosion inhibitors are extensively used in hybrid coatings and in particular sol-gel-derived organic–inorganic hybrid coatings (326). In early times, chromates were used as corrosion inhibitors in coatings. Since chromates are known to be toxic, alternative substitutes for chromates such as cerium salts (327) may be used. Cerium ions are incorporated in silica gel coatings, which are applied on aluminum alloys. The barrier effect of silica coatings is combined with the corrosion inhibitor effect of cerium ions. Chromate was used in early times and it has been replaced by cerium in view of the toxicity of chromate. Thin (1 nm thick) and transparent cerium-doped silica gel coatings were prepared by dipping 3005 aluminum alloys in sol-gel solutions. Spectroscopic evidence showed the presence of both Ce^{3+} and Ce^{4+} ions in the coatings. Single- and two-layer coatings prepared with Ce^{4+} salt showed good protection of aluminum, while two or multiple layers of coating were required when Ce^{3+} ions were present. The improvement in protection with immersion time indicates migration of cerium ions through the coating to the site of attack and then passivation. Sol-gel derived organosilicate hybrid coatings preloaded with organic corrosion inhibitors have been developed by Khramov et al. (328), and these coatings provide good corrosion protection when the integrity of the coating is compromised. The incorporation of the organic corrosion inhibitor in the coating is achieved by physical entrapment of the inhibitor in the coating material at the film formation stage and cross-linking. The entrapped corrosion inhibitor becomes active in corrosive electrolyte and slowly diffuses out of the host matrix. To ensure continuous delivery of the inhibitor to corrosion sites, a sustained release of the inhibitor is achieved by a reversible chemical equilibrium of either ion exchange of the inhibitor with the coating material or through cyclodextrin-assisted molecular encapsulation. Some of the inhibitors tested for this procedure include mercaptobenzothiazole, mercapto-benzimidazole, mercaptobenzimidazole sulfonate, and thiosalicylic acid. Lamaka et al. (329) proposed an approach for the development of a nanoporous reservoir for storing corrosion inhibitors on the metal/coating interface. The nanostructured porous TiO_2 layer was prepared on the aluminum alloy surface by the controlled hydrolysis of titanium alkoxide in the presence of a template agent. In contrast to direct embedding of the inhibitor into the sol-gel matrix, the use of a porous reservoir eliminated the negative effect of the inhibitor on the stability of the hybrid sol-gel matrix. Titania/inhibitor/sol-gel systems showed enhanced corrosion protection and self-healing ability.

A new protective system (330) with self-healing ability comprising hybrid sol-gel films doped with nanocontainers that release entrapped corrosion inhibitor in response to pH changes caused by the corrosion process has been developed. A silica-zirconia-based hybrid film was used as an anticorrosion coating deposited on 2024 aluminum alloy. Silica nanoparticles covered layer by layer with polyelectrolyte layers and layers of benzotriazole inhibitor were randomly introduced in the hybrid films. The hybrid film with nanocontainers gave long-term protection compared to undoped hybrid film. The technique involving a scanning vibrating electrode revealed the effective self-healing nature of the defects. The self-healing of the defects may be attributed to the regulated release of the corrosion inhibitor triggered and self-controlled by the pH feedback due to the corrosion processes in the cavities. The concept of feedback was studied further involving nanocontainers with the ability to release the encapsulated active materials in a controlled way, which ultimately help develop self-repairing multifunctional coatings that are not only passive but also possess rapid feedback activity in response to changes in local environment. Several methods of fabricating self-repairing coatings have been reviewed (331). The release of the active component was regulated to occur only when triggered by the specific nature of the corrosion process, which prevented the leakage of the active component out of the coating and thereby increased the durability of the coating. The self-repair properties of sol-gel films involving trivalent chromium, Ce(III), molybdate, and permanganate were further studied and were compared with the action of chromium(III). The morphology of sol-gel and the solubility of the additive appear to play an important role in the effectiveness of long-term corrosion protection. Additives such as molybdate and permanganate were too soluble to stabilize the sol-gel network, and cerium(III)—being soluble—was not efficient. Contrary to the findings of Pepe et al. (327), sol-gel films containing Cr(III) and Cr(VI) gave good protection due to their stability and low solubility. Titania containing organic–inorganic hybrid sol-gel films has been advanced as a replacement for the environmentally undesirable chromate-based coating for surface pretreatment of aluminum alloys. Stable hybrid sols were prepared by hydrolysis of 3-glycidoxypropyltrimethoxysilane and other different titanium organo compounds in the presence of small amounts of acidified water. Different diketones were used as complexing agents in the synthesis of controlled hydrolysis of organotitanium compounds. The performance of the resulting coatings was compared with that of zirconia-bearing films, using electrochemical impedance and salt spray tests to evaluate the performance of the hybrid coatings (333). The protective properties of the coatings depended on the nature of metalorganic precursors and complexing agents used in the preparation of sol-gel. The best corrosion protection of AA2024 alloy was observed when titania-containing sol-gels were prepared with titanium(IV) tetrapropoxide and acetyl acetone as starting materials. In the case of zirconia films, good protection was observed when ethylacetoacetate was used as a complexing agent. Yasakau et al. (334) used a number of corrosion inhibitors for AA2024 alloy to produce the sol-gel without affecting the coating. The additives such as 8-hydroxyquinoline, benzotriazole, and cerium nitrate have been used at different stages of the synthesis to gain an understanding of the interaction of inhibitors with the sol-gel components. 8-Hydroxyquinoline and cerium nitrate did not affect the stability of the hybrid sol-gel films but augmented corrosion protection. On the other hand, benzotriazole had an adverse effect on the corrosion protection of the sol-gel. Testing of green algae as an additive for paint formulation based on vinyl chloride copolymer for the protection of steel corrosion in seawater showed an extracted form of algae to perform better than a suspended form (335).

Mathematical and computer-aided models to analyze and organize experimental data in corrosion inhibition are beginning to be useful in understanding the corrosion inhibition mechanisms (336, 337). Computer modeling and mathematical analysis are beginning to prove useful in corrosion inhibition. A robust and comprehensive finite element model (338) for predicting corrosion rates of steel in concrete has been advanced. This model may be useful in the development of better concrete structures with respect to corrosive deterioration. Kubo et al. developed a mathematical model for the simulation of changes in the pore solution phase chemistry of carbonated hardened cement paste when aqueous organic base inhibitors are applied at the surface of the material. The model, based on the Nernst equation, is used to obtain concentration profiles of electrochemically injected inhibitors and the major ionic species present in the pore electrolyte. The corrosion inhibition model that takes into account changes in inhibitor concentration over steel by thermomechanical processing has been advanced (339). The model is also capable of accurately predicting current values in short simulation times. The flow-induced corrosion model for carbon steel in high-velocity flow seawater has been studied by Jingjun et al. (340) The corrosion rates calculated theoretically were in good agreement with the observed rates. The theoretically calculated corrosion rates were in fair agreement with the measured corrosion rates due to flow-induced corrosion.

Environmentally toxic inhibitors are sometimes referred to as gray inhibitors and environment-friendly inhibitors are termed as green inhibitors. These consist of (i) organic compounds, (ii) amino acids, (iii) plant extracts, and (iv) rare earth metal compounds. Khaled (308) synthesized a substituted guanidine compound and used it in the corrosion inhibition of mild steel in 1 M hydrochloric and 0.5 M sulfuric acid solutions. The adsorption of guanidine inhibitor was found to obey Langmuir's adsorption isotherm. Aminoalcohols such as 1,3-bis-dialkylaminoprane-2-ols were used as inhibitors and the degree of corrosion inhibition increased with an increase in the length of the alkyl chain and also an increase in the concentration of the inhibitor (341). Polarization curves were characteristic of anodic inhibitors and the adsorption of the inhibitors on the metal surface followed Langmuir's adsorption isotherm.

Amino acids-based green inhibitors and their corrosion inhibition effect on copper were studied by weight loss technique and the electrochemical polarization method (332). Valine and glycine increased the corrosion rate, while arginine, lysine, and cysteine inhibited the corrosion process, with cysteine being the best among the amino acids tested. The cysteine inhibited the corrosion of copper by 61%at a concentration of 0.001 M and 84% in 0.6 M NaCl solution (343). The electrochemical behavior of copper in the presence of cysteine showed the inhibitor acting as a cathodic inhibitor. The data showed the adsorption of the inhibitor corresponded to the Langmuir isotherm, with 25 kJ/mol as the free energy of adsorption.

Green inhibitors consisting of plant extracts such as rosemary oil were tested as corrosion inhibitors (344). The principal constituent of rosemary oil is 1,8-cineole. The electrochemical polarization studies showed rosemary oil as a cathodic inhibitor with corrosion inhibition of 73% at a concentration of 10 g/L. Rosemary oil can be effectively used in the cleaning and removing of scales such as pickling operations.

The corrosion inhibition of mild steel by black pepper extract was studied (309) in 1 M H_2SO_4 solution and the extract acted as a good inhibitor and the adsorption conformed to the Temkin adsorption isotherm. Electrochemical impedance and scanning electron microscopic studies provide further evidence for the protection. The corrosion inhibition of steel, nickel, and zinc in acidic, neutral, and alkaline solution by aqueous extract of henna leaves (*Lawsonia*) was studied (345), showing that the henna extract provided good corrosion inhibition of all three metals. The degree of inhibition depended on the nature of the metal and the medium. In the case of carbon steel and nickel, the degree of inhibition varied with the medium: acid $<$ alkaline $<$ neutral, which also showed that *Lawsonia* extract is a mixed type of inhibitor. The adsorption of *Lawsonia* extract obeyed the Langmuir isotherm. Complex formation between steel/nickel and Lawson extract has been postulated as a mechanism of corrosion inhibition (345).

The corrosion inhibition of aluminum by *Ocimum basilicum* extract in 2 M HCl and 2 M KOH solution was studied (346) and the basilicum extract was found to inhibit corrosion in both the solutions. The degree of inhibition increased with an increase in the concentration of the extract but decreased in inhibition with an increase in temperature. Halide ions provided a synergistic effect on the *O. basilicum* extract and improved the degree of corrosion inhibition.

Corrosion inhibition of tin by natural honey (chestnut and acacia) and a mixture of natural honey and black radish juice showed the degree of inhibition with acacia honey was lower than that with the chestnut honey, and the addition of black radish juice to both types of honey resulted in an increased degree of corrosion inhibition (347). The authors attributed the protection rendered by honey to the formation of a multilayer adsorbed film on the metal conforming to Langmuir adsorption isotherm.

The corrosion inhibition of mild steel by an acid extract of *D. metel* followed both Temkin and Langmuir isotherms (348). The authors of this investigation advanced the idea that the *D. metel* plant may be used as a green inhibitor for the corrosion inhibition of machinery made out of mild steel. Arghel herb extract was used as a green inhibitor in 1.0 M sulfuric acid solution. From the data obtained, the authors concluded that Arghel extract functions as a mixed-type inhibitor (349).

The corrosion inhibition of aluminum in 2 M sodium hydroxide and 0.5 M NaCl by Damsissa (*Ambrosia maritima* L.) extract was studied using potentiodynamic polarization and AC impedance techniques (350). Nyquist plots showed the presence of two capacitative semicircles. The degree of inhibition of corrosion was marked and the data showed Damsissa extract to be a very promising green inhibitor for aluminum in chloride solutions. The extract was stable for over 35 days and the extract was more effective in the presence of a chloride ion. The corrosion inhibition increased with an increase in the concentration of the Damsissa extract but decreased with increasing temperature.

The corrosion inhibition of aluminum with a mixture of Lupine and Damsissa plant extracts along with a surfactant was studied by potentiodynamic polarization and AC impedance techniques and the degree of inhibition was maximum at the critical micelle concentration of the surfactant.

The extract of Arghel has a marked effect in reducing the attack of steel by chloride in concrete. Arghel extract behaved as an anodic inhibitor in concrete. Olive extract (*Oleo europaea* L.) functions as both an antiscalant for calcium carbonate deposit and an inhibitor for steel in alkaline calcium chloride brine solution. The Arghel extract impedes calcium carbonate supersaturation and decreases the time of nucleation. The surface area occupied by the scale deposits decreased with an increase in Arghel extract concentration. The plant leaves extract inhibit the corrosion of steel by controlling the cathodic oxygen reduction process. Both olive and fig leaf extracts decrease the corrosion rate and scale buildup under the tested conditions (351).

Rare earth compounds such as cerium and lanthanum cinnamate compounds have been used as green inhibitors in the protection of iron from corrosion (352). The species FeO•OH normally present in the absence of rare earth cinnamate was not present on the iron sample treated with cerium cinnamate inhibitor (353).

7.8 BARBITURATES AS GREEN CORROSION INHIBITORS

Barbiturates such as barbituric acid (BA), ethyl barbituric acid (EBA), and 2-thiobarbituric acid (2-TEA) have been studied as green corrosion inhibitors in the corrosion of mild steel in 0.5 M HCl solution by linear polarization and electrochemical impedance spectroscopy (354). The degree of corrosion of mild steel in the presence of the inhibitor was found to depend on the structure and concentration of the inhibitor. The degree of inhibition was found to depend upon the concentration of the inhibition. The corrosion inhibition was in the order: 2-TBA > EBA > BA. The corrosion inhibition by these inhibitors increased with the increase in concentration of the inhibitors. Some data on the corrosion inhibition by 2-thiobarbituric acid are given in Table 7.21.

The degree of corrosion inhibition increases marginally with increase in inhibitor concentration by tenfold at a time interval of 168 h. The adsorption of 2-thiobarbituric acid on mild steel was in accordance with the Langmuir adsorption isotherm. The values of adsorption equilibrium constant and adsorption free energy suggest strong adsorption of 2-thiobarbituric acid on the surface of mild steel. The plot of polarization resistance against potential showed a maximum at -0.488 V, which can be termed the potential of zero charge (PZC) in 0.5 M HCl and 10 mM of 2-thiobarbituric acid. The open circuit potential of mild steel in the same conditions (i.e., 10 mM of 2-TBA) is -0.448 V, which is higher than the potential of zero charge. These data show that the surface charge of mild steel under these conditions is positive and hence initially negative chloride followed by adsorption of 2-thiobarbituric acid molecules.

7.9 CORROSION PREVENTION OF COPPER USING ULTRATHIN ORGANIC MONOLAYERS

Copper and its alloys are used in chemical and electronics industries for their excellent properties such as high thermal and electrical conductivities and their low cost. But copper and its alloys are easily prone to corrosion under the combined effect of

TABLE 7.21 Corrosion Inhibition by 2-Thiobarbituric Acid

Concentration of Inhibitor (mM)	Inhibition (%)
1.00	95.2
5.00	96.5
10.00	97.7

moisture, ionic dust, and gases such as sulfur dioxide, nitrogen dioxide, hydrogen chloride, and chlorine (355). Some methods available for surface treatment are sol-gel hybrid-derived coating (356), monolayer formation by self-assembly (357), anodic oxidation, and organic coating (358).

Self-assembled monolayers (SAMs) are ordered molecular assemblies formed by the adsorption of an active surfactant on a solid surface. Self-assembled monolayers are surfaces consisting of a single layer of molecules on substrate. Monolayer formation may be achieved by mere immersion of the metal substrate into a solution of the desired compound for optimum assembly duration and washing away the excess solution.

A familiar example of self-assembly is alkanethiol on copper, as shown in Fig. 7.3. The self-assembly approach of film formation is based on the chemical affinity between the substrate surface and the molecular species chosen for fixing on the surface. The bonding between the adsorbed species and the metal is strong (covalent or ionic) and van der Waals type of interactions between the self-assembled molecules contribute to order and coherence of the monolayer film. Copper has a high affinity to bond with organic molecules containing sulfur in the functional group. The organosulfur complex of copper has a high degree of resistance to corrosion. The lone pair of electrons on the sulfur in the thiol compound are donated to the copper metal, forming a coordinate bond. The factors involved in a SAM are (i) high bond energy between the functional group donor atom(s) and the metal (Cu); (ii) energy of van der Waals interaction; and (iii) entropy lost, thermal effect, solvent displacement, and solvent solution structuring all affecting self-assembled monolayer formation.

The self-assembled monolayer offers a simple method for achieving ultrathin films, with particular application in microelectronics. The potential advantages of SAMs are that they (i) are cost-effective and environment-friendly; (ii) do not affect the metallic properties, except for changing hydrophobicity; (iii) are thin films of nanometer range; (iv) prevent diffusion of electrolyte; (v) have minimal impact on heat transfer; (vi) do not need continuous addition of inhibitors; (vii) have a strong chemical bond with metal; and (viii) provide nanometric films suitable for the electronics industry.

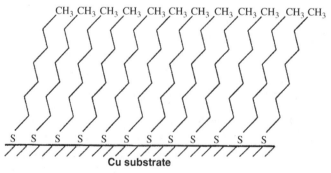

FIGURE 7.3   Self-assembly of alkanethiol on a copper surface.

Self-assembly of alkanethiol on copper occurs according to the reaction

$$Cu + C_nH_{2n+1}SH \rightarrow [Cu] - SC_nH_{2n+1} + H^+ \text{ or } 1/2H_2.$$

The resulting self-assembled monolayer inhibits the corrosion of copper substrate in an aerated aqueous corrosive solution and atmospheric environment. SAM also provided corrosion inhibition in both wet and dry environments. The chain length of alkanethiol $CH_3(CH_2)_{n-1}SH$ had an observable effect of protection. The longer the hydrocarbon chain, the better the corrosion protection:

$$C_{18}SH > C_{12}SH > C_6SH.$$

Modification of thiol SAM on copper through substitution by alkylchlorosilane resulted in two-dimensional film, which mitigated corrosion. Both the SAM and silane-modified SAM may be formed as detailed below:

$$Cu + HS(CH_2)_{11}OH \rightarrow [Cu]S(CH_2)_{11}OH + H^+ + e^-$$

$$[Cu]S(CH_2)_{11}OH + XC_nH_{2n+1}SiCl_3 + xH_2O \rightarrow \text{silane-modified SAM}.$$

Silane-modified SAM gave better atmospheric corrosion inhibition than unmodified SAM.

Modification of SAM with alkylisocyanate was found to be superior to that with alkylchlorosilane. Self-assembly film of Schiff bases with modified 1-dodecanthiol SAM with thiosulfates has been studied (359–361). The SAM may be used as a primer to prevent delamination of microelectronic packages.

7.10 CORROSION OF TITANIUM BIOMATERIALS

The relatively inert titanium bioalloys may corrode in the human body environment (362). Ionic and particulate matter from biomaterials may be involved in physiological interactions with human tissues followed by metabolical, bacteriological, and oncogenic effects, leading to inflammatory reaction or other adverse effects.

Dissolution of titanium and its alloys may occur in three different ways. It may be chemical or electrochemical dissolution of layers of simple and complex oxides. Dissolution may occur due to the transport of titanium and other elements through the oxide crystal lattice and its imperfections such as grain boundaries. Corrosion may manifest as cracks due to aging of oxide layers, exfoliation of the layer, and corrosion tunnels, which make liquid penetration to the bare metal and electrochemical reaction possible.

The direct reaction of human body liquids with the bare metal leads to corrosion. The structure and the thickness of the oxide layer, the interface layer, the material bulk strength, and the stresses imposed on the titanium bioalloy should be considered in predicting and improving the corrosion behavior of the alloy.

The basic issues associated with titanium bioalloys, such as the amount of metal released from the implant, the site to which the metal is transported along with the amount transported, the chemical form of the metal released, and the pathophysiological consequences of the metal release, are important. The techniques presently available cannot properly predict the behavior of implanted human organ implants over long duration, especially the corrosion rate and the corrosion fatigue limit.

Effective countermeasures against corrosion could involve change in structure of the oxide layer, increasing the thickness of the oxide layer by available methods such as laser oxidation, ion implantation, and creation of composite layers.

7.11 CORROSION CONTROL IN THE ELECTRONICS INDUSTRY

Corrosion of silver components in electronic devices such as television components due to the exposure to hydrogen sulfide was observed (363). The hydrogen sulfide tarnishes the silver components and produces silver sulfide. The weight loss method was used in obtaining corrosion rates and the surface film was identified as silver sulfide by scanning electron microscopy (SEM) and energy-dispersive X-ray analysis (EDX). Corrosion of the electronic components was mitigated by using VAPPRO vapor-phase corrosion inhibitor.

REFERENCES

1. CM Allen, LB Silbey, S Cosgrove, *14th Annual Meeting of the ASLE*, Buffalo, NY, Paper No. 59, April 1959.
2. S Rajendran, AJ Amalraj, MJ Joice, N Anthony, DC Trivedi, M Sundara-Vadivelu, *Corros Rev* **22**(3):233–248 (2004).
3. J Mathiyarasu, SS Pathak, V Yegnaraman, *Corros Rev* **24**:307 (2006).
4. SMA Shibli, V Anitha Kumary, *Anti-Corros Method Mater* **51**(4),(Aug 2004).
5. S Taj, A Siddekha, S Papavinasam, RW Revie, *Corrosion 2007*, NACE, Nashville, TN, Mar 11–15, 2007.
6. YI Kuznetsov, *Corrosion 2004*, New Orleans, LA, April 2004.
7. AM Alfuraij, *Corrosion & Prevention—2000, Auckland*, New Zealand, Nov 19–22, 2000.
8. HJ Chen, Y Chen, *Corrosion 2002*, Denver, CO, Apr 7–11, 2002.
9. MM Attar, JD Scantlebury, *J Corros Sci Eng* **1** (1995–2000).
10. PC Okafor, VI Osabor, EE Ebenso, *Pigm Resin Technol* **36**:299–305 (2007).
11. C Georges, E Rocca, C Caillet, *EUROCORR 2004: Long Term Prediction and Modelling of Corrosion*, Nice, France, Sept 12–16, 2004.
12. GR Sparrow, *Asia Pacific Interfinish 1994*, Australian Institute of Metal Finishing, North Melbourne, Victoria, Australia, Oct 2–6, 1994.
13. SA Verma, GN Mehta, *Trans SAEST (India)* **33**(4):160–162 (1998).
14. PC Okafor, EE Ebenso, *Pigm Resin Technol* **36**:134–140 (2007).

15. KN Tatleriou, KM Penn, ML Mellott, CA Sizemore, CT Lin, *221st ACS National Meeting*, San Diego, Apr 1–5, 2001.

16. E Juzeliunas, R Ramanauskas, *Electrochim Acta* **51**(27):6085–6090 (2006).

17. BV Appa Rao, S Srinivasa Rao, *Bull Electrochem* **21**(3):139–144 (2005).

18. E Morris, JO Stoffer, TJ O'Keefe, P Yu, X Lin, *Polym Mater Sci Eng* **78**:172–173 (1998).

19. IV Branzoi, I Camenita, F Branzoi, A Bondarev, *UPB Bull Stiint B: Chem Mater Sci* **69**(3):9–18 (2007).

20. Z Faska, L Majidi, R Fihi, A Bouyanzer, B Hammouti, *Pigm Resin Technol* **36**(5):293–298 (2007).

21. AA El Hosary, RM Saleh, *Progress in Understanding and Prevention of Corrosion*, Vol. II, Barcelona, Spain, 1993, pp. 911–915.

22. AE Jenkins, WY Mok, CG Gamble, SR Keenan, *Corrosion 2004*, New Orleans, LA, Mar 28–Apr 1, 2004.

23. IH Farooqi, MA Nasir, MA Quraishi, *Corros Prev Control* **44**(5):129–134 (1997).

24. L Miskovic, T Madzar, *Corrosion 2002*, Denver, CO, Apr 7–10, 2002.

25. BV Apa Rao, S Srinivasa Rao, M Venkateswara Rao, *Corros Eng Sci Technol*, **43**:46–53 (2008).

26. HA Videla, PS Guiamet, SG De Saravia, LK Herrara, C Gaylarde, *Corrosion 2004*, New Orleans, LA, Mar 28–Apr 1, 2004.

27. A Schiopescu, L Antonescu, M Moraru, I Camenita, *EUROCORR 2001: The European Corrosion Congress*, Lake Garda, Italy, Sept–Oct 2001.

28. TY Chen, CB Batton, *Proceedings of the 9th European Symposium on Corrosion Inhibitors*, Vol. 1, Ferrara, Italy 2000, pp. 53–64.

29. D Mukherjee, Ramprasad, Benchman, Sudarshan, Marthamuthu, *Tool Alloy Steels (India)* **31**(2):22–36 (1997).

30. B Miksic, *Tekhnologiya Legkikh Splavov*, University of Ferrara, Italy, pp. 569–588, 1995.

31. B Miksic, L Gelner, D Bjegovic, L Sipos, *Proceedings of the 8th European Symposium on Corrosion Inhibitors,* Vol. 1, Ferrara, 1995, pp. 569–588.

32. SM Powell, HN McMurray, DA Worsley, *Corrosion 99*, San Antonio, TX, Apr 25–30, 1999.

33. BV Appa Rao, M Venkateswara Rao, *EUROCORR 2004: Long Term Prediction and Modelling of Corrosion*, Nice, France, Sept 12–16, 2004.

34. A Schiopescu, L Antonescu, M Moraru, *EUROCORR '99: Eurocongress*, Aachen, Germany, Aug 30–Sept 2, 1999.

35. BRW Hinton, *Asia Pacific Interfinish 90: Growth Opportunities in the 1990s*, Singapore, Nov 19–22, 1990, pp. 301–3021.

36. SG Gomez de Saravia, PS Guiamet, HA Videla, *79th Annual Meeting and International Coatings Technology*, Edinburgh, UK, Sept 18–21, 2001 (published by Institute of Corrosion, Bedfordshire, UK).

37. D Mukerjee, J Berchman, A Rajasekar, N Sudarsanan, R Mahalingam, C Hari, J Thangarajan, *Tool Alloy Steels (India)* **31**:13–19 (1997).

38. DI Bain, G Fan, J Fan, H Brugman, K Enoch, *Corrosion 2003*, San Diego, CA, Mar 16–20, 2003.

39. G Schmitt, R Buschman, H Theunissen, *EUROCORR 2004: Long Term Prediction and Modelling Corrosion*, Nice, France, Sept 12–16, 2004.

40. O Lahodny–Sarc, F Kapor, *Mater Sci Forum* **289–292**(2):1205–1216 (1998).

41. CB Batton, TY Chen, DM Cicero, *Corrosion 99*, San Antonio, TX, Apr 25–30, 1999.

42. M Bendahou, M Benabdellah, B Hammouti, *Pigm Resin Technol* **35**(2):95–100 (2006).

43. D Labouche, M Traisnel, *J Oil Colour Chem Assoc* **77**(10):404, 406–408, 410, 411, 414–416, 418–419 (1994).

44. S Tebbal, RD Kane, BN Al-Shumaimri, PK Mukhopadyay, *Corrosion '98*, San Diego, CA, Mar 22–27, 1998.

45. H Akrout, L Bousselmi, E Triki, S Maximovitch, F Dalard, *J Mater Sci* **39**(24):7341–7350 (2004).

46. BV Apparao, K Christina, *EUROCORR 2004: Long Term Prediction and Modelling Corrosion*, Nice, France, Sept 12–16, 2004.

47. H Akrout, S Maximovitch, L Bousselmi, E Triki, F Dalard, *Mater Corros* **58**(3):202–206 (2007).

48. DI Bain, G Fan, J Fan, H Brugman, K Enoch, *Corrosion 2002*, Denver, CO, Apr 7–11 2002.

49. L Crotty, L Gelner, *Triservice Conference on Corrosion*, Wrightsville Beach, NC, Nov 17–21, 1997, pp. 121–127.

50. M Khobaib, FW Valdiek, CT Lynch, *Triservice Conference on Corrosion*, Vol. II, US Air Force Academy, Colorado, Nov 5–7, 1980, pp. 345–406.

51. A Schiopescu, M Moraru, M Negoiu, O Georgescu, M Rizea, *EUROCORR 1997*, Vol. I, Trondheim, Norway, Sept 22–25, 1997, pp. 207–211.

52. O' Lahodny-Sarc, I Gotic, E Levacic, *Proceedings of the 9th European Symposium on Corrosion Inhibitors*, Vol. 1, Ferrara, Sept 4–8 2000, pp. 93–104.

53. A Schiopescu, L Antonescu, M Moraru, *Proceedings of the 9th European Symposium on Corrosion Inhibitors*, Vol. 2, Ferrara 2000, pp. 1065–1074.

54. G Banerjee, A Banerjee, CS Shah, *Corrosion 99*, San Antonio, TX, Apr 25–30, 1999, pp. 1999.

55. Anonymous, *Anti-Corros Method Mater* **31**(4):4 (1984).

56. SA Verma, GN Mehta, *Trans SAEST (India)* **32**:89–93 (1997).

57. IH Farooqi, A Hussain, MA Quraishi, PA Saini, *Anti-Corros Method Mater* **46**(5):328–331 (1999).

58. MA Quraishi, D Jamal, *Corrosion* **56**:983–985 (2000).

59. G Gunasekaran, LR Chauhan, *Electrochim Acta* **49**:4387–4395 (2004).

60. MI Awad, *J Appl Electrochem* **36**:1163–1168 (2006).

61. AM Beccaria, *EUROCORR 1997*, Vol. 1, Trondheim, Norway, Sept 22–25, 1997, pp. 381–386.

62. WH Durnie, R De Marco, BJ Kinsella, A Jefferson, *Corrosion and Prevention 1998*, Hobart, Australia, Nov 23–25, 1998, pp. 208–213.

63. BK Failon, RG Gabriel, BL Downward, PR Fowler, R Talbot, *Corrosion 99, San Antonio, TX, Apr 25–30, 1999*, p. 10.

64. SM Powell, HN McMurray, DA Worsley, *Corrosion* **55**(11):1040–1051 (1999)

65. M Bethencourt, FJ Botana, M Marcos, MA Arenas, J De Damborenea, *9th European Symposium on Corrosion Inhibitors*, Vol. 1, Ferrara 2000, pp. 83–91.

66. G Schmidt, AO Saleh, *Corrosion 2000*, Orlando, FL, Mar 2000, pp. 003351–0033513.

67. G Schmidt, AO Saleh, *Mater Perform* **39**(8):62–65 (2000).

68. C Simpson, *ICE 2001, 79th Annual Meeting*, Atlanta, GA, Nov 5–7, 2001.

69. AJ McMahon, NJ Phillips, N Bretherton, C Sutherland, I Martin, *79th Annual Meeting and International Coatings Technology Conference*, Edinburgh, UK, Sept 18–20, 2001, p. 16.

70. C Chandler, *Corrosion 2001*, Houston, TX, Mar 11–16, 2001, p. 14.

71. B Miksic, RW Kramer, *Mater Perform* **40**:24–27 (2001).

72. BM Badran, HA Mohammad, HA Aglan, *J Appl Polym Sci* **85**:879–885 (2002).

73. C Chandler, M Kharshan, A Furman, *Corros Rev* **20**:379–390 (2002).

74. MA Arenas, JJ De Damborenea, *Electrochim Acta* **48**:3693–3698 (2003).

75. CJ D'Arcy, JR Still, *Corros Manag* (56):16–20 (2003).

76. JS Gill, *Corrosion NACE 2006, 61st Annual Conference*, San Diego, CA, Mar 12–16, 2006.

77. Y Gao, U Ana, GD Wilcox, *Trans Inst Met Finish* **84**(3):141–148 (2006).

78. MS Morad, *J Appl Electrochem* **37**(6):661–668 (2007).

79. P Narmada, MV Rao, G Venkatachari, BV Apparao, *Anti-Corros Method Mater* **53**(5):310–314 (2006).

80. C Shivane, P Puomi, WJ Ooij, *Proceedings of the 20th International Conference on Surface Modification Technologies*, Vienna, 2006, pp. 1–4 2007.

81. YI Kuznetsov, *Prot Met (Russia)* **35**(4):322–329 (1999).

82. YI Kuznetsov, *Proceedings of the 9th European Symposium on Corrosion Inhibitors*, Vol. 1, Ferrara, Italy, Sept 4–8 2000, pp. 375–386.

83. M Zubielewicz, W Gnot, *Prog Org Coating* **49**(4):358–371 (2004).

84. K Ravichandran, NM Kumar, K Subash, TSN Sankara Narayanan, *Corros Rev (Israel)* **19**(1):29–42 (2001).

85. H Wang, R Akid, *Corros Sci* **50**(4):1142–1148 (2008).

86. N Srisuwan, N Ochoa, N Pebere, B Tribollet, *Corros Sci* **50**:1245–1250 (2008).

87. H Ju, Y Li, *Corros Sci* **49**(11):4185–4201 (2007).

88. E Khamis, A Hefnawy, AM El-Demerdash, *Materialwiss Werkstofftech* **38**(3):227–232 (2007).

89. L Gelner, *Corrosion 98*, San Diego, CA, Mar 22–27, 1998, pp. 712/1–712/6.

90. A Rauscher, G Kutsan, R Bundula, T Szailer, *Proceedings of the 9th European Symposium on Corrosion Inhibitors*, Vol. 1, Ferrara, Italy, Sept 4–8 2000.

91. J Banas, B Mazurkiewicz, W Sclarski, KZA Marchut, A Kaezmarczyk, L Komblit, *EUROCORR 1998: Solutions to Corrosion Problems*, Utrecht, The Netherlands, Sept 28–Oct 1, 1998.

92. AM Beccaria, G Castello, MG Zampella, G Poggi, *EUROCORR 1997*, Vol. 1, Trondheim, Norway, Sept 22–25, 1997, pp. 381–386.

93. BA Miksic, RW Kramer, *Steel Times (UK)* **227**(5):170–172 (1999).

94. FP Galliano, D Landolt, *42nd Corrosion Symposium*, Swansea, UK, Sept 11–13 2001.

95. M Islam, H al-Mazeedi, AM Abdullah, *8th International Congress on Metallic Corrosion*, Vol. II, Mainz, West Germany, 1981, pp. 1233–1238.

96. R Athey, *J Coating Technol* **57**(726):71–77 (1985).

97. MM Osman, E Khamis, AF Hefny, A Michael, *Anti-Corros* **41**(2–3):7 (1994).

98. JC Oung, SK Chiu, HC Shih, *Corros Prev Control* **45**:156–162 (1998).

99. M Kalpana, GN Mehta, *Trans SAEST* **38**(1):40–42 (2003).

100. A Abdul Rahim, E Rocca, J Steinmetz, R Adnan, MJ Kassim, *EUROCORR 2004: Long Term Prediction and Modelling of Corrosion*, Nice, France, Sept 12–16, 2004.

101. N Ochoa, F Moran, N Pebere, B Tribollet, *EUROCORR 2004: Long Term Prediction and Modelling of Corrosion*, Nice, France, Sept 12–16, 2004.

102. N Ochoa, F Moran, N Pebere, B Tribollet, *Corros Sci*, **47**:593–604 (2005).

103. N Etteyeb, L Dhouibi, M Sanchez, C Alonso, C Andrade, E Triki, *J Mater Sci* **42**(13):4721–4730 (2007).

104. B Bozzini, V Romanello, C Mele, F Bogani, *Mater Corros* **58**:20–24 (2007).

105. M Salasi, T Shahrabi, E Roayaei, *Anti-Corros Method Mater* **54**:82 (2007).

106. G Pallos, G Wallwork, *Corros Rev*, **6**(3):237–278 (1985).

107. I Jiricek, J Vosta, J Macak, J Fikar, F Liska, *Proceedings of the 8th European Symposium on Corrosion Inhibitors*, Vol. 1, Ferrara, Italy, Sept 18–22 1995, pp. 235–244.

108. YI Kuznetsov, *Prot Met (Russia)* **37**(2):101–107 (2001).

109. S Cheng, S Chen, T Liu, X Chang, Y Yin, *Mater Lett* **61**:3276–3280 (2007).

110. Anonymous, *J Coating Technol* **53**:29–36 (1981).

111. MA Amin, SS Abd El-Rehim, EEF El-Sherbini, RS Bayoumi, *Electrochim Acta* **52**(11):3588–3600 (2007).

112. DJ Banerjee, *Corros Maint* **4**(3):193–196 (1981).

113. P Munn, *Corros Sci* **35**:1495–1501 (1993).

114. MB Lawson, LD Martin, WP Banks, *Corrosion 83*, Anaheim, CA, 1983.

115. AM Borshchevskii, AM Sukhotin, LI Esina, *Prot Met (Russia)* **26**(4):514–516 (1990).

116. GP Tishchenkov, SF Vazhenin, NU Moisenko, IG Tishchenko, *Prot Met (Russia)* **26**(5):657–658 (1990).

117. O Lahodny-Sarc, P Orlovic-Leko, H Zagar, *Progress in Understanding and Prevention of Corrosion*, Vol. II, Barcelona, Spain, 1993, pp. 886–890.

118. GV Zinchenko, YuI Kuznetsov, *Prot Met (Russia)* **41**(2):167–172 (2005).

119. MA Quraishi, IH Farooqi, PA Saini, *Corrosion 99*, San Antonio, TX, Apr 25–30, 1999.

120. MA Quraishi, D Jamal, *Bull Electrochem (India)* **18**(7):289–294.

121. PB Raja, MG Sethuraman, *Surf Rev Lett* **14**(6):1157–1164 (2007).

122. C Monticelli, A Frignani, G Brunoro, G Trabanelli, F Zucchi, M Tassineri, *Corros Sci* **35**(5–8):1483–1489 (1993).

123. B Del Amo, M Deya, G Blustein, VF Vitere, R Ramagnoli, *EUROCORR 2001: The European Corrosion Congress*, Lake Garda, Italy, 2001, p. 10.

124. G Parashar, M Bajpayee, PK Kamari, *Surf Coating Int B* **86**(B3):209–216 (2003).

125. VB Rodin, SK Zhigletsova, NA Zhirkova, IA Irkhina, V Chugnov, VP Kholodenko, *EUROCORR 2004: Long Term Prediction and Modelling of Corrosion*, Nice, France, Sept 12–16, 2004.

126. MA Finan, A Harris, A Marshall, *Corrosion 79*, Atlanta, GA, Mar 12–16, 1979, p. 11.

127. J Iwanow, Yu I Kuznetsov, K Setkowiczk, *Proceedings of the 7th European Symposium on Corrosion Inhibitors*, Vol. 2, 1990, pp. 795–806.

128. V Mircheva, S Rangalov, *Comptes rendus, Académie bulgare des sciences (Bulgaria)* **45**(6):55–58 (1992).

129. Yu I Kuznetsov, AF Raskol'nikov, *Prot Met (Russia)* **29**(1):59–64 (1993).

130. Yu I Kuznetsov, *Progress in the Understanding and Prevention of Corrosion*, Vol. II, Barcelona, Spain, 1993, pp. 853–859.

131. Yu M Loshkaryov, *Proceedings of the 8th European Symposium on Corrosion Inhibitors*, Vol. 2, Ferrara, Sept 18–22 1995, pp. 711–720.

132. F Chmilenko, L Sobol, *EUROCORR 1997*, Vol. II, Trondheim, Norway, Sept 22–25, 1997, pp. 121–123.

133. W Hater, B Mayer, HD Speckmann, *EUROCORR 1997*, Vol. I, Trondheim, Norway, Sept 22–25, 1997, pp. 695–700.

134. I Esih, T Soric, Z Pavlinic, *Br Corros J*, **33**(4):309–334 (1998).

135. J Iwanow, J Senatorski, J Tracikowski, *Ochrona Przedkorozja (Poland)* Special Issue, pp. 437–442 (2002).

136. YY Tur, *Mater Sci (Russia)* **38**(2):304–306 (2002).

137. BD Oakes, SR Wilson, CF Schrieber, Report W7603143, Office of Water Research and Technology, Washington, DC, June 1975.

138. M Khobaib, CT Lynch, *Material and Process Applications: Land, Sea, Air, Space*, Los Angeles, CA, Apr 28–30, 1981, pp. 52–64.

139. H Krasts, V Kadeck, S Klevina, *Proceedings of the 7th European Symposium on Corrosion Inhibitors*, Vol. 1, Ferrara, Italy, 1990, pp. 569–581.

140. TC Chevrot, YM Gunaltun, *Corrosion 2000*, Orlando, FL, Mar 26–31, 2000, pp. 1–006.

141. BS Fultz et al., SNAME Panel SP-3 Surface Preparation and Coatings, Report No. 210254, May 1980.

142. J Kilts, MW Joosten, PG Hamble, J Clapham, *Corrosion 98*, San Diego, CA, Mar 22–27, 1998, pp. 38/1–38/12.

143. Z Wei, P Duby, P Somasundaran, *Corrosion 2004*, New Orleans, LA, Mar 28–Apr 1, 2004, p. 14.

144. MG Sethuraman, P Bothi Raja, *Pigm Resin Technol* **34**(6):(2005)

145. EE Oguzie, *Pigm Resin Technol* **35**(6):334–340 (2006).

146. PB Raja, MG Sethuraman, *Mater Lett* **62**:2977–2979 (2008)

147. K Wilson, M Forsyth, GB Deacon, C Forsyth, J Cosgriff, *Proceedings of the 9th European Symposium on Corrosion Inhibitors*, Ferrara, Italy, Sept 2000, pp. 1125–1140 2000.

148. MA Arenas, A Conde, JJ Damborenea, *Corros Sci* **44**(3):511–520 (2002).

149. BRW Hinton, DR Arnott, NE Ryan, *Metals Forum* **7**(4):12–18 (1984).

150. BRW Hinton, *J Alloys Compd* **180**:15–25 (1992).

151. B Hinton, A Hughes, R Taylor, K Henderson, K Nelson, L Wilson, *ATB Metallurgie* **37** (1997).

152. M Forsyth, K Wilson, T Behrsing, C Forsyth, GB Deacon, A Phanasgoankar, *Corros Sci* **58**:953 (2002).

153. DR Arnolt, BRW Hinton, NE Ryan, *Mater Performance* **26**:42 (1987).

154. AE Hughes, SG Hardin, TG Harvey, T Nikour, B Hinton, A Galassi, G McAdam, A Stonham, SJ Harris, S Church, C Figgures, D Dixon, C Bowden, P Mogan, SK Toh, D McCulloch, J Du Plessis, *ATB Metallurgie* **43**(1–2):264 (2003).

155. WG Fahrenholtz, MJ O'Keefe, H Zhou, JT Grant, *Surf Coat Technol* **155**:208 (2002).

156. AE Hughes, KJH Nelson, PR Miller, *Mater Sci Technol* **15**:1124 (1999).

157. D Ho, M Forsyth, unpublished data.

158. GB Deacon, CM Forsyth, M Forsyth, *Zeitschrift für anorganische und allgemeine Chemie* **629**(9):1472–1474 (2003).

159. M Forsyth, K Wilson, C Forsyth, T Behrsing, GB Deacon, *Corrosion* **58**:953 (2002).

160. M Forsyth, CM Forsyth, K Wilson, T Behrsing, GB Deacon, *Corros Sci* **44**:2651–2656 (2002).

161. H Daniel, M Grant, R Bruce, W Hinton, M Forsyth, PC Junk, S Leary, GB Deacon, *Proceedings of Corrosion Control & NDT*, 2003, p. 102.

162. AE Hughes et al., *ATB Metallurgie* **43**(1–2):264 (2003).

163. A Mol, Filiform corrosion of aluminum alloys: the effect of microstructural variations in the substrate, PhD thesis, Delft University of Technology, The Netherlands, 2000.

164. M Stern, *J Electrochem Soc* **24**:787–806 (1958).

165. D Bienstock, H Field, *Corrosion* **17**:87–90 (1961).

166. NR Whitehouse, *Polym Paint Colour J* **178**:239 (1984).

167. J Boxall, *Polym Paint Colour J* **174**:382–384 (1984).

168. BRW Hinton, *Met Finish* **9**:15–20, 55–61 (1991).

169. HE Hager, CJ Johnson, KY Blohowiak, CM Wong, JH Jones, SR Taylor, RL Cook, Jr., US Patent 5,866,652, The Boeing Company, United States (1999).

170. RL Cook, SR Taylor, *Corrosion* **56**:321–333 (2000).

171. M Favre, D Landolt, *Proceedings of the 7th European Symposium on Corrosion Inhibitors*, Vol. 2, University of Ferrara, Italy, p. 787.

172. M Manimegalai, P Rajeswari, S Mohanan, S Maruthamuthu, N Palaniswamy, *10th National Congress on Corrosion Control* (10th NCCC), Madurai, India, 2000, p. 153.

173. GH Awad, *Proceedings of the 6th European Symposium on Corrosion Inhibitors*, Vol. 1, University of Ferrara, Italy 1985, p. 385.

174. VA Altekar, I Singh, MK Banerjee, MN Singh, TR Soni, *Proceedings of the 5th European Symposium on Corrosion Inhibitors*, Vol. 2, University of Ferrara, Italy, p. 367 (1980).

175. M Khullar, S Chakraborti, B Allard, *Proceedings of the 5th European Symposium on Corrosion Inhibitors*, Vol. 3, University of Ferrara, Italy, p. 815 (1980).

176. AA El-Hosary, RM Saleh, HA El-Dahan, *Proceedings of the 7th European Symposium on Corrosion Inhibitors*, Vol. 1, University of Ferrara, Italy, p. 725 (1990).

177. NC Subramanyam, BS Seshadri, SM Mayanna, *Proceedings of the 6th European Symposium on Corrosion Inhibitors*, Vol. 1, University of Ferrara, Italy, p. 317 (1985).

178. IH Farooqi, A Hussain, MA Quraishi, PA Saini, *Anti-Corros Method Mater* **46**:328 (1999).

179. A Minhaj, PA Saini, MA Quraishi, IH Farooqi, *Corros Prev Control* **32**:(1999).

180. IH Farooqi, MA Quraishi, PA Saini, *Corros Prev Control* **93**:(1999).

181. M Kliskic, J Radosevic, S Gudic, V Katalinic, *J Appl Electrochem* **30**:823 (2000).

182. SR Taylor, BD Chambers, *Corros Rev* **25**(5–6):571–590 (2007).

183. Y Feng, KS Siow, KT Teo, AK Hsich, *Corros Sci* **41**:829–852 (1999).

184. S Sayed Azim, S Muralidharan, SV Iyer, B Muralidharan, T Vasudevan, *Br Corros J* **33**:297 (1998).

185. M Mustafa, SM Shahinoor, I Dulal, *Br Corros J* **32**:133 (1997).

186. S Sayed Azim, S Muralidharan, S Venkat Krishna Iyer, *J Appl Electrochem* **25**:495–500 (1995).

187. DDN Singh, AK Dey, *Corrosion* **49**:594–600 (1993).

188. MA Quraishi, J Rawat, M Ajmal, *Corrosion* **55**:919–923 (1999).

189. GN Mu, TP Zhao, T Gu, *Corrosion* **52**:853–856 (1996).

190. MA Quraishi, S Ahmed, M Ansari, *Br Corros J* **32**:297–300 (1997).

191. JM Abd El Kader, AA Warraky, AM Abd El Aziz, *Br Corros J* **33**:152–157 (1998).

192. S Rajendran, BV Appa Rao, N Palaniswamy, *Electrochim Acta* **44**:533–537 (1998).

193. T Suzuki, H Nishihara, K Aramaki, *Corros Sci* **38**:1223–1234 (1996).

194. Y Gonzalez, MC Lafont, N Pebere, F Moran, *J Appl Electrochem* **26**:1259–1265 (1996).

195. Y Feng, KS Siow, WK Teo, KL Tan, AK Hsieh, *Corrosion* **53**:546 (1997).

196. S Gonzalez, MM Laz, RM Suoto, RC Salvarezza, AJ Arvia, *Corrosion* **49**:450–456 (1993).

197. GS Chen, M Gao, RP Weir, *Corrosion* **52**:8–15 (1996).

198. RG Bucheit, RP Grant, PF Hlava, B McKenzie, GL Zender, *J Electrochem Soc* **144**:8 (1997).

199. N Dimitrov, JA Mann, K Sieradski, *J Electrochem Soc* **146**:98 (1999).

200. SR Taylor, BD Chambers, US Provisional Patent 60,657,298.

201. Czechoslovakian Patent 116908.

202. S. Racz, T. Sziebert (1996) Hungarian Patent 726.

203. Shell Dev. (24 August 1948) UK Patent 607013; WGO 1803168.

204. *Kogyo Kagaku Zasshi*, **61**:69–72 (1958).

205. *Met Finish* **58**(8):40–436 (1960).

206. Japanese Patent 70/27692–27693.

207. G.L. Doelling, J.A. Siefker (27 Oct. 1964) US Patent 3,347,796.

208. (a) *J Oil Chem Soc (Japan)* **3**:149–154 (1954); (b) *Werkst und Korrosion* **10**:767–9 (1959); (c) French Patent 1329948.

209. US Patent 3,262,038.

210. J. Balvay, C. Jouandet (30 March 1951) French Patent 977348.

211. Canadian Patent 137,558.

212. Laporte Chemical (14 June 1963) French Patent 1329948.

213. Anon.

214. E.M. Hoffman (11 October 1966) US Patent 3,278,434.

215. J. Iaciofano (14 Nov. 1967) US Patent 3,352,695.

216. Buhler Fontaine SA (29 January 1965) French Patent 1387181.

217. J.B. Rust (21 May 1946) US Patent 2,400,784.

218. US Patent 2,665,995.

219. *Schmierstaffe Schmierungstek* **7**:69–79 (1996); US Patent 250,012,020.

220. *Official Digest* (1964, 1966), 1348/1358; G. Nottes, H. Otterbach (24 March 1967) French Patent 1474514.

221. A.H. Matuszak (7 Aug. 1951) US Patent 2,563,609.

222. A.H. Matuszak, J.W. Hand Jr. (25 Dec. 1951) US Patent 2,580,036.

223. Canadian Patent 17593.

224. J.W. Bishop (24 Oct. 1950) US Patent 2,527,296.

225. Canadian Patent 113,302.

226. *Abura Kagawu* **3**:210–215 (1954).

227. J. Arthur Sharp (30 May 1961) US Patent 2,986,482.

228. Japanese Patent 76/70216.

229. *J Appl Chem* **8**:339–345 (1963).

230. *J Soc Chem Ind* **65**:196–204 (1946).

231. G.W. Graham, G.R. Lusby (12 July 1960) Canadian Patent 601,668; 87061.

232. *J Oil Chem Soc (Japan)* **3**:149–154 (1954).

233. E.A. Deiman (04 May 1954) US Patent 2,677,618; Canadian Patent 207,531; Japanese Patent 70/34030.

234. Japanese Patent K77/126637.

235. Japanese Patent 69/23210.

236. *Kogyo Kagaku Zasshi* **60**:556–579 (1957).

237. *J Oil Chem Soc (Japan)* **3**:149 (1954).

238. Dutch Patent 6112851.

239. Polish Patent 92191 (31 March 1977).

240. Canadian Patent 119,034; 195,536.

241. J.B. Coates, A.H. Maxey (29 Aug. 1949) UK Patent 628419.

242. H.C. Thayer, jr (19 April 1960) US Patent 2,933,412.

243. Japanese Patent 77/114445.

244. US Patent 3,180,829.

245. *Izv Vyssh Ucheb Zaved Khim Teknol* **16**(2):240–243 (1973).

246. D.L. Klass, V. Brozowski (04 Aug. 1964) US Patent 3,143,552.

247. Japanese Patent 32416.

248. *Kogyo Kagaku Zasshi* **61**:69–72 (1958).

249. R. Leutz (29 Dec. 1960) West German Patent 1096145.

250. G.W. Duncan (14 Aug. 1945) US Patent 2,382,699; UK Patent 573623.

251. H. Marcinkowski, D. Haensel (03 April 1985) West German Patent 2200743.

252. R.A. Westlund Jr (14 Oct. 1958) US Patent 2,856,299.

253. W.A. Zisman, H.R. Baker (04 Aug 1953) US Patent 2,647,839; Canadian Patent 158,327.

254. G.W. Duncan (14 Aug. 1945) US Patent 2,382,699.

255. Indian Patent 46821.

256. W.A. Zisman, H.R. Baker (04 Aug 1953) US Patent 2,647,839; Shell Dev (24 Aug 1948) UK Patent 586679, 707162.

257. Fuji Iron & Steel (14 Jan. 1970) UK Patent 1177873.

258. L.W. Sproule, L.F. King (16 Aug 1949) US Patent 2,479,424; Shell Dev (24 Aug 1948) UK Patent 607013.

259. *Lubr Eng* **9**:19 (1947).

260. K.L. Houghton, P. G. N. Leonard (23 Aug. 1972) UK Patent 1286473.

261. Dutch Patent 6909920.

262. Canadian Patent 6740; Czech Patent 165655 (22 Nov. 1975).

263. D. Apikos (02 May 1972) US Patent 3,660,334; 3,714,094 (30 Jan. 1973); 3,843,574 (22 Oct. 1974).

264. *Ind J Technol* **2**:12, 402–415 (1964); *Defence Sci J (New Delhi)* **9**: 212–221 (1959).

265. A. Wachter, N. Stillman (24 Feb. 1953) US Patent 2,629,649; Polish Patent 93952 (30 July 1977).

266. A. Wachter, S. Thurston (21 June 1955) US Patent 2,711,360.

267. F. Ross, C. Mellick (25 Feb. 1958) US Patent 2,824,782.

268. UK 9197787; West German Patent 1172921.

269. J.B. Hinkamp (22 May 1962) US Patent 3,035,906; 3,294,705 G.J. Kautsky (27 Dec. 1966).

270. Canadian Patent 54600.

271. Hungarian Patent 140708.

272. Canadian Patent 125,000.

273. C. Fiquet (18 Mar. 1966) French Patent 1431718.

274. Olin Mathieson (26 Mar. 1969) UK Patent 1146710.

275. Canadian Patent 157,034; 230,558; 211,381.

276. US Patent 3,861,824.

277. Canadian Patent 97,818; 9752; UK Patent 1227189.

278. P.W. Solomon (04 Oct. 1960) US Patent 2,955,093.

279. D.L. Andersen (18 July 1961) US Patent 2,993,007.

280. US Patent 2,993,007.

281. T.O. Counts (27 Dec. 1960) US Patent 2,966,458.

282. Renault, Peugeot (13 Dec. 1973) West German Patent 2324532.

283. Canadian Patent 158,327.

284. J.W. Bishop (13 Nov 1951) US Patent 2,574,954; 24,482,581.

285. S.T. Polley (02 Dec. 1959) UK Patent 824,405.

286. *Morskoi Flot* **18**:16 (1958).

287. Japanese Patent K76/108680.

288. W.F. Dewey (12 May 1959) US Patent 2,886,531.

289. C.M. Blair Jr, W.F. Gross (26 May 1953) US Patent 2,640,029.

290. Canadian Patent 10759.

291. L.B. High, W.E. Hague (20 Aug 1957) US Patent 2,803,593.

292. H.W. Sigworth, M.R. Barusch (08 Aug 1961) US Patent 2,995,427.

293. *J Iron Steel Inst (London)* **188**:36–45 (1958).

294. Canadian Patent 130,299.

295. J.A. Ellis (15 Nov. 1966) US Patent 3,285,755.

296. Canadian Patent 111,681.

297. Canadian Patent 91,130.

298. Canadian Patent 196,175.

299. *J Appl Chem* **13**(8):339 (1963).

300. *Zestzyty Nauk Politek Slash Chem* **2**:24–27 (1957).

301. GH Brundtland, *Our Common Future*, Oxford University Press, New York, 1987.

302. O Hutzinger, *Environ Sci Poll Res* **6**:123 (1999).

303. L Desimone, F Popoff, *Ecoefficiency: The Business Link to Sustainable Development*, MIT Press, Cambridge, MA, 2000.

304. R Sanghi, *Corros Sci* **79**:1662 (2000).

305. SK Sharma, A Chaudhary, RV Singh, *Rasayan J Chem* **1**(1):68–92 (2008).

306. B Valdez, J Cheng, F Flores, M Schorr, L Yeleva, *Corros Rev* **21**:445–458 (2003).

307. SR Taylor, BD Chambers, *Corros Rev* **25**:571–590 (2007).

308. KF Khaled, *Int J Electrochem Sci* **3**:462–475 (2008).

309. P Bothi Raja, MG Sethuraman, *Mater Lett* **62**(1):113–116 (2008).

310. S Rajendran, AJ Amalraj, MJ Joice, N Anthony, DC Trivedi, M Sundaravadivelu, *Corros Rev* **22**(3):233–248 (2004).

311. J Mathiyarasu, SS Pathak, V Yegnaraman, *Corros Rev* **24**:307–322 (2006).

312. A Mesbah, C Jeurs, F Lacouture, S Mathieu, *Solid State Sci* **9**:322–328 (2007).

313. PC Okafor, VI Osabor, EE Ebenso, *Pigm Resin Technol* **36**:299 (2007).

314. M Lebrini, M Traisnel, M Lagren, B Mernari, F Bentiss, *Corros Sci* **50**:504 (2008).

315. K Radojei, S Berkovi, S Kovac, J Vorkapi-Furac, *Corros Sci* **50**:504 (2008).

316. SAM Refaey, AM Abdel Malak, F Taha, HTM Abdel-Fatah, *Int J Electrochem Sci* **3**:167–176 (2008).

317. EM Mansour, AM Abdel-Gaber, A Ludwick, *Corrosion* **59**:242 (2003).

318. E Hefnawy, *Werkstofftechnik* **38**:227 (2007).

319. AM Abdel-Gaber, E Khamis, H Abo-Eldahab, S Adeel, *Mater Chem Phys* **109**:297 (2008).

320. M Elayyachi, B Hammouti, A El Idrissi, *Appl Surf Sci* **249**(1–4):176–182 (2005).

321. GY Elewady, IA El-Said, AS Fouda, *Int J Electrochem Sci* **3**:177 (2008).

322. R Fuchs-Godec, *Colloids Surf* **280**:130–139 (2006).

323. M Lebrini, M Traisnel, M Lagren, B Mernari, F Bentiss, *Corros Sci* **50**(2):473 (2008).

324. A Yurt, S Ulutas, H Dal, *Appl Surf Sci* **253**:919–925 (2006).

325. ESH El-Ashry, A El Nemr, SA Essawy, S Ragab, *Prog Org Coating* **61**(1):1161 (2008).

326. SS Pathak, AS Khanna, TJM Sinha, *Corros Rev* **24**:281–306 (2006).

327. A Pepe, M Aparichio, S Cer, A Durin, *J Non-Cryst Solids* **348**:162–171 (2004).

328. AN Khramov, NN Voevodin, VN Balbyshev, MS Donley, *Thin Solid Films* **447**(8):549–557 (2004).

329. SV Lamaka, ML Zheludkevich, KA Yasakau, R Serra, SK Poznyak, *Prog Org Coating* **58**:127–135 (2007).

330. ML Zheludkevich, DG Schukin, KA Yasakau, H Mohwald, MGS Ferreira, *Chem Mater* **19**(3):402–411 (2007).

331. DG Shchukin, H Mhwald, *Small* **3**(6):926–943 (2007).

332. V Moutarlier, B Neveu, MP Gigardet, *Surf Coating Technol* **202**(10):2058 (2008).

333. SK Poznyak, ML Zheludkevich, D Raps, F Gammel, KA Yasakaei, MG Ferreira, *Prog Org Coating* **62**(2):226–235 (2008).

334. KA Yasakau, ML Zheludkevich, OV Karaval, MGS Ferreira, *Prog Org Coating*, 2008.

335. EME Mansour, AM Abdel-Gaber, BA Abd-El Nabey, N Khalil, E Khamis, A Tadros, H Aglan, A Ludwick, *Corrosion* **59**:242 (2003).

336. S Papavinasam, RW Revie, WI Friesen, A Doiron, T Paneerselvan, *Corros Rev* **24**(3–4):173 (2006).

337. A Zieliski, S Sobieszczyk, *Corros Rev* **26**:1–22 (2008).

338. OB Isgor, AG Razakpur, *Mater Struct* **39**:291–302 (2006).

339. D Colorado–Garrido, DM Ortega–Toledo, JA Hernandez, JG Gonzalez–Rodriguez, *Proceedings of Electronics, Robotics and Automotive Mechanics Conference*, CERMA, 2007, p. 213.

340. L Jingjun, L Yuzhen, L Xiaoyu, *Anti-Corros Method Mater* **55**:66 (2008).

341. G Gao, C Liang, *J Electrochem Soc* **154**:144–151 (2007).

342. K Barouni, L Bazzi, R Salghi, M Mihit, B Hammouti, A Albourine, S El Issami, *Mater Lett* **62**(19):3325–3327 (2008).

343. KM Ismail, *Electrochim Acta* **52**(28):7811–7819 (2007).

344. M Bendahou, M Benabdellah, B Hammouti, *Pigm Resin Technol* **35**(2):953 (2006).

345. AY El-Etre, M Abdallah, ZE El-Tantawy, *Corros Sci* **47**(2):385 (2005).

346. EE Oguzie, AI Onuchukwu, PC Okafor, EE Ebenso, *Pigm Resin Technol* **34**:327–331 (2005).

347. I Radojci, K Berkovi, S Kovac, J Yorkapi-Furac, *Corros Sci* **50**(5):504 (2008).

348. MG Sethuraman, P Bothi Raja, *Pigm Resin Technol* **34**:327 (2005).

349. A Khamis, E Hefnawy, *Werkstofftechnik* **38**:227 (2007).

350. AM Abdel-Gaber, E Khamis, H Abo-Eldahab, S Adeel, *Mater Chem Phys* **109**:297 (2008).

351. E Khamis, BA Abd-El-Nabey, AM Abdel-Gaber, DE Abd-El-Khalek, *26th Water Treatment Technology Conference*, Alexandria San Stefano, Egypt, June 7–9, 2008.

352. F Blin, SG Leary, GB Deacon, PC Junk, M Forsyth, *Corros Sci* **48**:419 (2006).

353. F Blin, P Kontsoukos, P Klepetsianis, M Forsyth, *Electrochim Acta* **52**:6212–6220 (2007).

354. G Kardas, R Solmaz, *Corros Rev* **24**:151–171 (2006).

355. RW Revie, *Uhlig's Corrosion Handbook*, 2nd edition, John Wiley & Sons, Inc., 2000.

356. SR Davis, AR Brough, A Atkinson, *J Non-Cryst Solids* **315**:197 (2003).

357. N Tsuji, K Nozawa, K Aramaki, *Corros Sci* **42**:1523 (2000).

358. S Sathiyanarayanan, S Muthukrishnan, G Venkatachari, *Electrochim Acta* **51**:6313 (2006).

359. Z Quan, S Chen, S Li, *Corros Sci* **43**:1071 (2001).

360. AT Lusk, GK Jennings, *Langmuir* **17**:7830 (2001).

361. C Wang, S Chen, S Zhao, *J Electrochem Soc* **151**:B11–B15 (2004).

362. A Zielinski, S Sobieszczyk, *Corros Rev* **26**:1–22 (2008).

363. B Valdez, J Cheng, F Flores, M Schorr, L Veleva, *Corros Rev* **21**:445 (2003).

INDEX

Green Corrosion Inhibitors: Theory and Practice, First Edition. V. S. Sastri.
© 2011 John Wiley & Sons, Inc. Published 2011 by John Wiley & Sons, Inc.